2020 IEEE 29th Conference on Electrical Performance of Electronic Packaging and Systems (EPEPS 2020)

San Jose, California, USA
5 – 7 October 2020

IEEE Catalog Number: CFP20EPP-POD
ISBN: 978-1-7281-6162-4

**Copyright © 2020 by the Institute of Electrical and Electronics Engineers, Inc.
All Rights Reserved**

Copyright and Reprint Permissions: Abstracting is permitted with credit to the source. Libraries are permitted to photocopy beyond the limit of U.S. copyright law for private use of patrons those articles in this volume that carry a code at the bottom of the first page, provided the per-copy fee indicated in the code is paid through Copyright Clearance Center, 222 Rosewood Drive, Danvers, MA 01923.

For other copying, reprint or republication permission, write to IEEE Copyrights Manager, IEEE Service Center, 445 Hoes Lane, Piscataway, NJ 08854. All rights reserved.

*** This is a print representation of what appears in the IEEE Digital Library. Some format issues inherent in the e-media version may also appear in this print version.

IEEE Catalog Number:	CFP20EPP-POD
ISBN (Print-On-Demand):	978-1-7281-6162-4
ISBN (Online):	978-1-7281-6161-7
ISSN:	2165-4107

Additional Copies of This Publication Are Available From:

Curran Associates, Inc
57 Morehouse Lane
Red Hook, NY 12571 USA
Phone: (845) 758-0400
Fax: (845) 758-2633
E-mail: curran@proceedings.com
Web: www.proceedings.com

2020 IEEE 29th Conference on Electrical Performance of Electronic Packaging and Systems (EPEPS 2020)

San Jose, California, USA
5 – 7 October 2020

IEEE Catalog Number: CFP20EPP-POD
ISBN: 978-1-7281-6162-4

TABLE OF CONTENTS

AUGMENTED PEEC FOR DIRECT TIME-DOMAIN THERMAL AND POWER ESTIMATION OF INTEGRATED VOLTAGE REGULATOR ARCHITECTURES ARISING IN HETEROGENEOUS INTEGRATION 1
Venkatesh Avula, Vanessa Smet, Yogendra Joshi, Madhavan Swaminathan

EXTRACTING THE DYNAMIC CURRENT OF A POWER DELIVERY NETWORK 4
Heidi Barnes, Steve Sandler, Jack Carrel

ON DISSIPATIVITY CONDITIONS FOR LINEARIZED MODELS OF LOCALLY ACTIVE CIRCUIT BLOCKS 7
T. Bradde, S. Grivet-Talocia, G. C. Calafiore, A. V. Proskurnikov, Z. Mahmood, L. Daniel

ENERGY-AREA AWARE CHANNEL DESIGN FOR MULTI-CHIP INTERFACES 10
Muhammad Waqas Chaudhary, Andy Heinig, Bhaskar Choubey

DUAL SIDED HIGH FREQUENCY MEASUREMENT OF MICROELECTRONIC PACKAGES 13
Sean R. Christ, Ahmet C. Durgun, Kemal Aygün, Michael J. Hill

POST-FEC BER PERFORMANCE ANALYSIS FOR MULTISTAGE PAM4 SYSTEMS 16
Xiaoqing Dong, Chunxing Huang

GAUSS-NEWTON METHOD FOR PERFORMANCE EVALUATION OF DECOUPLING CAPACITORS ON RESONANT PARALLEL PLATES 19
Ihsan Erdin, Ram Achar

MEASUREMENT UNCERTAINTY PROPAGATION IN THE VALIDATION OF HIGH-SPEED INTERCONNECTS 22
Cemil S. Geyik, Michael J. Hill, Zhichao Zhang, Kemal Aygün, James T. Aberle

A NON-RANDOM EXPLORATION BASED METHOD FOR THE OPTIMIZATION OF CAPACITORS IN POWER DELIVERY NETWORKS 25
Seunghyup Han, Madhavan Swaminathan

HIGH-DIMENSIONAL UNCERTAINTY QUANTIFICATION VIA ACTIVE AND RANK-ADAPTIVE TENSOR REGRESSION 28
Zichang He, Zheng Zhang

A PARALLEL-IN-TIME CIRCUIT SIMULATOR FOR POWER DELIVERY NETWORKS WITH NONLINEAR LOAD MODELS 31
Chung-Kuan Cheng, Chia-Tung Ho, Chao Jiao, Xinyuan Wang, Zhiyu Zen, Xin Zha

DESIGN, SIMULATION AND MEASUREMENT OF A FLEXIBLE VOLTAGE-CONTROLLED OSCILLATOR (VCO) CHIP WITH BENDING RADIUS 34
Seungtaek Jeong, Seongsoo Lee, Seokwoo Hong, Boogyo Sim, Hyunwook Park, Subin Kim, Youngwoo Kim, Keeyeong Son, Joungho Kim, Jaehak Lee, Junyeop Song

A COMPARISON OF FINITE VS. INFINITE PLANE MODELS OF REFERENCE CONDUCTORS IN ELECTRONIC PACKAGES 38
Yi-Ru Jeong, Ali E. Yilmaz

ANALYSIS OF POWER SUPPLY NOISE INDUCED JITTER OF I/O SUBSYSTEMS WITH MULTIPLE POWER DOMAINS 41
Hyo-Soon Kang, Ashkan Hashemi, Guang Chen, Xiaoping Liu, Wendemagegnehu Beyene

REINFORCEMENT LEARNING-BASED AUTO-ROUTER CONSIDERING SIGNAL INTEGRITY.. 44
Minsu Kim, Hyunwook Park, Seongguk Kim, Keeyoung Son, Subin Kim, Kyunjune Son, Seonguk Choi, Gapyeol Park, Joungho Kim

PCIE GEN-5 DESIGN CHALLENGES OF HIGH-SPEED SERVERS 47
Mallikarjun Vasa, Chun-Lin Liao, Sanjay Kumar, Ching-Huei Chen, Bhyrav Mutnury

A SHIELDED-BLOCK PRECONDITIONER FOR REDUCED-DOMAIN LAYERED-MEDIUM INTEGRAL-EQUATION METHODS.. 50
Chang Liu, Ali E. Yilmaz

ON THE ACCURACY OF CROSS-TALK MODELING IN HIGH-SPEED DIGITAL CIRCUITS USING THE ACCELERATED BOUNDARY ELEMENT METHOD 53
Dongwei Li, Giacomo Bianconi, Swagato Chakraborty

UNIFORMLY ACCURATE ELECTROSTATIC LAYERED MEDIUM GREEN'S FUNCTION APPROXIMATION VIA SCATTERED FIELD FORMULATION 56
Xinbo Li, Vladimir Okhmatovski

AN EFFICIENT AND PARALLEL ELECTROMAGNETIC SOLVER FOR COMPLEX INTERCONNECTS IN LAYERED MEDIA 59
Damian Marek, Shashwat Sharma, Piero Triverio

HIGH-SPEED LINK DESIGN OPTIMIZATION USING MACHINE LEARNING SVR-AS METHOD.. 62
Hanzhi Ma, Er-Ping Li, Andreas C. Cangellaris, Xu Chen

A TRANSMISSION LINE COUPLER COMPONENT FOR DIRECT B2B COMMUNICATIONS.............. 65
Reiji Miura, Tadahiro Kuroda, Mototsugu Hamada

A REVIEW OF 90 DEGREE CORNER DESIGN FOR HIGH-SPEED DIGITAL AND MMWAVE APPLICATIONS.. 68
Heidi Barnes, Giovanni Bianchi, Jose Moreira

A TUNABLE NEURAL NETWORK BASED DECISION FEED-BACK EQUALIZER MODEL FOR HIGH-SPEED LINK SIMULATION... 71
Thong Nguyen, Jose Schutt-Aine

VIA DESIGN OPTIMIZATION FOR HIGH SPEED DIFFERENTIAL INTERCONNECTS ON CIRCUIT BOARDS... 74
Armen Vardapetyan, Chong-Jin Ong

CAUSAL TRANSMISSION LINE GEOMETRY OPTIMIZATION FOR IMPEDANCE CONTROL IN PCBS 77
Zachariah M. Peterson

PREDICTOR-CORRECTOR ALGORITHM WITH EMBEDDED DIMENSION REDUCTION FOR UNCERTAINTY QUANTIFICATION OF MWCNT ON-CHIP INTERCONNECT NETWORKS.. 80
Surila Guglani, Sourajeet Roy

SI MODEL TO HARDWARE CORRELATION ON A 44 GB/S HLGA SOCKET CONNECTOR.............. 83
Pavel Roy Paladhi, Yanyan Zhang, Junyan Tang, Daniel Rodriguez, Jose Hejase, Sungjun Chun, Wiren Becker, Brian Beaman, Daniel Dreps

ANN PERFORMANCE FOR THE PREDICTION OF HIGH-SPEED DIGITAL
INTERCONNECTS OVER MULTIPLE PCBS 86
    Katharina Scharff, Christian Morten Schierholz, Cheng Yang, Christian Schuster

COST-EFFECTIVE IMPLEMENTATION OF AIR-FILLED WAVEGUIDES ON PRINTED
CIRCUIT BOARDS 89
    *Felix Sepaintner, Andreas Scharl, Johannes Jakob, Florian Keck, Kevin Kunze, Franz Xaver
    Röhrl, Werner Bogner, Stefan Zorn*

THERMAL SENSITIVITY OF DIELECTRIC MATERIALS IN HIGH-SPEED DESIGNS 92
    *Sunil Pathania, Bhyrav Mutnury, Mallikarjun Vasa, Vijender Kumar, Sukumar Muthusamy, P
    K Seema, Rohit Sharma*

AN INSPECTION BASED METHOD TO ANALYSE DETERMINISTIC NOISE IN N-PORT
CIRCUITS 95
    Vijender Kumar Sharma, Jai Narayan Tripathi, Hitesh Shrimali

ACCELERATED BOUNDARY ELEMENT MODELING OF LOSSY CONDUCTORS IN
LAYERED MEDIA WITH A SINGLE-SOURCE SURFACE IMPEDANCE OPERATOR 98
    Shashwat Sharma, Piero Triverio

ESTIMATING PER-UNIT-LENGTH RESISTANCE PARAMETER IN EMERGING COPPER-
GRAPHENE HYBRID INTERCONNECTS VIA PRIOR KNOWLEDGE BASED
ACCELERATED NEURAL NETWORKS 101
    *Rahul Kumar, S. S. Likith Narayan, Somesh Kumar, Sourajeet Roy, Brajesh K. Kaushik,
    Ramachandra Achar, Rohit Sharma*

VARIATIONAL INFERENCE APPROACH TO JITTER DECOMPOSITION IN HIGH-SPEED
LINK 104
    Bobi Shi, Thong Nguyen, Jose Schutt-Aine

ASSESSMENT OF 2X THRU DE-EMBEDDING ACCURACY FOR PACKAGE
TRANSMISSION LINE DUTS 107
    Stephen A. Smith, Zhichao Zhang, Kemal Aygün

ACCURATE BGA PACKAGE SOLDER JOINT MODELING FOR HIGH SPEED SERDES
INTERFACES 110
    Jiwei Sun, Zhiguo Qian, Cemil S. Geyik, Kemal Aygün

3D INTEGRATION OF KA-BAND RFIC BY INDUCTIVE INTERCHIP WIRELESS
COMMUNICATION USING FIGURE-8 COILS 113
    Masairo Usui, Kota Shiba, Mototsugu Hamada, Tadahiro Kuroda

ANALYSIS OF THE INFLUENCE OF ROUGHNESS ON THE PROPAGATION CONSTANT
OF A WAVEGUIDE VIA TWO SPARSE STOCHASTIC METHODS 116
    *Ruben Waeytens, Dries Bosman, Martijn Huynen, Michiel Gossye, Hendrik Rogier, Dries
    Vande Ginste*

DETERMINE SOCKET'S INDUCTANCE AND CONTACT RESISTANCE BY USING PRF
METHOD 119
    Tao Wang, Jun Fan

RX EQUALIZATION FOR A HIGH-SPEED CHANNEL BASED ON BAYESIAN ACTIVE
LEARNING USING DROPOUT 122
    *Xianbo Yang, Junyan Tang, Hakki M. Torun, Wiren D. Becker, Jose A. Hejase, Madhavan
    Swaminathan*

HYPERPARAMETER DETERMINATION IN MULTIVARIATE MACROMODELING BASED ON RADIAL BASIS FUNCTIONS... 125

Alessandro Zanco, Stefano Grivet-Talocia

SIGNAL INTEGRITY CHARACTERIZATION OF CHANNELS WITH ASYMMETRIC VIA STUBS 128

Yanyan Zhang, Mahesh Bohra, Nam Pham, Pavel R. Paladhi, Wiren D. Becker, Daniel M. Dreps

Author Index

EPEPS 2020

IEEE 29TH CONFERENCE ON ELECTRICAL PERFORMANCE OF ELECTRONIC PACKAGING AND SYSTEMS

OCT 5-7, 2020

SPONSORED BY
IEEE ELECTRONICS PACKAGING SOCIETY,
MICROWAVE THEORY & TECHNIQUES SOCIETY
& ANTENNAS AND PROPAGATION SOCIETY

IEEE 29th Conference on Electrical Performance of Electronic Packaging and Systems
Welcome Message
from the
Conference Chairs
EPEPS 2020

Greetings EPEPS Attendees,

Welcome to the 29th Electrical Performance of Electronic Packaging and Systems (EPEPS) conference, held virtually for the first time! This conference provides a forum for the presentation and discussion of the latest advances in the electrical design, analysis, modeling and characterization of interconnections and packaging structures of electronic systems covering all the application families and frequency ranges namely, digital, RF, microwave and mm-wave applications. One of the key objectives of this meeting is to bring together researchers and practicing engineers from industry, universities, and government laboratories from around the world to address current and future issues affecting the electrical performance of high-speed electronic systems.

The Technical Program Committee (TPC) is proud to present a diverse technical program of 43 paper presentations. These contributions cover the latest advances and emerging technologies in electrical modeling and analysis of packaging and systems. Our Paper Review Committee (PRC) continued assisting the TPC in reviewing and selecting the papers at this year's conference. Their contributions are very much appreciated and acknowledged.

Our two keynote speakers, James P. Held of Intel and Daniel Dreps of IBM; and our invited presenter, Vaishnav Srinivas of Qualcomm, will present their insight into our technical challenges and our opportunity for innovation. We have a special presentation this year from IEEE Electronics Packaging Society, Technical Committee - Electrical Design, Modeling & Simulation, on package benchmarking. We also offer excellent educational opportunities with 5 tutorials from experts in industry and academia.

This year, EPEPS will offer two best paper awards, a best conference paper and a best student paper. A total of 18 student papers will be competing for the best student paper award. We are also very excited to announce a logo competition for this year's conference. The winning logo will be used for future EPEPS conferences.

A further objective of the meeting is the encouragement of interaction among the participants. In our experience, such interaction has resulted in the most productive industry-academia collaboration and leads to significant advances in the area of electronic packaging and systems. Our corporate sponsorship enables the TPC to provide you with an exciting EPEPS program. Please visit the virtual sponsor showcase area to see what they have to offer. We greatly appreciate the support of our Platinum Sponsor, Qualcomm; Gold Sponsor, Intel; Silver Sponsors, Keysight and Xpeedic; and Exhibitors, Cadence and Amphenol.

Last but not least, the chairs of the 29th EPEPS conference wish to thank the invited speakers, authors, presenters, tutorial instructors, and the members of the TPC and PRC for their contributions in creating this year's outstanding technical program. The sponsorship of the IEEE Microwave Theory and Techniques Society, the IEEE Electronics Packaging Society and the IEEE Antennas and Propagation Society is acknowledged and greatly appreciated. The support of the IEEE MCE Digital Events Team has been tremendous to enable the digital platform for the virtual conference, for which we are truly appreciative. We also acknowledge the continuing effort and dedication of Prof. José E. Schutt-Ainé in maintaining the EPEPS web site and review portals. And finally, we thank the session chairs, conference organizing volunteers, and our paper award committee members for their contributions.

Kemal Aygün, Intel   Jose Hejase, Nvidia
EPEPS 2020 Conference Chairs

Keynote I

Title: Meeting the Challenge of Building a Scalable Quantum Computer

Speaker: James P. Held
Intel Fellow & Director of Emerging Technologies Research
Intel

Abstract: The potential of quantum computing to deliver compelling performance to a wide range of demanding applications is generating tremendous excitement. This talk will review the promise and the challenges to achieving large-scale quantum computers capable of real-world applications. Intel's research to address the challenges spans the entire quantum system, from qubit devices to the hardware and software required to control these devices and the quantum algorithms that will harness the power of quantum technologies.

Jim Held leads a team conducting research in new technologies for Intel's future, ranging from quantum computing to neuromorphic computing. Since joining Intel in 1990, Held has served in a variety of positions leading research on platform technology such as operating system support for real-time media processing, extensible processor architecture, and multi-core processor architecture. He has led a variety of labs conducting research in platform interconnect, microarchitecture, parallel computing and programming systems. Before coming to Intel, Held worked in research and teaching capacities in the Medical School and Department of Computer Science at the University of Minnesota. He is a Member of the IEEE Computer Society and the Association for Computer Machinery (ACM).

Keynote II

Title: High Speed and Large Bandwidth Server Computer Bus Links: Past Milestones, Current State of The Art and Future Directions

Speaker: Daniel Dreps
Distinguished Engineer
IBM

Abstract: The presentation will focus on key aspects of high-speed interface design including memory, socket to socket and on-module interfaces. Case study details on memory, on module DCM, on module MCM and cable direct links are shown. System challenges of the transition from NRZ to PAM4 for low latency links are discussed. Moreover, organic approaches for channel design improvement are described. Finally, future outlooks of possibilities as we begin this new decade are laid out.

Daniel Dreps is a Distinguished Engineer working in the IBM Systems Group. He received his B.S.E.E. degree in 1983 from Michigan State University. During his IBM career, he has designed and developed: transistor models, fiber optic links, ASIC technology custom elements, high-speed serial links and clocking systems for IBM servers. His interests currently focus on ultra-low power high-speed link development and applications for the entire range of IBM servers. He has published multiple papers and holds over 200 worldwide patents in broad areas of interconnect and server design.

Invited Presentation

Title: Closing the Loop from Architecture to Post-Silicon for Signal and Power Integrity

Speaker: Vaishnav Srinivas
Senior Director
Qualcomm

Abstract: Signal and power integrity has an impact across many layers of abstraction in the IC design flow. This talk covers the impact during (1) early architecture definition, particularly relating to interconnect technology choices; (2) the design phase driving various aspects of the IO circuit, PKG and PCB; and finally (3) during post-silicon bring-up, debug and optimization. It focuses on the need for early exploration using exploratory frameworks able to study feasibility from a signal and power integrity perspective, and shows the benefit in closing the loop with key learnings from post-silicon measurements and debug. To illustrate this, examples including die-to-die exploration, memory interfaces, high-speed SERDES interfaces and PDN exploration are highlighted.

Vaishnav Srinivas is a Senior Director at Qualcomm, where he leads a mixed-signal systems team that works on various circuits-to-systems efforts in the HW design organization. This includes signal and power integrity for all interfaces; PDN design for all cores; training algorithms, firmware and debug relating to mixed-signal circuits; and circuits-to-system exploration relating to IO architecture, PDN mitigation and other mixed-signal design. He received his PhD from UCSD, his MS from UCLA and BTech from IIT Madras in Electrical Engineering. His interests include circuits-and-system architecture, design, analysis and validation, including HW and SW techniques to optimize mixed-signal circuits and systems.

Tutorial I

Title: How to Find and Validate Power Rail Resonances

Speakers: Heidi Barnes, Keysight / Steve Sandler, Picotest / Jack Carrel, Xilinx

Abstract: Power delivery is not a DC problem. Digital loads generate extremely wide spectral content ranging from low frequency cycling of power save modes to the higher frequencies of bursted data and continuous data transmission. This AC demand for power can easily excite resonances in the power delivery network creating excessive power rail ripple and extreme dynamic currents. This tutorial demonstrates how to use the PDN impedance peaks to find worst-case dynamic load both in simulation and in measurement. The magnitude and resonant Q of these impedance peaks directly correlate with the potential for power rail noise. Learn how to use EM models combined with measurement-based component models to minimize both power rail voltage ripple and excessive dynamic currents. See how minimizing the PDN impedance can actually increase the impedance peak caused by the package inductance and nicknamed the "Bandini Mountain" by industry. This often-unavoidable impedance peak can be reduced using flat impedance design techniques.

Heidi Barnes is a Senior Application Engineer for High Speed Digital applications in the EEsof EDA Group of Keysight Technologies. Her recent activities include the application of electromagnetic, transient, and channel simulators to solve signal and power integrity challenges. Author of over 20 papers on SI and PI and recipient of the DesignCon 2017 Engineer of the Year. Experience includes 6 years designing ATE test fixtures for Verigy, 6 years in RF/Microwave microcircuit packaging for Agilent Technologies, and 10 years with NASA in the aerospace industry. Heidi graduated from the California Institute of Technology in 1986 with a bachelor's degree in electrical engineering. She has been with Keysight EEsof since 2012.

Steve Sandler has been involved with power system engineering for more than 40 years. Steve is the founder of PICOTEST.com, a company specializing in power integrity solutions including measurement products, services and training. He frequently lectures and leads workshops internationally on the topics of power, PDN and distributed systems and is a Keysight certified expert for EDA software. Steve frequently writes articles and books related to power supply and PDN performance and his latest book, Power Integrity Using ADS was published by Faraday Press in 2019. Steve founded AEi Systems, a well-established leader in worst case circuit analysis and troubleshooting of high reliability systems.

Jack Carrel is an Applications Engineer at Xilinx. He has over 25 years of experience in product development and design in the fields of Instrumentation, Test and Measurement, and Telecommunications. His background includes development of electro-optic modules, Multi-gigabit transceiver boards, high speed and high-resolution data acquisition systems for government and commercial applications. Most recently he has been involved in product design using multi-gigabit transceivers with specific focus on PCB design issues. He has published in several professional publications. Jack received his Bachelor of Science degree in Electrical Engineering from the University of Oklahoma. Jack has been with Xilinx since 2006.

Tutorial II

Title: Challenges and solutions for efficient 3D EM analysis of IC-Package-Board problems

Speaker: Amir Ahmed Asif, Cadence

Abstract: As industry and academia pushes the limits on the speed of data transmission and incorporate more high-speed channels in the products, the challenges keep emerging from all the frontiers of design and simulation. The philosophy of 'divide and conquer' has been in the simulation workflows for a long time. However, to capture the bigger picture and reduce stitching of results, it is desirable that more of the design can stay under the scope of one simulation. At the same time, complex structures are showing the need to shift from 2.5D or hybrid-solvers to the full-wave 3D FEM solvers. These poses a big challenge for 3D EM solvers in terms of capacity, computing resources and performance. In addition to that, preparing a design with multiple technology (IC-Package-Board) to be under one simulation can become a difficult undertaking. In this tutorial, some challenges would be discussed for such scenarios and corresponding solutions with Clarity 3D EM solver and assembly of IC-Package-Board for a simulation would be presented.

Amir A. Asif is actively working with electromagnetic extraction and modeling of electrical designs. He is currently serving as a Lead Product Engineer, focusing on Finite Element Method and Method of Moment tools, at Cadence Design Systems. Previously, he served at Ansys, Inc. and Lorentz Solutions, Inc. Throughout his career, he has worked on EM simulations involving different technologies like IC, Packages, PCB, etc. As a member of simulation tool provider companies, he has collaborated and worked with designers of leading companies in electronics industry. Dr. Asif received his PhD from Clemson University, South Carolina, USA, and Bachelor from Bangladesh University of Engineering and Technology (BUET), Bangladesh.

Tutorial III

Title: Power Delivery Architectures for Next Generation Microprocessors

Speaker: Kaladhar Radhakrishnan, Intel

Abstract: The tutorial will start by covering the basic fundamentals of power delivery. We will then provide an overview of how microprocessor power delivery has evolved over time. The power delivery requirements for the early microprocessors were fairly rudimentary due to the relatively low power levels. However, several decades of exponential scaling powered by Moore's law has greatly increased the power requirements and the complexity of the power delivery scheme. Furthermore, there has been a segmentation of microprocessor design based on their end application from the really low power handheld devices to higher power desktops to power hungry datacenter CPUs and GPUs. Even though the power levels are vastly different, each segment has its own unique power delivery challenges. Microprocessors used to power handheld devices and laptops need to have a very compact power delivery footprint and provide excellent light load efficiency to enhance battery life. General purpose datacenter CPUs need to support high power levels and still maintain a per core granularity in their power supplies to ensure maximum power utilization efficiency for heterogenous workloads. On the other hand GPUs and TPUs designed for massively parallel homogenous workloads require even higher power levels but can often get by without the need for per core granularity in their power supply.The tutorial will cover all the different power delivery architectures that are currently being used in the industry to power the microprocessors. These range from fairly simple PMIC based systems that are popular in smaller handheld devices all the way to a more complex integrated switching regulator implemented on the SoC. There are several power delivery schemes that fall somewhere in between such as moving the VR on to the package or using a simpler linear regulator on die. The tutorial will also look to the future to see where the power delivery requirements are headed and how some of the advanced packaging technologies are being used to come up with novel PD architectures.

Kaladhar Radhakrishnan is a Senior Principal Engineer with the Technology and Manufacturing Group at Intel Corporation. He joined Intel in 2000 after completing his Ph.D. in Electrical Engineering from the University of Illinois at Urbana Champaign. His area of specialization is Power Delivery with an emphasis on Integrated Voltage Regulation, and Magnetics for Power Conversion.

Tutorial IV

Title: Analysis of Direct-to-Package 112G Channels

Speaker: Michael Rowlands, Amphenol

Abstract: The electrical properties of printed circuit board (PCB) materials have a vast range of values. The properties for each material are given on the associated data sheet at one frequency. This is excellent for cross-sectional impedance calculations and first order approximations. However, signal integrity engineers want to know the broadband performance of these materials. Contemporary techniques to calculate PCB material properties use two or more transmission lines of different lengths, forcing engineers to waste valuable time preparing for an outcome which may not even be feasible. We questioned, "Can you do it with one?" It turns out, you can!

This presentation expands the usefulness of an already useful de-embedding technology to extract material properties with just one measurement. The de-embedding technology used in this article was created in association with the IEEE P370 effort and has been adapted to this application. This paper explains how to apply the de-embedding technology to isolate a transmission line of a known length. The methodology here utilizes the work of Jose Moraiea, Heidi Barnes, Marina Kolendensteva and Jason Ellison to extract the relative permittivity of the printed circuit board (PCB) dielectric material and the effective surface roughness with surprising accuracy.

Michael Rowlands is an SI Engineering Manager at Amphenol designing 112G connectors. He specializes in signal integrity at multi-gigahertz frequencies. He received a Bachelor's and Master's degree in Electrical Engineering from MIT in 1998. In his career, he's worked on a variety of system components, inluding cable assemblies, circuit boards, optical chips, chip packaging, twin-axial cable, and connectors. In 2015, he led a team developing carbon-nanotube-based electronics, which won the Molex Innovation Challenge business competition. He holds multiple patents in multi-gigahertz signal design. He has authored and presented technical papers at ECTC, DesignCon, IMAPS, IPC-APEX and PCB East; winning a DesignCon Best Paper Award in 2014 for "Quantitative EMI Analysis of Electrical Connectors Using Simulation Models."

Tutorial V

Title: Graphene-Based Emerging Interconnects - From Physics-Based Deterministic SPICE Models to Uncertainty Quantification

Speakers: Sourajeet Roy, IIT, Roorkee and Rohit Sharma, IIT, Ropar

Abstract: The aggressive scaling of VLSI technology into the sub-22nm levels has rendered conventional copper on-chip interconnects highly susceptible to surface roughness effects, grain boundary scattering, thermal breakdown, and electromigration. Therefore, copper on-chip interconnects are nearing their performance and reliability limits within the 22nm technology node. This has motivated the investigation into new 2D materials such as graphene which can potentially overcome the limitations faced by copper interconnects at sub-22nm nodes. Unfortunately, the coupled quantum physics-EM behavior of such interconnects leads to massive SPICE equivalent circuits whose solution is prohibitively expensive. This problem is further compounded by the fact that in order to study fabrication process variations and perform statistical analysis, we need to perform often tens of thousands of repeated SPICE simulations.

Sourajeet Roy received the Bachelor of Technology degree in electrical engineering from Sikkim Manipal University, Gangtok, India, in 2006, and the M. E. Sc. and Ph.D. degrees in electrical engineering from the University of Western Ontario, London, ON, Canada, in 2009 and 2013, respectively. From 2013 to 2019, he was an Assistant Professor with the Department of Electrical and Computer Engineering, Colorado State University, Fort Collins, CO, USA. Since 2019, he has been working as an Assistant Professor with the Department of Electrical and Communications Engineering, IIT Roorkee, Roorkee, India where he leads the Center for Advanced EDA. His current research interests include signal and power integrity analysis of integrated circuits, machine learning based EDA of electronic packaging, and uncertainty quantification of microwave/RF circuits. Dr. Roy was a recipient of the Vice-Chancellors Gold Medal at the undergraduate level in 2006, the Queen Elizabeth II Graduate Scholarship in Science and Technology in 2012, the Ontario Graduate Scholarship in 2012, and the Graduate Thesis Research Award from University of Western Ontario in 2012. He also received the Early Career Research Award from the Science and Engineering Research Board, Department of Science and Technology India in 2019. He currently serves as a reviewer for various IEEE transaction journals. He is a current Associate Editor for the IEEE Transactions on Components, Packaging, and Manufacturing Technology. He is also a member of the Technical Program Committee for the IEEE Electrical Performance of Electronic Packaging and Integrated Systems Conference and the IEEE Workshop on Signal and Power Integrity.

Rohit Sharma received the B.E. degree in electronics and telecommunication engineering from North Maharashtra University, India, in 2000, the M. Tech. degree in systems engineering from Dayalbagh Educational Institutes, India, in 2003 and the Ph.D. degree in electronics and communication engineering from Jaypee University of Information Technology, India, in 2009. He worked as a Post-Doctoral Fellow at the Design Automation Lab at Seoul National University,

Seoul, Korea from Jan 2010 to Dec 2010. He was a Post-Doctoral Fellow at the Interconnect Focus Centre at Georgia Institute of Technology, Atlanta, USA from Jan 2011 to Jun 2012. Dr. Sharma joined the department of electrical engineering at the Indian Institute of Technology Ropar in 2012, where he is currently an Associate Professor. All along his tenure at IIT Ropar, he has initiated activities in the area of Electronic Packaging. His current research interests include design of high-speed chip-chip and on-chip interconnects, Graphene based nanoelectronic devices and interconnects, Signal and Thermal integrity in high-speed interconnects and 3D ICs/packages and application of Machine Learning in advanced packaging and systems. He is also the coordinator of the Indo-Taiwan Joint Research Centre on Artificial Intelligence and Machine Learning at IIT Ropar. He is an Associate Editor of the IEEE Transactions on Components, Packaging and Manufacturing Technology and a Program Committee member in IEEE EPEPS and IEEE EDAPS. He has been the General Co-Chair of the IEEE EDAPS in 2018. He is the Co-Chair of the IEEE EPS Technical Committee on Electrical Design, Modeling, and Simulation and is a Senior Member of the IEEE.

EPEPS 2020 Technical Program Committee (TPC)

- **Ramachandra Achar,** *Carleton University*
- **Kemal Aygun,** *Intel*
- **Heidi Barnes,** *Keysight*
- **Wendem Beyene,** *Intel*
- **Henning Braunisch,** *Intel*
- **Xu Chen,** *University of Illinois, Urbana*
- **Swagato Chakraborty,** *Mentor Graphics*
- **Paul Franzon,** *North Carolina State University*
- **Dipanjan Gope,** *Indian Institute of Science*
- **Stefano Grivet-Talocia,** *Politecnico di Torino*
- **Xiaoxiong (Kevin) Gu,** *IBM*
- **Jose Hejase,** *Nvidia*
- **Lijun Jiang,** *University of Hong Kong*
- **Roni Khazaka,** *McGill University*
- **Joungho Kim,** *KAIST*
- **Zhen Peng,** *University of Illinois, Urbana*
- **Sourajeet Roy,** *IIT, Roorkee*
- **Albert Ruehli,** *Emeritus IBM; MST*
- **Rohit Sharma,** *IIT, Ropar*
- **Jai Narayan Tripathi,** *IIT Jodhpur*
- **Piero Triverio,** *University of Toronto*
- **Dries Vande Ginste,** *Ghent University*
- **Andreas Weisshaar,** *Oregon State University*
- **Thomas-Michael Winkel,** *IBM*
- **Tzong-Lin Wu,** *National Taiwan University*
- **Ali Yilmaz,** *University of Texas, Austin*
- **Tingdong Zhou,** *Freescale*
- **Yaping Zhou,** *Nvidia*

EPEPS 2020 Paper Review Committee (PRC)

- **Giulio Antonini,** *University of L'Aquila*
- **CK Cheng,** *University of California at San Diego*
- **Paul Dahlen,** *IBM*
- **Daniel DeAraujo,** *Mentor Graphics*
- **Dirk Deschrijver,** *Ghent University*
- **Matt Doyle,** *IBM*
- **Ege Engin,** *San Diego State University*
- **Francesco Ferranti,** *Vrije Universiteit Brussel*
- **Sunil Gupta,** *Qualcomm*
- **Anand Haridass,** *Intel*
- **Shaowu Huang,** *Marvell*
- **Jingook Kim,** *UNIST*
- **Gokul Kumar,** *Sandisk*
- **Naiguang Lei,** *Synopsys*
- **Bowen Li,** *North Carolina State University*
- **Tianjian Lu,** *Google*
- **Antonio Maffucci,** *University of Cassino*
- **Ivan Maio,** *Politecnico di Torino*
- **Paolo Manfredi,** *Politecnico di Torino*
- **Mosin Mondal,** *Mentor Graphics*
- **Zhen Mu,** *Cadence*
- **Rajen Murugan,** *Texas Instruments*
- **Bhyrav Mutnury,** *Dell*
- **Behzad Nouri,** *Carleton University*
- **Vladimir Okhmatovsky,** *University of Manitoba*
- **Pavel Roy Paladhi,** *IBM*
- **Zhiguo Qian,** *Intel*
- **Sameer Shekhar,** *Intel*
- **Eakhwan Song,** *Kwangwoon University*
- **Jairam Sukumar,** *Qualcomm*
- **Junyan Tang,** *IBM*
- **Tao Wang,** *Qualcomm*
- **Boping Wu ,***Huawei*
- **Biancun Xie,** *Qualcomm*
- **Zhuo Yan,** *Apple*
- **Ming Yi,** *Nvidia*
- **Zhen Zhou,** *Intel*

Thank you to all the EPEPS 2020 Sponsors!

PLATINUM SPONSORS

Qualcomm

GOLD SPONSORS

SILVER SPONSORS

EXHIBITORS

SOCIETY SPONSORS

IEEE ELECTRONICS PACKAGING SOCIETY

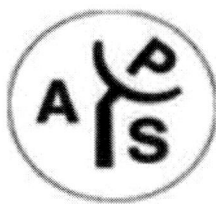

Conference Program for Day 1

7:50 - 8:00 am Pacific, Conference Kickoff

8:00 - 9:00 am Pacific

Keynote I - Meeting the Challenge of Building a Scalable Quantum Computer
(Live Q&A - 8:50 - 9:00 am Pacific)
Jim Held

Intel Corporation

9:00 - 10:20 am Pacific
Session M1A - High-Speed Links I
Session Chair:
Henning Braunisch, *Intel*

- 9:00 - 9:15 am Pacific - **M1A.1. PCIe Gen-5 Design Challenges of High-Speed Servers [115]**
  Mallikarjun Vasa, Chen Ching-Huei, Kumar Sanjay, Ching-Huei Chen, Mutnury Bhyrav
  Dell

- 9:15 - 9:30 am Pacific - **M1A.2. Measurement Uncertainty Propagation in the Validation of High-Speed Interconnects [24]**
  Cemil Geyik*, Michael Hill*, Zhichao Zhang*, Kemal Aygun*, Aberle James+
  *Intel,
  +ASU*

- 9:30 - 9:45 am Pacific - **M1A.3. Rx Equalization for a High-Speed Channel Based on Bayesian Active Learning Using Dropout [128]**
  Xianbo Yang**, Junyan Tang**, Hakki Torun*, Wiren Becker**, , Jose Hejase+, Madhavan Swaminathan*
  *Georgia Tech,
  +Nvidia,
  **IBM*

- 9:45 - 10:00 am Pacific - **M1A.4. Post-Fec BER Performance Analysis for Multi-Stage PAM4 Systems [34]**
  Xiaoqing Dong+, Chunxing Huang*,
  *Zhongzeling Electronics,
  +Xilinx*

- 10:00 - 10:20 am Pacific - **Live Q&A1**
  Panel of Presenters

9:00 - 10:20 am Pacific
Session M1B - Advanced CAD I
Session Chair:
Stefano Grivet-Talocia, *Politecnico di Torino*

- 9:00 - 9:15 am Pacific - **M1B.1. Hyperparameter determination in multivariate macromodeling based on radial basis functions [72]** (Student competition)
  Alessandro Zanco, Stefano Grivet-Talocia
  Politecnico di Torino

- 9:15 - 9:30 am Pacific - **M1B.2. Augmented PEEC for direct time domain thermal and power estimation of Integrated Voltage Regulator Architectures arising in Heterogeneous Integration [2]** (Student competition)
  Venkatesh Avula, Vanessa Smet, Yogendra Joshi, Madhavan Swaminathan
  Georgia Tech

- 9:30 - 9:45 am Pacific - **M1B.3. Accelerated Boundary Element Modeling of Lossy Conductors in Layered Media with a Single-Source Surface Impedance Operator [80]** (Student competition)
  Shashwat Sharma, Piero Triverio
  University of Toronto

- 9:45 - 10:00 am Pacific - **M1B.4. Predictor-Corrector Algorithm with Embedded Dimension Reduction for Uncertainty Quantification of MWCNT On-Chip Interconnect Networks [87]** (Student competition)
  Surila Guglani, Sourajeet Roy
  IIT Roorkee

- 10:00 - 10:20 am Pacific - **Live Q&A2**
  Panel of Presenters

10:20 - 10:30 am Pacific	Break

10:30 - 11:20 am Pacific   Tutorial I
Moderator:
Swagato Chakraborty, *Mentor*
How to Find and Validate Power Rail Resonances
Heidi Barnes**, Steve Sandler**, and Jack Carrel+
*Keysight
**Picotest
+Xilinx*

11:20 - 11:30 am Pacific   Live Q&A1

11:30 am - 12:50 pm Pacific
Session M2A - Power Integrity
Session Chair:
Matt Doyle, *IBM*

- 11:30 - 11:45 am Pacific - **M2A.1. Extracting the Dynamic Current of a Power Delivery Network [65]**
  Heidi Barnes*, Steve Sandler**, Jack Carrel*
  *Keysight Technologies,
  **Picotest
  +Xilinx*

- 11:45 am - 12:00 pm Pacific - **M2A.2. A Novel Multipin Gauss-Newton Method for Performance Evaluation of Decoupling Capacitors [92]**
  Ihsan Erdin*, Ram Achar**
  *Celestica,
  **Carleton University*

- 12:00 - 12:15 pm Pacific - **M2A.3. A Non-Random Exploration based Method for the Optimization of Capacitors in Power Delivery Networks [41]**
  Seunghyup Han, Madhavan Swaminathan
  Georgia Tech

- 12:15 - 12:30 pm Pacific - **M2A.4. A Parallel-in-Time Circuit Simulator for Power Delivery Networks with Nonlinear Load Models [47]**
  Chung-Kuan Cheng*, Chia-Tung Ho*, Chao Jiao+, Xinyuan Wang*, Zhiyu Zeng+, Xin Zhan+,
  *UCSD,
  +Cadence*

- 12:30 - 12:50 pm Pacific - **Live Q&A1**
  Panel of Presenters

10:30 - 11:20 am Pacific   Tutorial II
Moderator:
Zhiguo Qian, *Intel*
Challenges and solutions for efficient 3D EM analysis of IC-Package-Board problems
Amir Ahmed Asif
Cadence

11:20 - 11:30 am Pacific   Live Q&A2

11:30 am - 12:50 pm Pacific
Session M2B - Applied Electromagnetics
Session Chair:
Ram Achar, *Carleton University*

- 11:30 - 11:45 am Pacific - **M2B.1. High-Dimensional Uncertainty Quantification via Active and Rank-Adaptive Tensor Regression [30]** (Student competition)
  Zichang He, Zheng Zhang
  UCSB

- 11:45 am - 12:00 pm Pacific - **M2B.2. Analysis of the Influence of Roughness on the Propagation Constant of a Waveguide via Two Sparse Stochastic Methods [64]** (Student competition)
  Ruben Waeytens, Dries Bosman, Martijn Huynen, Michiel Gossye, Hendrik Rogier, Dries Vande Ginste+
  Ghent University/imec,

- 12:00 - 12:15 pm Pacific - **M2B.3. On the Accuracy of Cross-Talk Modeling in High-Speed Digital Circuits Using the Accelerated Boundary Element Method [108]**
  Dongwei Li, Giacomo Bianconi , Swagato Chakraborty
  Mentor

- 12:15 - 12:30 pm Pacific - **M2B.4. A Comparison of Finite vs. Infinite Plane Models of Reference Conductors in Electronic Packages [105]**
  Yiru Jeong, Ali Yilmaz
  UT Austin

- 12:30 - 12:50 am Pacific - **Live Q&A2**
  Panel of Presenters

12:50 - 1:00 pm Pacific   Day 1 Wrap Up

Conference Program for Day 2

7:50 - 8:00 am Pacific, Welcome

8:00 - 9:00 am Pacific

Keynote II - High Speed and Large Bandwidth Server Computer Bus Links: Past Milestones, Current State of The Art and Future Directions
(Live Q&A - 8:50 - 9:00 am Pacific)
Daniel Dreps

IBM

9:00 - 10:20 am Pacific
Session T1A - High-Speed Links II
Session Chair:
Kevin Gu, *IBM*

- 9:00 - 9:15 am Pacific - **T1A.1. Via Design Optimization for High Speed Differential Interconnects on Circuit Boards** [46]
  Armen Vardapetyan, Ong Chong-Jin
  Intel Corporation

- 9:15 - 9:30 am Pacific - **T1A.2. Signal Integrity Characterization of Channels With Asymmetric Via Stubs** [44]
  Yanyan Zhang, Mahesh Bohra, Nam Pham , Pavel Paladhi, Dale Becker, Daniel Dreps
  IBM

- 9:30 - 9:45 am Pacific - **T1A.3. High-Speed Link Design Optimization Using Machine Learning SVR-AS Method** [93] (Student competition)
  Hanzhi Ma+, Andreas C. Cangellaris*, Er-Ping Li+, Xu Chen*
  *UIUC,
  +Zhejiang University

- 9:45 - 10:00 am Pacific - **T1A.4. ANN Performance for the Prediction of High-Speed Digital Interconnects over Multiple PCBs** [38] (Student competition)
  Katharina Scharff, Christian Morten Schierholz, Cheng Yang, Christian Schuster
  Hamburg University of Technology

- 10:00 - 10:20 am Pacific - **Live Q&A1**
  Panel of Presenters

9:00 - 10:20 am Pacific
Session T1B - Advanced CAD II
Session Chair:
Xu Chen, *UIUC*

- 9:00 - 9:15 am Pacific - **T1B.1. A Shielded-Block Preconditioner for Reduced-Domain Layered-Medium Integral-Equation Methods** [131]
  Chang Liu*, Ali Yilmaz+
  *Cadence
  +UT Austin

- 9:15 -9:30 am Pacific - **T1B.2. An Efficient and Parallel Electromagnetic Solver for Complex Interconnects in Layered Media** [84] (Student competition)
  Damian Marek, Shashwat Sharma, Piero Triverio
  University of Toronto

- 9:30 - 9:45 am Pacific - **T1B.3. On Dissipativity Conditions for Linearized Models of Locally Active Circuit Blocks** [35] (Student competition)
  Tommaso Bradde*, Stefano Grivet-Talocia*, Giuseppe Carlo Calafiore*, Anton Proskurnikov*, Zohaib Mahmood+, Luca Daniel**
  *Politecnico di Torino,
  +NanoSemi, Inc.,
  **MIT

- 9:45 - 10:00 am Pacific - **T1B.4. Uniformly Accurate Electrostatic Layered Medium Green's Function Approximation via Scattered Field Formulation** [127] (Student competition)
  Xinbo Li, Vladimir Okhmatovski
  University of Manitoba

- 10:00 -10:20 am Pacific - **Live Q&A2**
  Panel of Presenters

10:20 - 10:30 am Pacific | **Break**

10:30 - 11:20 am Pacific Tutorial III
Moderator:
Sweagato Chakraborty, *Mentor*
Power Delivery Architectures for Next Generation Microprocessors
  Kaladhar Radhakrishnan
  Intel

11:20 - 11:30 am Pacific Live Q&A1

10:30 - 11:20 am Pacific Tutorial IV
Moderator:
Junyan Tang, *IBM*
Analysis of Direct-to-Package 112G Channels
  Michael Rowlands
  Amphenol

11:20 - 11:30 am Pacific Live Q&A2

11:30 am - 12:00 pm Pacific IEEE EPS TC-EDMS Presentation
Moderator:
Dale Becker, *IBM*

- TC-EDMS Packaging Benchmarks Committee Progress Report 2020
  Fei Guo*, Ali Yilmaz+
  *AMD, +UT Austin

12:00 - 12:50 pm Pacific
Session T2A - 3D Interconnects
Session Chair:
Marco Kassis, *Cadence*

- 12:00 - 12:15 pm Pacific - **T2A.1. 3D Integration of Ka-band RFIC by Inductive Inter-chip Wireless Communication Using Figure-8 Coils** [60] (Student competition)
  Masahiro Usui+, Kota Shiba*, Mototsugu Hamada*, Tadahiro Kuroda*
  *The University of Tokyo,
  +Keio University

- 12:15 - 12:30 pm Pacific - **T2A.2. Estimating Per-Unit-Length Resistance Parameter in Emerging Copper-Graphene Hybrid Interconnects via Prior Knowledge based Accelerated Neural Networks** [98] (Student competition)
  Somesh Kumar*, Sourajeet Roy+, B K Kaushik +, Ramachandra Achar**, Rohit Sharma++, Rahul Kumar++, Likith Narayan S S+
  *ABV-Indian Institute of Information Technology & Management,
  +IIT Roorkee,
  **Carleton University,
  ++IIT Ropar

- 12:30 - 12:50 pm Pacific - **Live Q&A1**
  Panel of Presenters

12:00 - 12:50 pm Pacific
Session T2B - Measurements I
Session Chair:
Ming Yi, *Nvidia*

- 12:00 - 12:15 pm Pacific - **T2B.1. A Review of 90 Degree Corner Design for High-Speed Digital and mmWave Applications** [29]
  Heidi Barnes+, Giovanni Bianchi*, Jose Moreira
  *Advantest,
  +Keysight Technologies

- 12:15 - 12:30 pm Pacific - **T2B.2. Assessment of 2x Thru De-embedding Accuracy for Package Transmission Line DUTs** [77]
  Stephen Smith, Zhichao Zhang, Kemal Aygun
  Intel Corporation

- 12:30 - 12:50 pm Pacific - **Live Q&A2**
  Panel of Presenters

12:50 - 1:00 pm Pacific Day 2 Wrap Up

Conference Program for Day 3

8:00 - 9:00 am Pacific

Invited Presentation - Closing the Loop from Architecture to Post-Silicon for Signal and Power Integrity
(Live Q&A - 8:50 - 9:00 am Pacific)

Vaishnav Srinivas

Qualcomm

9:00 - 10:20 am Pacific
Session W1A - Signal and Thermal Integrity
Session Chair:
Pavel Roy Paladhi, *IBM*

- 9:00 - 9:15 am Pacific - **W1A.1. Accurate BGA Package Solder Joint Modeling for High Speed SerDes Interfaces [71]**
  Jiwei Sun, Zhiguo Qian, Cemil S. Geyik, Kemal Aygun
  Intel Corporation

- 9:15 - 9:30 am Pacific - **W1A.2. Reinforcement Learning-based Auto-router considering Signal Integrity [110]** (Student competition)
  Minsu Kim, Hyunwook Park, Seongguk Kim, Keeyoung Son, Subin Kim, Kyungjune Son, Seonguk Choi, Gapyeol Park, Joungho Kim
  KAIST

- 9:30 - 9:45 am Pacific - **W1A.3. Thermal Sensitivity of Dielectric Materials in High-Speed Designs [104]** (Student competition)
  Sunil Pathania*, Bhyrav Mutnury+, Mallikarjun Vasa**, Vijender Kumar**, Sukumar Muthusamy**, Seema P K**, Rohit Sharma*
  *IIT-Ropar.
  +Dell
  **DellEMC

- 9:45 - 10:00 am Pacific - **W1A.4. A Tunable Neural Network based Decision Feed-back Equalizer model for High-speed Link Simulation [12]** (Student competition)
  Thong Nguyen, Jose Schutt-Aine
  UIUC

- 10:00 - 10:20 am Pacific - **Live Q&A1**
  Panel of Presenters

9:00 - 10:20 am Pacific
Session W1B - Novel Interconnects
Session Chair:
Rohit Sharma, *IIT, Ropar*

- 9:00 - 9:15 am Pacific - **W1B.1. Design, Simulation and Measurement of a Flexible Voltage-controlled Oscillator (VCO) Chip with Bending Radius [102]**
  Seungtaek Jeong*, Seongsoo Lee*, Seokwoo Hong*, Boogyo Sim*, Hyunwook Park*, Subin Kim*, Youngwoo Kim*, Keeyeong Son*, Joungho Ki*, Jaehak Lee+, Junyeop Son+
  *KAIST
  +Korea Institute of Machinery & Materials

- 9:15 - 9:30 am Pacific - **W1B.2. Cost-Effective Implementation of Air Filled Waveguides on Printed Circuit Boards [52]** (Student competition)
  Felix Sepaintner*, Andreas Scharl*, Johannes Jakob*, Florian Keck*, Kevin Kunze*, Franz Roehrl*, Werner Bogner*, Stefan Zorn+
  *Technische Hochschule Deggendorf,
  +Rohde & Schwarz

- 9:30 - 9:45 am Pacific - **W1B.3. A Transmission Line Coupler Component for direct B2B communications [66]**
  Reiji Miura, Tadahiro Kuroda, Mototsugu Hamada
  The University of Tokyo

- 9:45 - 10:00 am Pacific - **W1B.4. Causal Transmission Line Geometry Optimization for Impedance Control in PCBs [85]**
  Zachariah Peterson
  Northwest Engineering Solutions

- 10:00 - 10:20 am Pacific - **Live Q&A2**
  Panel of Presenters

10:20 - 10:30 am Pacific   Break

10:30 - 11:20 am Pacific   **Tutorial V**
Moderator:
Swagato Chakraborty, *Mentor*
Graphene-Based Emerging Interconnects - From Physics-Based Deterministic SPICE Models to Uncertainty Quantification
Sourajeet Roy*, Rohit Sharma+
*IIT, Roorkee
+IIT, Ropar

11:20 - 11:30 am Pacific   Live Q&A

11:30 am - 12:50 pm Pacific
Session W2A - Jitter Noise in High-Speed Links
Session Chair:
Mohiuddin Mazumder, *Intel*

- 11:30 - 11:45 am Pacific - **W2A.1. Analysis of Power Supply Noise Induced Jitter of I/O Subsystems with Multiple Power Domains [114]**
  Hyo-Soon Kang, Ashkan Hashemi, Guang Chen, Xiaoping Liu, Wendemagegnehu Beyene
  Intel

- 11:45 am - 12:00 pm Pacific - **W2A.2. An Inspection Based Method to Analyse Deterministic Noise in N-port Circuits [73]** (Student competition)
  Vijender Kumar Sharma+, Jai Narayan Tripathi*, Hitesh Shrimali+
  *IIT Jodhpur,
  +IIT Mandi

- 12:00 - 12:15 pm Pacific - **W2A.3. Variational Inference approach to Jitter decomposition in High-speed Link [81]** (Student competition)
  Bobi Shi, Thong Nguyen, Jose Schutt-Aine
  UIUC

- 12:15 -12:30 pm Pacific - **W2A.4. Energy-Area Aware Channel Design for Multi-Chip Interfaces [123]**
  Muhammad Waqas Chaudhary*, Andy Heinig*, Bhaskar Choubey+
  *Fraunhofer Institute,
  +Siegen University

- 12:30 - 12:50 pm Pacific - **Live Q&A1**
  Panel of Presenters

10:30 - 11:30 am Pacific   **Live Virtual Booths**

11:30 am - 12:50 pm Pacific
Session W2B - Measurements II
Session Chair:
Heidi Barnes, *Keysight*

- 11:30 - 11:45 am Pacific - **W2B.1. Determine Socket's Inductance and Contact Resistance by Using PRF Method [63]**
  Tao Wang*, Jun Fan+
  *Qualcomm
  +MST

- 11:45 am - 12:00 pm Pacific - **W2B.2. Dual Sided High Frequency Measurement of Microelectronic Packages [48]**
  Sean Christ, Ahmet Durgun, Kemal Aygun, Michael Hill
  Intel

- 12:00 - 12:15 pm Pacific - **W2B.3. SI Model to Hardware Correlation on a 44Gb/s HLGA Socket Connector [129]**
  Pavel Roy Paladhi+, Yanyan Zhang+, Junyan Tang+, Daniel Rodriguez+, Jose Hejase*, Sungjun Chun+, Wiren Becker+, Brian Beaman+, Daniel Dreps+,
  *Nvidia,
  +IBM

- 12:30 - 12:50 pm Pacific - **Live Q&A2**
  Panel of Presenters

12:50 - 1:00 pm Pacific   Conference Wrap-up (Live)

Augmented PEEC for direct time-domain thermal and power estimation of Integrated Voltage Regulator architectures arising in Heterogeneous Integration

Venkatesh Avula*, Vanessa Smet†, Yogendra Joshi† and Madhavan Swaminathan*

*School of Electrical and Computer Engineering
†George W. Woodruff School of Mechanical Engineering
Georgia Institute of Technology
Atlanta, GA 30332, USA
Email: vavula6@gatech.edu

Abstract—An enhanced PEEC model for periodic steady-state analysis is presented. It interprets a dynamical domain as a network of linear and switching elements and represents them with their augmented spectral equivalents. Its computational efficiency is verified on the thermal and power delivery analysis of integrated voltage regulator architectures.

Keywords—Partial Element Equivalent Circuit (PEEC), time-periodic steady-state analysis, heterogeneous integration

I. Introduction

Heterogeneous integration has become an avenue to reach beyond-Moore's law performance. But the performance comes at a cost of multi-physics challenges. Integrated voltage regulator (IVR), for example, has disparate components like embedded passives and fast-switching GaN devices [1] in close proximity. In implementing such a functionally dense architecture, its multi-physics time-domain characterization is necessary. For example, steady-state temperature response [2] under periodic power cycle loading condition indicates how long the architecture can be run continuously at its peak performance before being overheated and subsequently thermal gated [2]. Similarly, voltage droop response [3] of a power delivery network (PDN) under a switching circuit's step load current determines the power integrity of the system-on-chip (SOC). But in obtaining such time-domain responses, traditional transient time-domain methods are prone to long computation time, instability [4], and integration errors [5].

To circumvent such limitations, an enhanced PEEC model capable of periodic steady-state time-domain analysis is presented in this paper. It combines the PEEC method [5] with the augmented spectral equivalent circuit technique [6]. The augmented PEEC method is particularly advantageous for the aforementioned power and thermal computational problems. For the dynamical thermal domain, it directly solves the steady-state temperature response, while the traditional

This work was supported in part by ASCENT, one of six centers in JUMP, a Semiconductor Research Corporation (SRC) program sponsored by DARPA.

transient methods suffer from long computation time due to their time-stepping nature and large thermal time-constant of the domain. On the other hand, for the dynamic power delivery analysis, the augmented PEEC method can model the PDN end-to-end as a single black-box, including its DC voltage source, passive interconnects and load switching circuits of SOC, and directly solve the voltage droop response in a single-step, while the traditional methods need separate black-box models, S-parameter model for the interconnects and piece-wise linear (PWL) model for the load switching circuits. Also, the augmented PEEC model is stable, as the model's system stiffness matrix is sparse, full rank and guaranteed to have an inverse due to the Toeplitz nature of its building blocks.

II. Formulation

Extending the traditional PEEC method's linear time-invariant (LTI) circuit interpretation, the augmented PEEC method considers a given dynamical computational domain as a periodic switched linear (PSL) system. Such a system can be characterized by a Fourier series [7] based bi-frequency transfer function [8]. So we use the Fourier series based augmented spectral equivalent [6] representation for each partial element/variable of the computational domain, assemble the global stiffness matrix from the mesh data of the computational problem, and solve it to determine the unknown periodic steady-state responses of the domain. For the dynamical power and thermal domains considered in this paper, their equivalent PSL systems are DC or '0' frequency excited, which simplifies their augmented representations. Note that the simplified framework given below is analogous to and extends [9] the harmonic balance method for the class of PSL systems.

A. Augmented PEEC computational framework

The augmented framework uses a complex exponential basis set. For a dynamical computational domain or PSL system, the switching frequency of its switch or time-varying source, ω_s, is used as the fundamental frequency of the basis.

The number of positive harmonics of the fundamental, N, determine the fidelity of the steady-state response. For a time-varying variable or switching element of the PSL system, the coefficients of the basis are the Fourier coefficients of its respective time-varying characteristic. The basis expansion for such a variable and switch is given in eqs. 1 and 2 respectively.

$$X(t) = \sum_{n=-N}^{+N} X_n e^{-jn\omega_s t} \quad (1)$$

where $X(t) \in \mathbb{R}$ and $X_n \in \mathbb{C}$ are its Fourier coefficients.

$$Y_{sw}(t) = \sum_{n=-N}^{+N} Y_n e^{-jn\omega_s t} \quad (2)$$

where $Y_{sw}(t) \in \mathbb{R}$ is the switch's time-varying periodic conductance profile, and $Y_n \in \mathbb{C}$ are its Fourier coefficients.

Based on the basis above, the augmented spectral equivalents for the variables, elements and excitation sources of a dynamical domain are listed in Table I. Classic LTI elements-resistors, conductance, capacitors and inductors- become diagonal matrices, while time-varying switches become Toeplitz matrices. Similarly, the time-varying variables and sources of the domain, like voltages and currents, become vectors.

TABLE I
AUGMENTED BLACKBOX REPRESENTATIONS [a]

Element/variable	Augmented spectral equivalent representation
Laplace, s	$\tilde{s} = j\omega_s.diag([-N,\ldots,0,\ldots,N]) \in \mathbb{R}^{D \times D}$
Voltage, V	$\tilde{V} = [V_{-N},\ldots,V_0,\ldots,V_N]^T \in \mathbb{R}^{D \times 1}$
Temperature, T	$\tilde{T} = [T_{-N},\ldots,T_0,\ldots,T_N]^T \in \mathbb{R}^{D \times 1}$
Current, I	$\tilde{I} = [I_{-N},\ldots,I_0,\ldots,I_N]^T \in \mathbb{R}^{D \times 1}$
Heat flux, Q_{sw}	$\tilde{Q}_{sw} = [Q_{-N},\ldots,Q_0,\ldots,Q_N]^T \in \mathbb{R}^{D \times 1}$
DC source, E_{DC}	$\tilde{E}_{DC} = [0,\ldots,E_{DC},\ldots,0]^T \in \mathbb{R}^{D \times 1}$
Scalar, k	$k = k.diag([1,\ldots,1,\ldots,1]) \in \mathbb{R}^{D \times D}$
Resistor, R	$\tilde{R} = R.diag([1,\ldots,1,\ldots,1]) \in \mathbb{R}^{D \times D}$
Inductor, L	$\tilde{L} = L.diag([1,\ldots,1,\ldots,1]) \in \mathbb{R}^{D \times D}$
Capacitor, C	$\tilde{C} = C.diag([1,\ldots,1,\ldots,1]) \in \mathbb{R}^{D \times D}$
Conductance, G	$\tilde{G} = G.diag([1,\ldots,1,\ldots,1]) \in \mathbb{R}^{D \times D}$
Switch S	$\tilde{Y}_{sw} = \begin{bmatrix} Y_0 & Y_{-1} & \cdots & Y_{-2N} \\ Y_1 & Y_0 & \cdots & Y_{-2N+1} \\ \vdots & \vdots & \ddots & \vdots \\ Y_{2N} & Y_{2N-1} & \cdots & Y_0 \end{bmatrix}_{D \times D}$

[a] augmented basis dimension, $D = 2N + 1$ for N positive harmonics

B. Augmented Thermal PEEC

Representing the finite-difference method's (FDM) [10] thermal cells using the augmented matrix forms from Table I and applying Kirchhoff's current law (KCL), the following augmented circuit eq. 3 can be written.

$$[\tilde{s}\tilde{C} + \tilde{G}][\tilde{T}] = [\tilde{Q}_{sw}] \quad (3)$$

where $\tilde{Q}_{sw}$ is the time-varying heat flux load. $\tilde{C}$ and $\tilde{G}$ are the augmented capacitance and conductance.

The solved nodal temperature, $\tilde{T}$, of eq. 3 can then be converted to time-domain profiles, T(t), using eq. 1.

C. Augmented PDN PEEC

The dynamical PDN can be modeled as an equivalent circuit consisting of RLCG elements [11], DC source and a switching load. Using the augmented matrix representation from Table I and applying KCL and Kirchhoff's voltage law (KVL), the following augmented circuit eq. 4 can be written.

$$\begin{bmatrix} \tilde{s}\tilde{C} + \tilde{G} + \tilde{Y}_{sw} & \tilde{A} \\ \tilde{A}^T & -(\tilde{R} + \tilde{s}\tilde{L}) \end{bmatrix} \begin{bmatrix} \tilde{V} \\ \tilde{I} \end{bmatrix} = \begin{bmatrix} \tilde{I}_S = \vec{0} \\ \tilde{E}_{DC} \end{bmatrix} \quad (4)$$

where $\tilde{A}$ is augmented incidence matrix, $\tilde{Y}_{sw}$ is switching load device, $\tilde{I}_S$ is current source, a null vector for our case, $\tilde{E}_{DC}$ is DC excitation voltage source and $\tilde{I}$ is unknown current profile.

The solved nodal voltage, $\tilde{V}$, of eq. 4 can then be converted to time-domain voltage droop profile, V(t), using eq. 1.

III. RESULTS

IVR architectures, shown in Fig. 1, are used as test cases to validate the proposed method and evaluate their performance. For thermal validation, a 500 quad element 2D mesh of the 2D IVR case is run in both the Wolfram Mathematica FEM package having 561 nodal unknowns and the proposed augmented PEEC method having 500×21 or 10500 unknowns with the boundary and loading conditions shown in Figs.1a and 2a. Similarly, for the power domain, 3D IVR package PDN is modeled with Ansys HFSS and is run in Keysight ADS based transient under the 1A current step loading and equivalent circuit conditions shown in Figs. 3a and 3b. 3D IVR's PDN include via and c4 arrays, while the 2d IVR additionally include parallel planes. SOC die is modeled as C_{die} of 50nF with parasitic resistances of 2mΩ. The validation study results are shown in Figs. 2b, 3c, and Table II, while the evaluation study results are in Figs. 2c, 3d and Table III.

IV. DISCUSSION

The validation results, in Figs. 2b and 3c, show perfect match and thus validate the proposed method. Table II shows that it is 2500x faster. On the other hand, typical PDN's time constant is much smaller, so its steady-state response is identical to the transient right from the start, as shown in Fig. 3c. So the method addresses the time-scale challenges in multi-physics. Also, multiple tools like ADS and HFSS were needed to generate the transient, while the augmented PEEC did it seamlessly in a single-step. Multi-physics evaluation results in Table III show that the 3D IVR performs better in power delivery but is thermally-limited. Short and direct current path, as highlighted in Fig. 1b, results in lower voltage droop for the 3D IVR, while the lateral power planes in the 2D IVR, as in Fig. 1a, cause additional loop inductance, resulting in its higher voltage droop. Also, comparing temperature maps, in Figs. 2d and 2e, indicate that the 3D IVR's embedded inductors are in the hotspot, therefore needing advanced thermal management solutions. As part of future work, Numerical inverse Laplace transform (NILT) [9] technique can be explored to capture the initial transient response from 0 to 20s in Fig. 2b. Joule heating caused by electro-thermal coupling can also be explored for a more accurate temperature profile.

978-1-7281-6162-4/20 $31.00 © 2020 IEEE

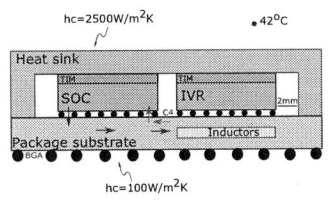 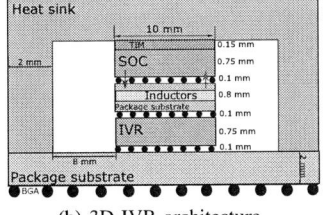

(a) 2D IVR architecture   (b) 3D IVR architecture

Fig. 1. Heterogeneously integrated voltage regulator (IVR) architectures

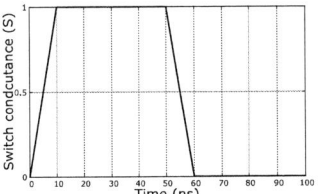

(a) SOC's switching load profile

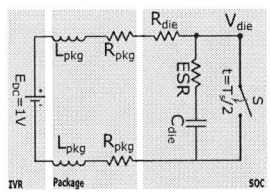

(b) Equivalent circuit representation of the dynamical PDN

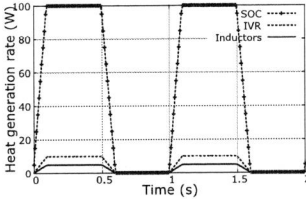 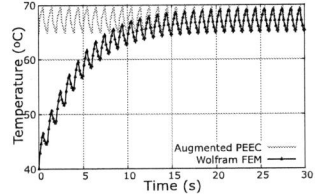

(a) Periodic heat flux loading profiles with time period, T_s=1s

(b) Inductor's response for the 2D IVR case: comparison with reference

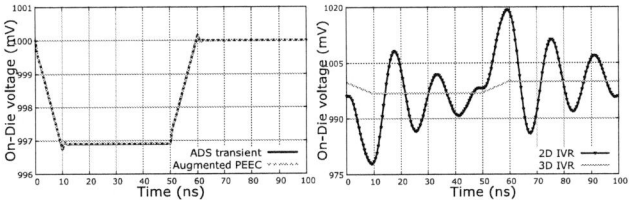

(c) Droop response for the 3D IVR (d) Droop response: 2D vs. 3D IVR case: comparison with reference

Fig. 3. Augmented PEEC based power delivery analysis

V. CONCLUSION

An extended PEEC model capable of direct time-periodic steady-state analysis is presented. From the test cases involving power and thermal evaluation of heterogeneously integrated voltage regulators, this paper shows that the proposed method is seamless and agile for dynamic multi-physics modeling.

REFERENCES

[1] V. Avula and S. Sandler, "Evaluation of Gallium Nitride HEMTs for VRM designs," presented at the DesignCon, 2016.

[2] S. K. Khatamifard, L. Wang, W. Yu, S. Köse, and U. R. Karpuzcu, "Thermogater: Thermally-aware on-chip voltage regulation," in *2017 ACM/IEEE 44th Annual International Symposium on Computer Architecture (ISCA)*, 2017, pp. 120–132.

[3] K. Aygün, M. J. Hill, K. Eilert, K. Radhakrishnan, and A. Levin, "Power delivery for high-performance microprocessors." *Intel Technology Journal*, vol. 9, no. 4, pp. 273 – 283, 2005.

[4] V. Avula and A. Zadehgol, "A novel method for equivalent circuit synthesis from frequency response of multi-port networks," in *2016 International Symposium on Electromagnetic Compatibility - EMC EUROPE*, 2016, pp. 79–84.

[5] A. Ruehli, "Partial element equivalent circuit (PEEC) method and its application in the frequency and time domain," in *Proceedings of Symposium on Electromagnetic Compatibility*, 1996, pp. 128–133.

[6] R. Trinchero, I. S. Stievano, and F. G. Canavero, "Steady-state analysis of switching power converters via augmented time-invariant equivalents," *IEEE Transactions on Power Electronics*, vol. 29, no. 11, pp. 5657–5661, 2014.

[7] T. Strom and S. Signell, "Analysis of periodically switched linear circuits," *IEEE Transactions on Circuits and Systems*, vol. 24, no. 10, pp. 531–541, 1977.

[8] L. A. Zadeh, "Frequency analysis of variable networks," *Proceedings of the IRE*, vol. 38, no. 3, pp. 291–299, March 1950.

[9] R. Trinchero, I. S. Stievano, and F. G. Canavero, "Simulation of buck converters via numerical inverse Laplace transform," in *2017 IEEE 21st Workshop on Signal and Power Integrity (SPI)*, 2017, pp. 1–4.

[10] L. Lombardi, R. Raimondo, and G. Antonini, "Electrothermal formulation of the partial element equivalent circuit method," *International Journal of Numerical Modelling: Electronic Networks, Devices and Fields*, vol. 31, no. 4, p. e2253, 2018, e2253 jnm.2253.

[11] Y. Li et al., "Analysis of multilayer structure near- and far-field radiation by the coupled PP-PEEC and field-equivalence principle method," *IEEE Transactions on Electromagnetic Compatibility*, vol. 61, no. 2, pp. 495–503, 2019.

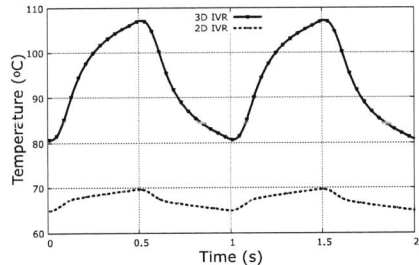

(c) Inductor's response: 2D vs. 3D IVR

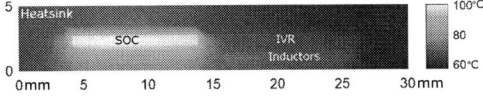

(d) Temperature map of 2D IVR at 0.5s

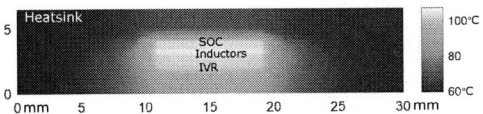

(e) Temperature map of 3D IVR at 0.5s

Fig. 2. Augmented PEEC based thermal analysis

TABLE II
STEADY-STATE RESPONSE COMPUTATION TIME COMPARISON[a]

Test case	Wolfram FEM	Augmented PEEC with N=10
2D IVR	173s	0.07s

[a] run on Intel Core i7 system with 16GB RAM.

TABLE III
IVR ARCHITECTURE EVALUATION USING THE AUGMENTED PEEC [a]

Test case	Voltage droop (mV)	$T_{inductor}$ (°C)
2D IVR	22	70
3D IVR	3.25	107

[a] with number of positive harmonics, N=10 for thermal and 20 for PDN.

978-1-7281-6162-4/20 $31.00 © 2020 IEEE

Extracting the Dynamic Current of a Power Delivery Network

Heidi Barnes
Keysight Technologies
Santa Rosa,California
heidi_barnes@keysight.com

Steve Sandler
Picotest.com
Phoenix, Arizona
steve@picotest.com

Jack Carrel
Xilinx
Rockwell, Texas
jackc@xilinx.com

Abstract—**Measuring dynamic current at a multi-pin load, such as an FPGA, creates significant challenges. There is no practical way of installing current loop probes, or sense resistors in every power and ground pin pair that connect to the load. This paper demonstrates a practical approach for extracting the dynamic current going into a multipin load component on a PCB. The method uses the creation of an accurate EM model with components of the distributed power delivery network (PDN) in combination with measured voltages at the VRM input to the PDN and at the load on the output of the PDN. This method enables power integrity engineers to see the hidden impact of dynamic current on the power rail noise.**

Keywords—Power Integrity, Power Distribution Network, Dynamic Current, Power Rail Noise, Target Impedance

I. INTRODUCTION

Power integrity simulations should be able to predict the dynamic time domain ripple on the power rail. However, this prediction is limited if one does not know the actual dynamic load current. The model in Fig. 1 shows the key elements for a simple power integrity simulation. A voltage regulator module (VRM) supplies the voltage on one side of a PCB power delivery network (PDN), and on the other side is the load represented by the package/die impedance with a dynamic current sink. The challenge is the lack of a good model for the FPGA di/dt dynamic current sink. It is not possible for the consumer of an FPGA component to measure the dynamic current at the die, and even direct measurements of the current going into the FPGA package is problematic with so many power and ground pin connections.

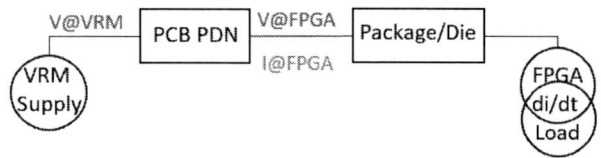

Fig. 1. Key elements of a power integrity simulation: Supply + PCB PDN + Package/Die + Dynamic Load Current.

This paper proposes the use of the easily measured dynamic voltage at the VRM and at the FPGA package pins to synthesize the actual dynamic current at the FPGA pins.

Dynamic voltage measurements can be made with high impedance probes to capture the high frequency transient behavior of the power rail voltage noise ripple [1]. This method of synthesizing dynamic current is shown in Fig. 2 and relies on an accurate model of the PCB PDN and simultaneous measurements of the dynamic voltage at the input and output of the PCB PDN.

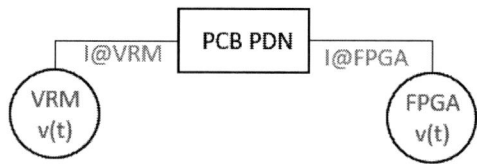

Fig. 2. Dynamic load current at the FPGA is synthesized by simulating the PCB PDN model with measured dynamic voltages at the input VRM side and at the output going to the FPGA load.

The PCB PDN can be modeled as a z-parameter impedance matrix behavioral model. Mathematically this system can be defined by two state equations [2]:

$$I_{VRM} = \frac{V_{VRM}}{Z_{11}} + \frac{V_{VRM}}{Z_{21}}$$

$$I_{FPGA} = \frac{V_{FPGA}}{Z_{21}} + \frac{V_{FPGA}}{Z_{22}} \tag{1}$$

This method of synthesizing dynamic current is demonstrated using the Xilinx Zynq™ ZCU104 FPGA characterization board [3]. The success of the method enables correlation between simulation and measurement to predict worst case noise. The result also highlights that the dynamic voltage ripple is not an accurate predictor of the dynamic noise power and that both dynamic current and voltage are needed.

II. PDN MODEL

The power delivery network consists of three key parts, the power supply, the PCB power delivery network, and the

978-1-7281-6162-4/20 $31.00 © 2020 IEEE

packaged FPGA load. Practical models are needed for each element to predict the time and frequency domain behavior.

A bench top supply with a remote VICOR™ current multiplier module was used as the power supply. This provided a stable low impedance delivery of power to the VCCINT power rail on the ZCU104 board. The output impedance of the power supply was measured and to first order can be represented by a simple series R – L model where R is 3 mOhms measured at low frequency, and the L is 3.33 nH measured in the inductive region at higher frequencies.

The PCB PDN was simulated using Keysight PathWave PIPro FEM simulator to create an EM model of the bare PCB in the form of a multiport s-parameter. This net-based FEM simulator places ports at the pins of the components attached to the VCCINT power rail and then solves for the multiport s-parameter over the frequency of interest. The model of the bare PCB PDN without capacitors was verified with measurements, and then PDN capacitor models attached. It is worth noting that each capacitor value was measured and in every case the vendor model included too much of the mounting inductance that is already a part of the EM model. For this reason, measurements under bias were used to create user models with the mounting inductance removed.

The package/die model can also be represented by an S-parameter supplied by the vendor. To simplify the initial test case and minimize the port count the power pins are grouped together and the ground pins are grouped together to look at the total dynamic current going into the FPGA. The dynamic current load at the die is also grouped to a single output pin to allow a total dynamic load to be applied at the die location. Connecting the package/die s-parameter to an open, short, and matched 5 milliohm load on the PCB input side shows how the PCB impedance can impact the impedance that the die sees, Fig. 3. Matched impedance provides the lowest overall impedance profile, while extremely low impedance increases the impedance peak of the package/die inductance resonating with the die capacitance. It is the frequency of this impedance peak that will be shown to be the worst case noise [4].

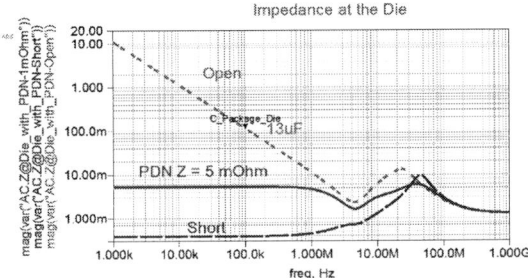

Fig. 3. Connecting the package die model to a PCB open, short, and matched 5 mOhm load shows how the impedance seen by the die is influenced by attaching the PCB PDN. There is no influence above 100MHz.

Connecting these three parts of the power integrity ecosystem together, power supply, PCB PDN, and package/die, shows an impedance profile in Fig. 4 that has an impedance peak at 30 MHz. Measurements actually miss seeing this impedance peak since it is internal to the package/die and measurement locations suffer from the inductance of vias and interconnects.

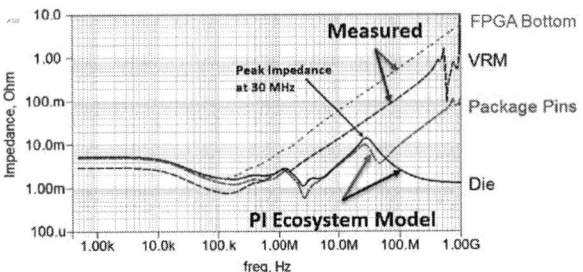

Fig. 4. The PI Ecosystem Model uncovers the impedance peak at 30 MHz that was not seen with measurement. The impedance peak indicates the potential for worst case noise.

This full model of the VCCINT PDN highlights the potential for worst case noise when the impedance peak at 30 MHz is excited with a forcing function at that frequency. The Zynq FPGA has the ability to be programmed to switch all 400,000 flipflops on and off at various clock rates to create a dynamic current load. Two specific clock rates were compared 30 MHz with an average current of 2.5 Amps, and 200 MHz with an average current of 13.5 Amps.

III. VALIDATION WITH SIMULATION

To validate the method of synthesizing dynamic current, the method was simulated by attaching a simulated 30 MHz, 2.5 Amp load at the die. The simulation schematic test bench is shown in Fig. 5. This full model of the VCCINT PDN with input 0.87 volts and dynamic current load then generated the dynamic voltages at the input and output of the PCB PDN EM model.

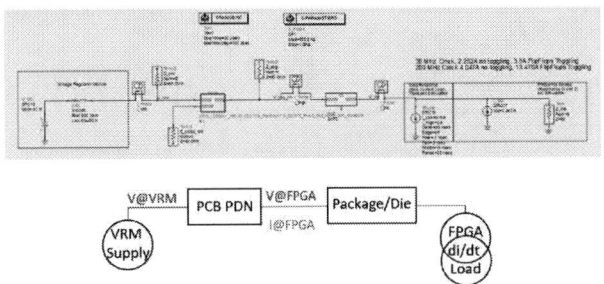

Fig. 5. Test bench for validating the methodology for synthesizing current. A simulated dynamic load is used to simulate the dynamic voltages across the PCB PDN.

The dynamic voltages were then used in a simulation to excite the PCB PDN EM model without the power supply and package/die model. The dynamic current at the input to the FPGA was synthesized from this simulation and then compared with the simulated currents from initial full VCCINT PDN model with dynamic load. The results in Fig. 6 show excellent correlation, even capturing the start-up behavior as the load starts switching.

978-1-7281-6162-4/20 $31.00 © 2020 IEEE

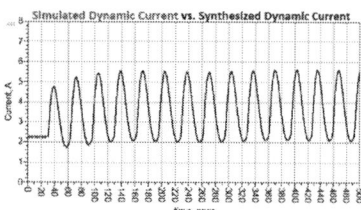

Fig. 6. Simulated currents using the test bench with known simulated currents in red matches well with the synthesized currents in blue.

IV. MEASUREMENT RESULTS

After validating the method with simulated dynamic voltages, the next step is to apply the methodology using measured dynamic voltages. Although it is not possible to measure the dynamic current at the FPGA, it is possible to measure dynamic current coming from the power supply. Measuring the current at the VRM can be used as an additional cross check with the synthesized currents. An AC coupled Rogowski current loop was chosen with its small size and minimal impact on the inductance of the power supply connection.

The results in Fig. 7 of the cross check with measured total current at the VRM supply with that of the synthesized total current at the VRM are limited by the performance of the Rogowski probe [5]. However, to first order it shows the similarity of measured current to synthesized current for the step load response when turning on 400,000 flip flops with a 200 MHz toggling rate.

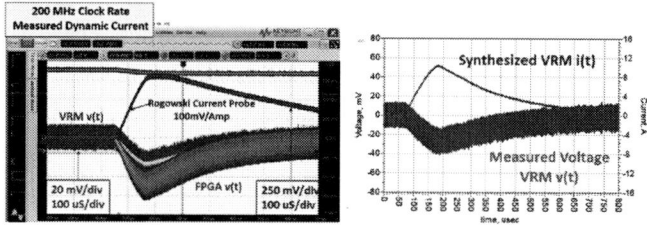

Fig. 7. Total current from the VRM supply was measured with an AC coupled Rogowski probe and compared with the synthesized current from the measured voltages across the PCB PDN.

The VRM current isn't nearly as interesting as what is going on at the FPGA pins. To highlight the benefit of being able to synthesize the dynamic current at the FPGA pins, two toggling rates for the load are used, 200 MHz with 13.5 Amps, and 30 MHz with 2.5 Amps. The results shown in Fig. 8 confirm the predicted worst-case measured voltage ripple is at 30 MHz, not 200 MHz even though the average current is 5 times less.

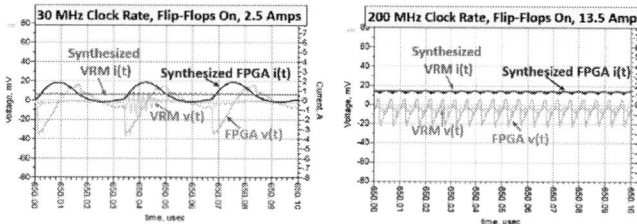

Fig. 8. Results showing synthesized dynamic current at the FPGA in blue for 30 MHz and 200 MHz. The impedance peak at 30 MHz correctly predicts worst case noise for both dynamic current and voltage.

Looking closely at the delta change in noise ripple going from 200 MHz to 30 MHz, as shown in Fig. 9, reveals that the measured voltage ripple is on the order of 2 times worse, but the synthesized current ripple is more than a factor of 10. This means that the AC noise power increased by a larger percentage then what the voltage ripple alone would indicate. Designers worried about EMI/EMC compatibility will benefit from this technique of looking at the noise power and not just step load voltage ripple.

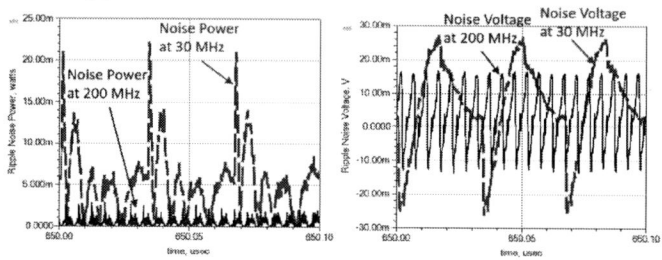

Fig. 9. Multiplying the voltage ripple times the current ripple to get AC noise power shows a factor of 10 increase at 30 MHz compared to 200 MHz, while the voltage is only twice as large.

V. CONCLUSION

The method of synthesizing the total dynamic current going into the FPGA from two dynamic voltage measurements across the PCB PDN works. The method relies on an accurate model of the full PDN which includes an EM model of the PCB and measured models of the capacitors. This full model has the additional benefit of making it easy to identify the frequency of impedance peaks with the potential for worst case noise. Comparing dynamic voltage ripple and dynamic current ripple at two different load frequencies shows that dynamic voltage alone does not indicate the full magnitude of the noise power.

ACKNOWLEDGMENT

The authors would like to acknowledge Darshan Moodgal Devendrappa and the Xilinx Power Integrity lab's assistance in the measurements of the ZCU104 characterization board.

REFERENCES

[1] S. M. Sandler, *Power Integrity*, McGraw-Hill Education, 2014.

[2] J. Fasig, C. White, B. Gilbert, C. Haider, "Introduction to Non-Invasive Current Estimation (NICE)" DesignCon, 2018.

[3] H. Barnes, J. Carrel, S. Sandler, "A Method for Dynamic Load Current Testing with a Bench Power Supply" DesignCon, 2020.

[4] S. M. Sandler, "Designing Power for Sensitive Circuits", EDICon, 2017.

[5] S. Sandler, "Faster-Switching GaN : Presenting a number of interesting measurement challenges," in IEEE Power Electronics Magazine, vol. 2, no. 2, pp. 24-31, June 2015, doi: 10.1109/MPEL.2015.2420232.

On Dissipativity Conditions for Linearized Models of Locally Active Circuit Blocks

T. Bradde[*], S. Grivet-Talocia[*], G. C. Calafiore[*], A. V. Proskurnikov[*], Z. Mahmood[†], L. Daniel[‡]

[*] Dept. Electronics and Telecommunications, Politecnico di Torino, Italy

[†] NanoSemi, Inc., USA

[‡] Dept. Electrical Engineering and Computer Science, Massachusetts Institute of Technology, USA

Abstract—This paper generalizes the concept of dissipativity of linear models to linearized models of nonlinear circuit blocks that may also include locally active behavior. We show that such models can be guaranteed to behave as dissipative, provided they are subjected to certain bounds on the small-signal input amplitude, and provided that total voltage and current signals (including bias) are considered in the energy balance. Potentially severe incorrect dynamic behaviour can result from the violation of such bounds, as illustrated through a linearized reduced-order model of a low dropout voltage regulator.

I. Introduction

Many nonlinear circuit blocks, such as amplifiers or low-dropout voltage regulators, are designed to operate under small-signal conditions at well-defined bias points. In such conditions, linearized small-signal models can safely replace full transistor-level descriptions during system-level verification via numeric simulations. Standard linear Model Order Reduction procedures can be applied to compress such models (while controlling the approximation error), hence enabling drastic reduction in transient simulation runtime. Speedup of up to three orders of magnitude have been reported in [1]–[4].

However, the above numerical simulation flow can be safely adopted only when the amplitude of the small-signal inputs is below a given threshold δ, which defines a trust region for accuracy. For larger inputs, the accuracy of the numerical results cannot be guaranteed, mainly due to the linearization error in the small-signal model extraction.

In this work, we consider a second and more subtle limitation arising from energy balance considerations. We show that even if small-signal models behave as locally active (e.g. power amplifiers), they are in fact dissipative when their dynamics are described using total voltages and currents (including bias terms) at all ports (including power supply ports). Such models are henceforth called *affine linearized models*. This is explained by the observation that any circuit block that does not include independent sources is dissipative, since energy is only provided from the external environment through power supply. A locally active behavior arises due to internal energy redistribution from the DC bias to the small-signal components.

Following the theory in [5], we provide a characterization of the energy behavior of affine linearized models. It is shown that dissipativity holds only when the amplitude of the small-signal input components is bounded by some threshold γ, generally larger than the accuracy threshold δ. We refer to this behavior as *Bounded Input Dissipativity (BID)*. For small-signal inputs larger than γ, dissipativity no longer holds. In such case, a linearized model-based simulation may lead to completely wrong results with self-sustained oscillations or exploding signals, due to the ability of the linearized model to provide an indefinite amount of power or energy. Therefore, the qualitative behavior of true transistor-level circuit and its linearized model may be dramatically different when BID does not hold. We illustrate this concept through a simple example of a low dropout voltage regulator, whose linearized model is driven to instability by a small-signal input with an amplitude that violates the limits imposed by proposed BID conditions.

II. Background: Affine Linearized Models

Let us consider a generic nonlinear dynamic M-port system, driven by inputs $u \in \mathbb{R}^M$ with resulting outputs $y \in \mathbb{R}^M$. Without loss of generality, we consider a voltage-controlled setting, so that u collects the total port voltages and y the corresponding port currents. We assume normal sign reference at all ports, so that the instantaneous power entering the M-port is $p(t) = u(t)^\top y(t)$.

We assume that the circuit block is biased by an external power supply, set at a stable operating point. The corresponding DC values of input and output are denoted as U_0 and Y_0, respectively. Assuming small-signal operation, we split total voltages and currents as

$$u(t) = U_0 + \tilde{u}(t), \quad y(t) = Y_0 + \tilde{y}(t), \tag{1}$$

where $\tilde{u}(t)$ and $\tilde{y}(t)$ are small-signal components. After a linearization process followed by a model order reduction, one obtains the following behavioral state-space representation

$$\begin{aligned} \dot{\tilde{x}}(t) &= A\,\tilde{x}(t) + B\,\tilde{u}(t), \quad \tilde{x}(0) = 0 \\ \tilde{y}(t) &= C\,\tilde{x}(t) + D\,\tilde{u}(t), \end{aligned} \tag{2}$$

where the small-signal states $\tilde{x}(t)$ are induced by the particular algorithm used to derive the model. In this work, we use for instance a data-driven approach [1] where the circuit block is first characterized through its small-signal AC responses via SPICE runs, which are then subjected to a rational approximation based on Vector Fitting [6] followed by a state-space realization [7]. The small-signal transfer function of (2) is

$$\tilde{H}(s) = D + C(sI - A)^{-1}B. \tag{3}$$

978-1-7281-6162-4/20 $31.00 © 2020 IEEE

Model (2) is intended to replace the complete transistor-level circuit block in numerical simulations. Therefore, not only the small-signal dynamics, but also the DC bias levels must be correctly included. This is possible by augmenting (2) by an affine term in the output equation

$$\dot{x}(t) = Ax(t) + Bu(t), \qquad x(0) = X_0$$
$$y(t) = Cx(t) + Du(t) + Y_C, \tag{4}$$

where

$$Y_C = Y_0 - (CX_0 + DU_0) \quad \text{and} \quad X_0 = -A^{-1}BU_0. \tag{5}$$

We call (4) the *affine linearized model*. Unlike system (2), this model operates on total voltages and currents, with the initial condition $x(0)$ defined by the operating point through (5). This is the standard setting employed in all circuit simulation environments, in particular SPICE.

III. BOUNDED INPUT DISSIPATIVITY

The total input power flow into model (4) is

$$p(t) = u(t)^\top y(t) = (U_0 + \tilde{u}(t))^\top (Y_0 + \tilde{y}(t)), \tag{6}$$

where the DC power component is $P_0 = U_0^\top Y_0$. For all cases of practical interest $P_0 > 0$, so that the model receives DC power from its environment when small-signals are zero.

The pure small-signal model (2) may be locally active, so that the transfer function (3) is *not* Positive Real (this is the case, e.g., for power amplifier circuits). It is then conceivable that the small-signal component of the power is $\tilde{p}(t) = \tilde{u}(t)^\top \tilde{y}(t) < 0$, as an indication of being locally active. Increasing the amplitude of $\tilde{u}$ leads to a (quadratic) increase of the small-signal power. Therefore, it is to be expected that the full model (4) based on total signals (DC + small signals) exhibits a transition in behaviour from dissipative when small signals are turned off, to non-dissipative when small-signals exceed an amplitude bound γ. This is precisely the concept of *Bounded Input Dissipativity* that we discuss in this work.

The BID characterization requires a number of technical arguments [5], we recall here only the main points in view of the illustrative example of Sec. IV. Model (4) is *dissipative* according to the classical definition [8] if there exists a storage function of the state $E(x) \geq 0$ (here assumed differentiable for simplicity) such that for each solution of (4) one has

$$\frac{d}{dt}E(x(t)) \leq p(t) \quad \forall t \geq 0. \tag{7}$$

Equivalently, the rate of increase of stored energy cannot exceed the total power received from the environment.

Let us consider the class of one-sided input signals with bounded amplitude and superimposed bias,

$$u(t) = U_0 + \tilde{u}(t), \quad \|\tilde{u}(t)\|_2 \leq \gamma \; \forall t, \; \tilde{u}(t) \equiv 0 \; \forall t \leq 0. \tag{8}$$

Subject to such inputs, the states x of (4) are driven by the system dynamics within a region that is usually denoted as *reachability set* R_γ, so that $x(t) \in R_\gamma$ for all $t \geq 0$. It can be shown that this set can be expressed as $R_\gamma = X_0 + R_\gamma^0$, where R_γ^0 is the reachability set of the small-signal model (2) subject

to small-signal inputs $\tilde{u}$. To enable further derivations, this set is here overbounded by an ellipsoid $R_\gamma^0 \subseteq \{\tilde{x} : \tilde{x}^\top W \tilde{x} \leq \gamma^2\}$, with matrix $W \succ 0$ determined below.

In order to check dissipativity when inputs are restricted through (8), it is sufficient to produce a storage function $E(x)$ in the domain R_γ and enforce (7). Inspired by the linearity of (4) (see also [5]) we consider as possible candidates, quadratic storage functions

$$E(x) = \frac{1}{2}x^\top P x + q^\top x \tag{9}$$

where $P = P^\top$ is a symmetric matrix (not necessarily sign definite), and q is a vector. This choice enables a direct and convenient algebraic characterization of BID conditions, generalizing the celebrated Kalman-Yakubovich-Popov (KYP) Lemma [9]–[11]. Our main result is expressed as the following

Theorem 1: [5] The affine linearized system (4) is BID with input amplitude level $\gamma > 0$, if a matrix P and a vector q exist such that the following inequality holds

$$\tilde{z}^\top \Sigma_0(P)\tilde{z} + 2\theta_0(P,q)^\top \tilde{z} + \phi_0 \leq 0,$$
$$\forall \tilde{x}, \tilde{u}: \quad \tilde{x}^\top W \tilde{x} \leq \gamma^2, \; \tilde{u}^\top \tilde{u} \leq \gamma^2 \tag{10}$$

where

$$\Sigma_0 = \begin{bmatrix} A^\top P + PA & PB - C^\top \\ B^\top P - C & -D - D^\top \end{bmatrix}, \quad \tilde{z} = \begin{bmatrix} \tilde{x} \\ \tilde{u} \end{bmatrix}$$
$$\theta_0 = \begin{bmatrix} A^\top(PX_0 + q) - C^\top U_0 \\ B^\top(PX_0 + q) - D^\top U_0 - Y_0 \end{bmatrix}, \tag{11}$$
$$\phi_0 = -2U_0^\top Y_0,$$

and $W \succ 0$ is a matrix that obeys the matrix inequality for some scalar $\alpha \geq 0$,

$$\begin{bmatrix} A^\top W + WA + \alpha W & WB \\ B^\top W & -\alpha I \end{bmatrix} \preceq 0. \tag{12}$$

We remark that setting $U_0 = 0$ and $q = 0$ reduces (10) to the standard KYP Lemma, which deals with $\gamma = \infty$, i.e. the standard notion of dissipativity for linear systems. The advocated generalization includes additional non-homogeneous terms that represent the DC power contribution (term ϕ_0) and the coupling between DC and AC powers (term θ_0).

Theorem 1 provides purely algebraic conditions expressed in terms of Linear and Bilinear Matrix Inequalities, whose verification can be performed through dedicated solvers. Three different algorithms are presented and discussed in [5]. In practice, one can first fix a value for γ and then verify whether the conditions in Theorem 1 are feasible for that value. An iterative search on γ can then be performed to find the largest $\gamma_{\max}$ for which BID can be established. When exceeding this amplitude, the system may not behave as dissipative and numerical simulation issues could very easily be expected. An illustrative example follows.

IV. AN EXAMPLE

We consider a basic system simulation involving a Low Drop-Out (LDO) regulator circuit based on the design presented in [12] and implemented in a 40 nm CMOS process. A

978-1-7281-6162-4/20 $31.00 © 2020 IEEE

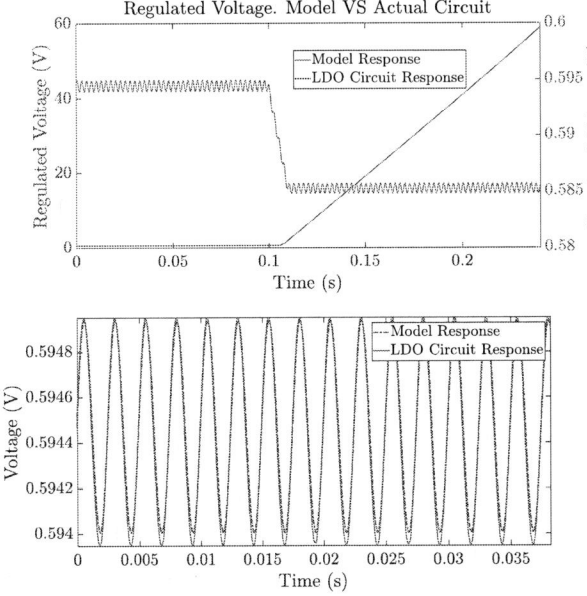

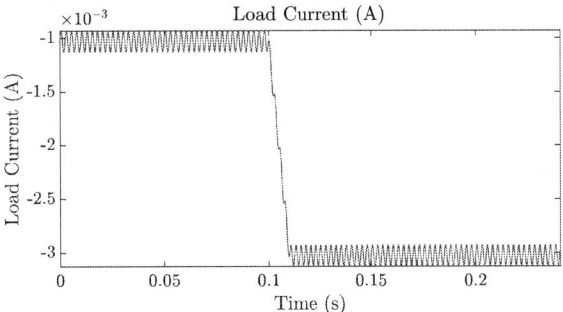

Fig. 1. Transient voltage of the LDO obtained with transistor-level circuit (red) and reduced model (blue). Top panel: model response diverges when Bounded-Input Dissipativity is lost (around $t = 0.1\,\text{s}$). Bottom panel: zoomed view for early time.

Fig. 2. Transient load current applied to the LDO. The amplitude of the small-signal current $\tilde{\imath}_L$ undergoes a jump at $t = 0.1\,\text{s}$ such that its amplitude exceeds the maximum allowed BID bound $\gamma_{\max}$.

physically-consistent energy behavior of the linearized models. We suggest that such conditions should be embedded in next-generation circuit and system solvers that employ behavioral reduced-order models of active circuit blocks, in order to provide automated self-consistency checks at runtime. This might avoid running models outside their validity limits, thus preventing designers to draw incorrect conclusions about the behavior of their systems.

linearized reduced order model of order 9 was generated from small-signal AC responses of a full layout extraction including parasitics, considering the operating point

$$U_0 = \begin{bmatrix} V_{\text{DD}} \\ -I_{\text{L}} \end{bmatrix} = \begin{bmatrix} 1.1\,\text{V} \\ -1.02\,\text{mA} \end{bmatrix}, \quad Y_0 = \begin{bmatrix} I_{\text{DD}} \\ V_{\text{L}} \end{bmatrix} = \begin{bmatrix} 1\,\text{mA} \\ 0.594\,\text{V} \end{bmatrix},$$

where V_{DD} is the DC input voltage and I_{L} is the DC load current. The allowed small-signal bound for model dissipativity, $\gamma_{\max} = 0.062$, is computed in 3 s. Both model (few kB netlist size) and transistor-level circuit (30 MB netlist size) were used in a system-level transient simulation, that included a diode in combination with an input capacitor in order to prevent possible reverse currents that could damage the regulating device [13]. As the load current fluctuations increase beyond $\gamma_{\max}$, the model starts to inject energy into its terminations (Fig. 2). This energy flow causes an uncontrolled increase of the input voltage, and consequently, of the regulated voltage at Port 2 (Fig. 1, top panel, blue line). This phenomenon is not observable when the actual transistor-level circuit is employed in the simulation (Fig. 1, top panel, red line), despite the response of the model being extremely accurate with respect to the transistor-level circuit (Fig. 1, bottom panel).

V. CONCLUSIONS

Linearized models of active circuit blocks can be safely employed in system-level transient simulations only when the amplitude of the small-signal inputs are constrained to a well-defined trust region. Beyond accuracy considerations, we have shown that the concept of Bounded Input Dissipativity and the associated conditions may lead to an a priori determination of the maximum input bound $\gamma_{\max}$ that guarantees a

REFERENCES

[1] S. B. Olivadese, G. Signorini, S. Grivet-Talocia, and P. Brenner, "Parameterized and DC-compliant small-signal macromodels of RF circuit blocks," *IEEE Trans. Components, Packaging and Manufacturing Technology*, vol. 5, no. 4, pp. 508–522, 2015.

[2] T. Bradde, P. Toledo, M. De Stefano, A. Zanco, S. Grivet-Talocia, and P. Crovetti, "Enabling fast power integrity transient analysis through parameterized small-signal macromodels," in *International Symposium on Electromagnetic Compatibility*, Barcelona, Spain, 2019, pp. 759–764.

[3] B. Nouri, M. S. Nakhla, and X. Deng, "Stable model-order reduction of active circuits," *IEEE Transactions on Components, Packaging and Manufacturing Technology*, vol. 7, no. 5, pp. 710–719, May 2017.

[4] B. Nouri and M. S. Nakhla, "Efficient time-domain sensitivity analysis of active networks," *IEEE Transactions on Components, Packaging and Manufacturing Technology*, vol. 9, no. 9, pp. 1721–1729, Sep. 2019.

[5] T. Bradde, S. Grivet-Talocia, G. C. Calafiore, A. V. Proskurnikov, Z. Mahmood, and L. Daniel, "Bounded input dissipativity of linearized circuit models," *IEEE Trans. Circuits and Systems I*, in press.

[6] B. Gustavsen and A. Semlyen, "Rational approximation of frequency domain responses by vector fitting," *IEEE Trans. Power Delivery*, vol. 14, no. 3, pp. 1052–1061, 1999.

[7] S. Grivet-Talocia and B. Gustavsen, *Passive Macromodeling: Theory and Applications*. John Wiley & Sons, 2015.

[8] J. C. Willems, "Dissipative dynamical systems part I: General theory," *Archive for Rational Mechanics and Analysis*, vol. 45, no. 5, pp. 321–351, 1972.

[9] B. D. O. Anderson and S. Vongpanitlerd, *Network Analysis and Synthesis*. Prentice-Hall, 1973.

[10] V. Popov, "Absolute stability of nonlinear systems of automatic control," *Autom. Remote Control*, vol. 22, no. 8, pp. 857–875, 1962.

[11] V. Yakubovich, "Solution of some matrix inequalities encountered in the nonlinear control theory," in *Dokl. Akad. Nauk SSSR*, vol. 156, no. 2, 1964, pp. 278–281.

[12] T. Y. Man, P. K. Mok, and M. Chan, "A high slew-rate push–pull output amplifier for low-quiescent current low-dropout regulators with transient-response improvement," *IEEE Trans. Circuits Syst. II: Express Briefs*, vol. 54, no. 9, pp. 755–759, 2007.

[13] M. Sellers, "LDO basics: Preventing reverse current." [Online]. Available: https://e2e.ti.com/blogs_/b/powerhouse/archive/2018/07/25/ldo-basics-preventing-reverse-current-in-ldos

978-1-7281-6162-4/20 $31.00 © 2020 IEEE

Energy-Area Aware Channel Design for Multi-Chip Interfaces

Muhammad Waqas Chaudhary, Andy Heinig
Fraunhofer Institute for Integrated Circuits IIS
Division Engineering of Adaptive Systems EAS
Zeunerstr. 38, 01069 Dresden, Germany
{muhammad.chaudhary, andy.heinig}@eas.iis.fraunhofer.de

Bhaskar Choubey
Chair of Analogue Circuits
and Image Sensors Siegen University
Hölderlinstr. 3, 57076 Siegen, Germany
bhaskar.choubey@uni-siegen.de

Abstract—**Multi-chip communication interfaces on an interposer or a package substrate must consume minimum routing area while consuming low power in the transceiver blocks. This paper presents an algorithm to design this channel in view of energy and area metrics for a given transceiver topology. It is then show-cased using an example of silicon interposers.**

Index Terms—**2.5D/3D interconnects and packages, electronic packages and microsystems, high-speed channels**

I. INTRODUCTION

Moore's law is reaching a communication bottleneck in 2D systems, which has led to development of multi-chip systems to further enhance the system performance [1]. Such a memory-processor system on an interposer is shown in Figure 1. These chips must transfer high speed data between each other which requires high speed chip-to-chip interfaces [2]. These transceivers are, however, designed for a specific channel represented by scattering (S) parameters. They are then optimized at circuit level to achieve minimum power consumption for given interconnect at required data rate [3]. However, for optimal space utilisation in multi-chip systems, the routing area is an important constraint which should be co-optimized with the transmitter or at least optimized for a given transceiver architecture.

A co-design of area and current mode logic driver was previously presented [4]. However, it does not consider the equalization needs of the transceiver and the required power consumption. Lho et al. describe an optimization approach for high speed channel, but do not consider relationship with technology node, equalization requirements, and combined energy-area performance [5]. This paper meets these needs through an algorithm for combined optimization of transceiver and channel for minimum energy-area costs.

II. DESIGN FLOW AND ALGORITHM

An overview diagram of design flow of the communication channel is shown in Figure 2. It consists of an extensive interconnect characterization, which is then used to derive the transceiver design constraints, especially with regards to drive strength, impedance matching and equalization. The energy consumption of transceiver with various interconnects is used to develop a combined performance metric of routing area and energy consumption. One can then derive the minimum energy-area measured by the performance metric

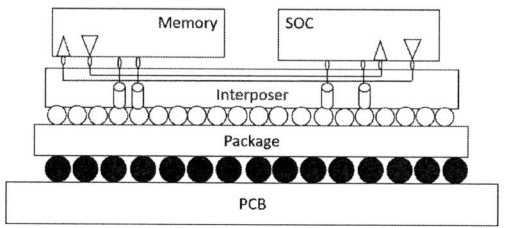

Fig. 1. Multi-chip interposer system model

of pJ/bit · μm (product of energy efficiency pJ/bit and signalling pitch μm) for given data rate, type of transceiver, substrate material and interconnect length.

This design flow is guided by the fact that while increasing the width of the interconnect leads to lower interconnect insertion loss, it also increases the signal routing pitch ρ, measured in μm. The very first step in the flow is hence, to characterize interconnects with various widths (W) and spacings (S) for a given length and substrate material. The interconnect s-parameters are then evaluated for a given data rate per wire (GSG) in single ended systems and per two-wires (GSSG) in differential signalling transceiver architectures.

The decrease in interconnect width leads to higher transceiver energy consumption, while an increase leads to higher signalling pitch. This flow of Figure 2 hence requires detailed analysis in each step and therefore an optimisation algorithm. This algorithmic approach with this flow would lead to an optimal channel design for a given substrate, bandwidth and transceiver topology. Such algorithm is derived in this paper. The algorithm is designed to be holistic and keeps the fixed constraints to as minimum as possible. In addition to the design space discussed earlier, it also considers different kinds of signalling topologies, and their correlation with channel area consumption along with total interface power consumption. This should therefore, provide an overall system level optimization.

To derive the algorithm, let us consider T as the set of possible transceiver topologies. This would include source series terminated signalling (SST), low voltage swing terminated logic (LVSTL), high swing push-pull signalling (CMOS) $T \in \{SST, LVSTL, CMOS\}$. The power consumption ϕ for a given transceiver topology $T_i \in T$ is a function of signalling pitch ρ defined by interconnect width, spacing and ground

978-1-7281-6162-4/20 $31.00 © 2020 IEEE

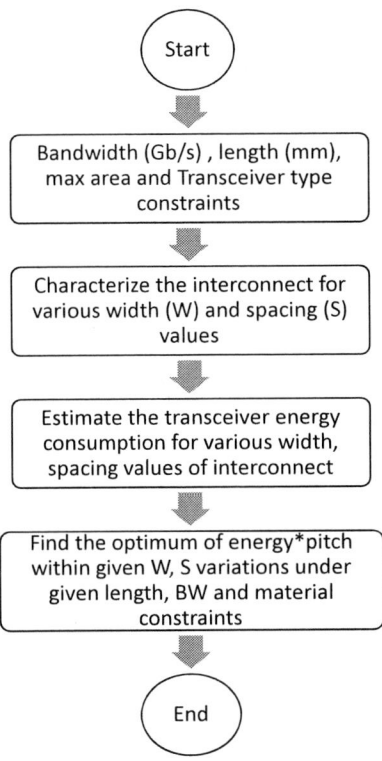

Fig. 2. Channel design flow

Algorithm 1: Channel design

Result: Optimum solution $T_{opt}, W_{opt}, S_{opt}$;
define Width range: $W = \{W_{min}, \ldots, W_{max}\}$;
define Spacing range: $S = \{S_{min}, \ldots, S_{max}\}$;
define Transceiver types: $T_i \in T$;
define Data bit rate: f_b;
define Interconnect average length: L;
initialize ψ_{old};
while $T_i \in T$ **do**
  for $W \leq W_{max}$ **do**
    for $S \leq S_{max}$ **do**
      find S-parameters for given W, S;
      find pulse response for given f_b;
      find required number of Taps for Tx;
      find required number of DFE Taps for Rx;
      calculate power consumption in Tx, Rx as
        $\phi_{Tx} = [\phi_{Drv} + \phi_{Eq} + \phi_{Ser} + \phi_{Ckbuf}]$
        $\phi_{Rx} = [\phi_{buf} + \phi_{Eq} + \phi_{DeSer} + \phi_{Ckbuf}]$
      calculate signalling pitch as
        $\rho = W + W_{min} + 2S$;
      calculate interface energy-area cost as
        $\psi = \frac{\phi}{f_b}(W + W_{min} + 2S)$;
      if $\psi < \psi_{old}$ **then**
        $T_{opt} = T_i, W_{opt} = W, S_{opt} = S$;
      end
      update $\psi_{old} = \psi$;
    end
  end
end

---

width. This would be a sum of the transmitter and receiver power consumption as $\phi_{T_i} = \phi_{Tx} + \phi_{Rx}$ where

$$\phi_{Tx} = [\phi_{Drv} + \phi_{Eq} + \phi_{Ser} + \phi_{Ckbuf}]$$
$$\phi_{Rx} = [\phi_{buf} + \phi_{Eq} + \phi_{DeSer} + \phi_{Ckbuf}]$$

Here ϕ_{Drv}, ϕ_{Eq}, ϕ_{Ser} and ϕ_{DeSer} represent the driver power, equalization, serialization and de-serialization blocks, respectively; while ϕ_{Ckbuf} denotes the clock buffering and distribution block. The back-end blocks in transmitter and receiver like the serializer, de-serializer, clock buffers and samplers are indirectly influenced by the interconnect width and spacing variations. They are rather defined by the transmitter and receiver front-ends, i.e. driver, receiver amplifier and equalization. The energy-area metric ψ is hence, given as $\phi/f_b * \rho$ in $pJ/bit \cdot \mu m$ where f_b is the data bit rate in Gb/s.

The range of width W is defined with a minimum W_{min} and maximum W_{max} values in given interconnect technology. The spacing between interconnects is restricted by the minimum value S_{min} and generally does not go above a few times of the width of the signal line, e.g. $3 \times W$. For single ended GSG signalling using minimum width ground interconnect, the signalling pitch is given as $\rho = W + W_{min} + 2S$. The final energy-area performance metric ψ is then given as

$$\psi(T_i, \rho) = \frac{\phi}{f_b}(W + W_{min} + 2S)$$

The algorithm iterates exhaustively through all possible combinations of width, spacing and transceiver topologies to find the minimum energy-area cost combination of (W, S, T).

III. CASE STUDY: SILICON SUBSTRATE CHANNEL

To understand the algorithm, a silicon interposer chip to chip interface is being presented as a case study. The stackup for this system is shown in Figure 3, where two metal layers in silicon-dioxide are placed on a silicon substrate. The tangent loss ($tan\delta$), here is dependent upon the resistivity, which for typical $100\,\Omega \cdot cm$ is chosen to be 0.1 for data rates around 5-10 GHz [6]. The length of the interconnect is selected as 10 mm. The impact of width variation on the channel insertion loss S21 from 1 to 2 μm is shown in Figure 4. The data rate for this study is chosen as $10\,Gb/s$ which has the Nyquist frequency of 5 GHz, at which 2 μm wide line has frequency dependent loss of only $-2\,dB$ while 1 μm has insertion loss of $-7\,dB$. It should be noted that there is 6 dB higher DC loss in 1 μm wide line which leads to a reduced voltage swing at the Rx input.

To study the equalization and voltage swing requirements, the channel is excited at the transmitter side with a $10\,Gb/s$ pulse with ideal rise time (1 ps) and unit interval (UI) of 0.1 ns. The received pulse response after channel is shown in Figure 5. As expected due to high resistivity of interconnect and DC loss, the voltage swing is just 0.2 V for 1 μm wide line.

From the pulse response in Figure 5, it can be observed that there is no pre-cursor inter symbol interference (ISI) for both lines. The signal rises within 1UI completely, as depicted

978-1-7281-6162-4/20 $31.00 © 2020 IEEE

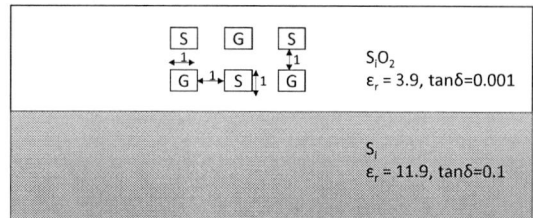

Fig. 3. Stackup for silicon interposer based multi-chip system

by the dotted blue line at 1UI tick of x-axis. However, both interconnects show some post-cursor ISI, as shown by the red dashed lines. The behavior is similar to an RC exponential voltage drop, especially significant in $1\,\mu m$ wide line. In order to completely cancel the post-cursor ISI, a high continuous time linear equalization (CTLE) or a number of decision feedback equalization (DFE) taps will be required, which shall impact the power consumption of the transceiver. For $1\,\mu m$ wide line, at least two DFE taps for 2nd and 3rd UI ISI cancellation are required. For $2\,\mu m$ line, however, only 1-tap DFE equalization for 2nd UI ISI cancellation is enough.

The power consumption estimate for different data rates and equalization requirement is based upon the work by Palaniappan et al. in [3]. A Continuous-Time Linear Equalizer (CTLE) based post-cursor ISI cancellation is generally used up to $12\,dB$ insertion loss. This is due to the fact that CTLE is a part of Rx input amplifier and hence, increases the power equally for signal and the noise. Therefore, for significantly higher losses than $12\,dB$, DFE taps are used which are calculated directly from the impulse response shown in Figure 5.

By using CTLE for equalization and $0.1\,mW/Gb/s$ power for every $6\,dB$ bandwidth peaking [3] at $10\,Gb/s$ in $90\,nm$ technology node, extra ϕ_{Eq} of $1\,mW$ is added to the total power consumption ϕ_{Rx} of $1\,\mu m$ wide wire interface as compared to the $2\,\mu m$ wire interface. The equalization constraints in CTLE and DFE will directly impact other design parameters for driver, samplers and clock buffers in Tx and Rx. However, if we ignore them for quick comparison of energy-area metric and just consider CTLE requirements, ψ

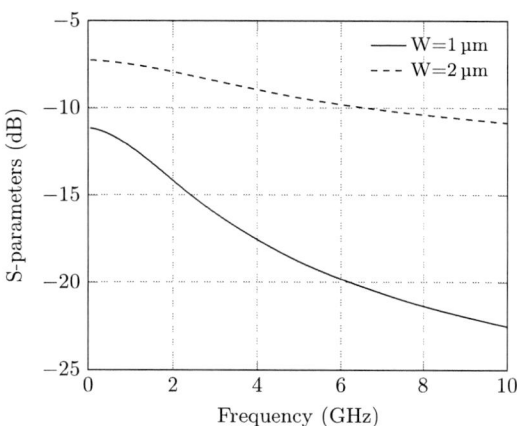

Fig. 4. S-parameters extracted using HSPICE 2D field solver

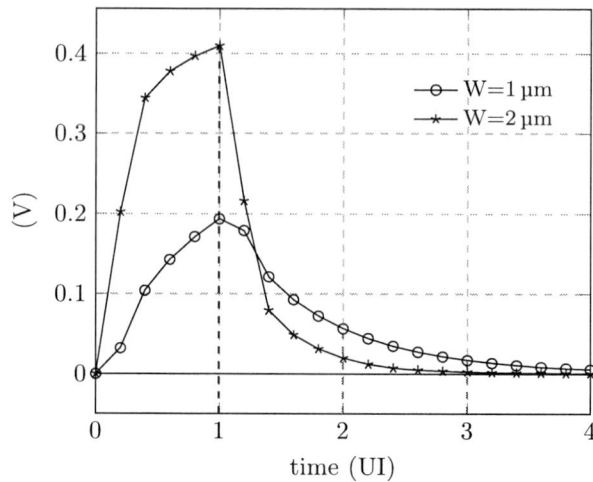

Fig. 5. Response for $10\,Gb/s$ input pulse with $1\,ps$ rise time

would be 0.7 and $0.8\,pJ/bit \cdot \mu m$ for 1 and $2\,\mu m$ wide wire interface respectively.

This shows that at specific data rates and equalization requirements, higher power consumption is not that detrimental if overall energy-area cost metric is used. However, noting the CTLE power consumption being constant for 6 and $12\,dB$ peaking in $45\,nm$ technology [3], the ψ for wider $2\,\mu m$ interconnect interface would be an even worse choice than in $90\,nm$ node. This leads to a conclusion that wider lines for high speed chip to chip links are useful from energy-area perspective in older technology nodes. But for newer nodes in the range of 45, 20 and $14\,nm$, thin wires with high receiver side equalization requirements are a better choice.

IV. Conclusion

A design flow for energy-area aware channel design for high speed chip to chip links is presented using silicon interposer interface case study. The flow shows that energy-area trade-offs can lead to an optimized interconnect width and spacing for given data rate, transceiver type and technology node.

References

[1] P. Vivet, E. Guthmuller et al., "2.3 a 220gops 96-core processor with 6 chiplets 3d-stacked on an active interposer offering 0.6ns/mm latency, 3tb/s/mm 2 inter-chiplet interconnects and 156mw/mm 2 @ 82%-peak-efficiency dc-dc converters," in *2020 IEEE International Solid- State Circuits Conference - (ISSCC)*. IEEE, 2/16/2020 - 2/20/2020, pp. 46–48.

[2] T. O. Dickson, Y. Liu et al., "A 1.4 pj/bit, power-scalable 16×12 gb/s source-synchronous i/o with dfe receiver in 32 nm soi cmos technology," *IEEE Journal of Solid-State Circuits*, vol. 50, no. 8, pp. 1917–1931, 2015.

[3] A. Palaniappan and S. Palermo, "A design methodology for power efficiency optimization of high-speed equalized-electrical i/o architectures," *IEEE Transactions on Very Large Scale Integration (VLSI) Systems*, vol. 21, no. 8, pp. 1421–1431, 2013.

[4] M. W. Chaudhary and A. Heinig, "Co-design of cml io and interposer channel for low area and power signaling," in *Formal proceedings of the 2016 IEEE 19th International Symposium on Design and Diagnostics of Electronic Circuits & Systems (DDECS)*, J. Brenkuš and Stopjaková, Eds. IEEE, 2016, pp. 1–6.

[5] D. Lho, J. Park et al., "Bayesian optimization of high-speed channel for signal integrity analysis," in *EPEPS 2019*. IEEE, 2019, pp. 1–3.

[6] R.-Y. Yang, C.-Y. Hung et al., "Loss characteristics of silicon substrate with different resistivities," *Microwave and Optical Technology Letters*, vol. 48, no. 9, pp. 1773–1776, 2006.

978-1-7281-6162-4/20 $31.00 © 2020 IEEE

Dual Sided High Frequency Measurement of Microelectronic Packages

Sean R. Christ
Intel Corporation
Chandler, AZ, USA
sean.r.christ@intel.com

Ahmet C. Durgun
Intel Corporation
Chandler, AZ, USA
ahmet.c.durgun@intel.com

Kemal Aygün
Intel Corporation
Chandler, AZ, USA
kemal.aygun@intel.com

Michael J. Hill
Intel Corporation
Chandler, AZ, USA
michael.j.hill@intel.com

Abstract—Traditionally, validation of simulation data of microelectronic packages has been done by high frequency measurement of coplanar transmission line structures. Although this metrology has matured over the years, it is not directly applicable to product packages since they also include vertical interconnects. To address this issue, we developed a dual-sided measurement metrology, by utilizing a special fixture design and employing short-open-load-reciprocal calibration to maintain measurement stability at high frequencies. Metrology capability analysis shows that there is minimal operator and environmental condition dependency and a very good correlation with simulations up to 40 GHz for differential transmission lines.

Keywords—dual-sided measurement, package measurement, S-parameter characterization, calibration techniques

I. INTRODUCTION

Methods for S-parameter characterization of packaged coplanar transmission line structures, using a Vector Network Analyzer (VNA) and radio frequency (RF) microprobes, are well-established. Turnkey probing solutions for test vehicles with coplanar devices, which include probe station, probe arms and manipulators, and vacuum fixtures for the device under test (DUT), are widely available from several manufacturers. In the case of a product package though, where function and cost dictate the layout of data transmission interfaces, vertical interconnects also play a fundamental role in addition to coplanar interconnects. Therefore, the ability to characterize these vertical interconnects provides valuable insight into the real performance of processor packaging for input/output interfaces such as dual data rate memory and peripheral component interconnect express links. Moreover, S-parameter data measured from the VNA can be plugged into 3D, frequency-dependent computer models to predict various parameters of an electrical system's performance.

In contrary to the coplanar high frequency measurements, measurement of top-to-bottom, through-package structures presents a unique set of physical and logistical challenges that have yet to be thoroughly addressed and quantified. A stable platform that securely supports the DUT, yet allows unimpeded probe access to both top and bottom must be implemented. Additionally, a calibration algorithm that makes use of an unknown thru standard must be used to calibrate the VNA, and must be studied to quantify the error contributions from calibration, multiple operators, and environmental factors. In the literature, there have been a few attempts to address these issues. For instance, in [1], the calibration is performed on a horizontal platform and then the structure is rotated for dual-sided measurement. This approach not only puts the quality of calibration in question, but also suffers from potential instability of the fixture during landing. Other attempts in [2-4] successfully handle the calibration and stability problems by utilizing a short-open-load-reciprocal (SOLR) calibration [5] on vacuum based horizontal platform. However, the provided data are only for single ended (SE) frequency domain measurements of vertical structures such as plated through hole (PTH) vias, pogo pins, and through silicon vias (TSV). Besides, the simulation to measurement correlation is sub-optimal, particularly for frequencies higher than 20 GHz.

In this paper, we extend the previous studies by providing a thorough metrology capability analysis (MCA) [6] for top-to-bottom measurement of differential ended (DE) product-like structures, comprised of probe pads, microvias, PTH vias, and horizontal routing. Moreover, a measurement-to-model correlation, which is supported by detailed cross-sectioning of the DUT, is performed to validate the accuracy of these dual-sided S-parameter measurements up to 40 GHz.

II. METROLOGY

A typical probe station serves as the basis of the physical measurement setup as shown in Fig. 1. The probe station has been modified to accept a bottom-mounted microscope for viewing the bottom-side probes. A probe positioner, whose nosepiece has been flipped and machined for clearance, manipulates the bottom-side probe. The top-side probe is attached to an unmodified positioner. Differential microprobes in GSGSG configuration with a pitch of 150 microns were used to probe the DUT. The measurement device is a 4-port VNA with a maximum frequency of 67 GHz. Four coaxial cables with 1.85mm connectors at each end connect the VNA ports to the probes. Commercially available software is used to calibrate the VNA and download the measured data.

The dual-sided measurement fixture shown in Fig. 2 was designed in-house and uses a combination of off-the-shelf and custom pieces. Its X-Y-Z stage attaches to the probe station with a vacuum base, and the front of the fixture is supported by mechanically-adjustable leveling feet. Spring-loaded steel bars provide stability and support for the DUT, while allowing easy adjustment and flexibility in positioning.

To ensure the most accurate measurement results, the probes and cables must remain in the same position during measurement as they were during calibration. To accomplish

978-1-7281-6162-4/20 $31.00 © 2020 IEEE

Fig. 1.   Dual-sided measurement in progress.

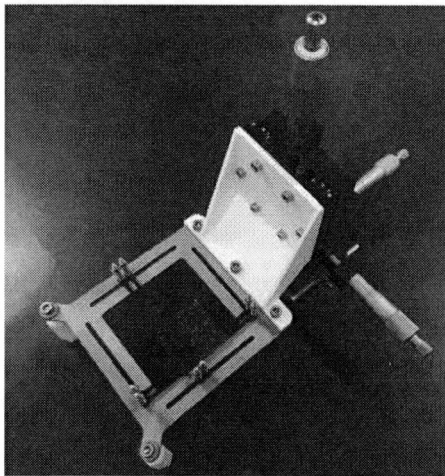

Fig. 2.   Dual-sided measurement fixture detail.

this, one would need a well-defined dual-sided thru structure (which is not commonly available) and use a line-reflect-reflect-match or a short-open-load-thru algorithm. Alternatively, a calibration algorithm that does not require a well-defined thru standard can be used. SOLR algorithm allows the use of the DUT as the reciprocal standard for top-to-bottom transmission paths, while using the normal short, open, load, and loopback structures found on a calibration substrate as the remaining calibration standards. To allow full access to all of the standards on the calibration substrate, and to ensure that it was able to be supported by all 4 support bars of the fixture, the calibration substrate was mounted to a larger carrier substrate. This also allowed easy handling and flipping of the substrate throughout the calibration process.

To quantify the performance of the measurement system, an MCA was performed. The study consists of a repeatability study to explore the variability from the measurement device and the probe landing, and a reproducibility study that explores the variability due to differences between operators and environmental differences over multiple days. The repeatability study involves one operator performing a single setup and calibration of the measurement system followed by 30 dynamic measurements (land, measure, lift, repeat) on the DUT in a single session. The reproducibility study involves 3 operators, each performing his own system calibration and measurement of 3 different DUTs, for each of 3 days.

III. RESULTS AND DISCUSSIONS

Fig.3 illustrates the test structure, which is composed of horizontal and vertical interconnects as in a product package, with solder ball pads replaced by probe pads.

Analysis of the repeatability data showed very little variability due to the measurement device and the operator's probe landing technique. Viewing the differential ended insertion loss (DEIL) and differential ended return loss (DERL) plots from the repeatability study showed very little difference across the 30 measurements in the data set. This gives great confidence in the stability of the measurement system and the ability of a well-trained operator to consistently achieve the same result.

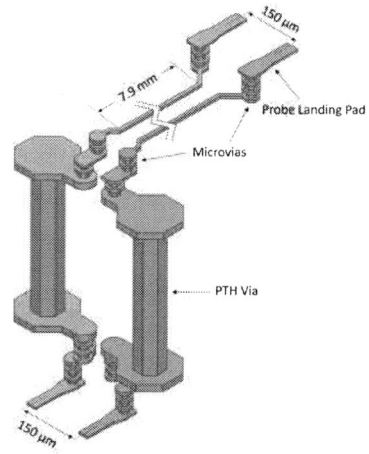

Fig. 3.   Detailed geometry of the test structure.

Analysis of the reproducibility study data gives a more realistic view of the tolerance that the measurement system and a group of operators can achieve. The DERL and DEIL plots are shown in Figs. 4 and 5. Variation in the results generally increases with frequency. A post-MCA study suggests that the variation in return loss is the result of differences in probing technique between operators. Differences in initial and final probe position, and probe skate distance all seem to have a noticeable impact on DERL, but have very little impact on DEIL.

While S-parameter measurement provides a full characterization of a device from probe to probe, all aspects of the routing (probe pad, trace, PTH) are lumped into one data set, making it difficult to see the electrical contribution of each piece of the routing. Therefore, plotting the device's time domain impedance is a useful method of intuitively viewing the contributions of each piece of a device's routing. Using a commercial circuit simulator, the S-parameters from the reproducibility study were converted to a TDR impedance response using a source with a rise time of 35 ps. Fig. 6 shows the processed TDR curves. One can intuitively map the top probe pad region, horizontal routing, and PTH to the curves shown.

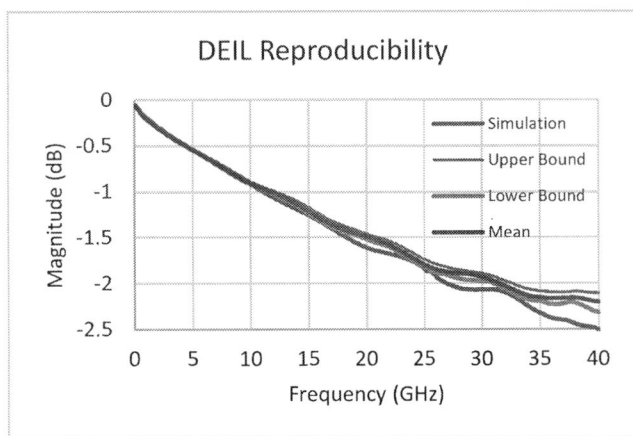

Fig. 4. DEIL reproducibility results.

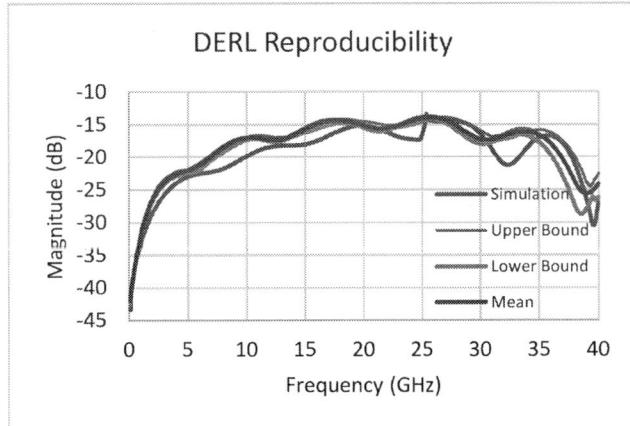

Fig. 5. DERL reproducibility results.

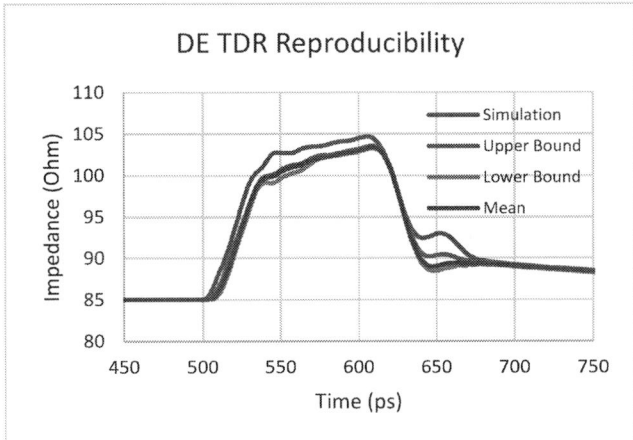

Fig. 6. DE TDR reproducibility results.

Figs. 4-6 also show the comparison of the measurements with simulations where cross-section images were utilized for accurate dimensioning of the design features. The measured dimensions are then loaded into a 3D model of the package, and a high-frequency solver simulates the S-parameters of the model. The simulated S-parameters are then compared to the measured S-parameters. It can be seen that the simulation slightly overpredicts IL compared to the measurements. This is

desired to cover the worst case scenario in the performance estimations. The measured and simulated RL are also in a good agreement. Finally, the TDR impedance comparison between simulations and measurements show less than 2 Ω difference for the horizontal routing part.

IV. CONCLUSIONS

It is possible to characterize top-to-bottom packaged interfaces by performing modifications to commercially available probe stations and probe positioners, and designing a custom fixture to allow dual-sided probing of the DUT. A VNA calibration algorithm that uses an unknown thru standard (SOLR) is well-suited to allow the probes to remain in the same position during calibration and measurement, ensuring that the VNA calibration remains accurate throughout the measurement process. Results of an MCA study show that the measurement system is capable of excellent measurement repeatability. Examining the reproducibility results of the MCA, it can be seen that DEIL variation is very small and not influenced by operator or environmental factors, while DERL shows more variation due mostly to operator probing technique. Measurement-to-model agreement for DEIL, DERL, and TDR impedance is achieved, further supporting the quality of measurement with the dual-sided probing system.

ACKNOWLEDGMENT

The authors would like to thank Kuang Liu, Raul Marquez, Leigh Wojewoda, and Gaurang Choksi for their contributions.

This research was developed with funding from the Defense Advanced Research Projects Agency (DARPA) under CHIPS program (Contract Number: HR00111790020). The views, opinions and/or findings expressed are those of the author and should not be interpreted as representing the official views or policies of the Department of Defense or the U.S. Government.

REFERENCES

[1] T. Burcham, P. McCann, and R. Jones, "Double sided probing structures," U.S. Patent No. 8,013,623. Sep. 6, 2011.

[2] K. Lu et al., "Vertical interconnect measurement techniques based on double-sided probing system and short-open-load-reciprocal calibration," 2011 IEEE 61st Electronic Components and Technology Conference (ECTC), Lake Buena Vista, FL, 2011, pp. 2130-2133.

[3] C. H. Lin, C. Liu, H.-K. Huang, K.-C. Fan and H. H. Lee, "Electrical model analysis of RF/high-speed performance for different designed TSV patterns by wideband double side measurement techniques," 2012 7th International Microsystems, Packaging, Assembly and Circuits Technology Conference (IMPACT), Taipei, 2012, pp. 72-75.

[4] J. Park et al., "High-Frequency Electrical Characterization of a New Coaxial Silicone Rubber Socket for High-Bandwidth and High-Density Package Test," in IEEE Transactions on Components, Packaging and Manufacturing Technology, vol. 8, no. 12, pp. 2152-2162, Dec. 2018.

[5] A. Ferrero and U. Pisani, "Two-port network analyzer calibration using an unknown 'thru'," in IEEE Microwave and Guided Wave Letters, vol. 2, no. 12, pp. 505-507, Dec. 1992.

[6] M. J. Hill and L. E. Wojewoda, "A Study of Permittivity Measurement Reproducibility Utilizing the Agilent 4291B," in IEEE Transactions on Advanced Packaging, vol. 29, no. 4, pp. 714-718, Nov. 2006

Post-FEC BER Performance Analysis for Multi-stage PAM4 Systems

Xiaoqing Dong
Xilinx
Singapore
adong@xilinx.com

Chunxing Huang
Zhongzeling Electronics
Shenzhen, China
nickhuang168@163.com

Abstract— In a multi-stage link system where forward error correction (FEC) encoding/decoding is only used in the end devices of the end-to-end link, mixed mode errors (random and burst errors) from each sub-link stage collectively impact the end-to-end link performance. This paper introduces an analytical method for evaluating RS-FEC performance in multi-stage PAM4 systems. Typical application scenarios include links with retimer(s), and electrical-optical-electrical (E-O-E) links such as 200GAUI-n/400GAUI-n systems. Simulations are performed to study the impact of sub-link stages on the overall system post-FEC BER performance.

Keywords— *RS-FEC, multi-stage link, burst error, BER, SER, end-to-end performance*

I. Introduction

Forward error correction (FEC) has been widely used in high speed 4-level PAM signaling (PAM4) system links. Both the 56G and 112G PAM4 standards adopt Reed-Solomon codes for error correction in long reach (LR), middle reach (MR) and very short reach (VSR) applications [1]. Unlike in single-stage links where FEC error correction is performed within single-stage closed systems, a multi-stage link system usually consists of multiple active components along the signal path, FEC encoding/decoding is only performed by the devices on both ends of the end-to-end link, the middle active components do not terminate errors in each sub-link stage for power and latency reasons.

For end-to-end FEC performance evaluation for such multi-stage systems, two aspects need to be considered: (1) the active components such as SerDes, the system level impairments such as insertion loss (related to inter symbol interference) or crosstalk (related to noise), are usually quite different in each sub-link stage, this leads to different error distribution signatures seen by the final stage FEC and largely determines error correction performances; (2) Coding techniques such as Reed-Solomon (RS) symbol distribution, FEC codewords interleaving and PMA bit-level multiplexing as defined in 200Gb/s and 400Gb/s Ethernet [1], play an important role in such multi-stage link systems and have great impact on the post-FEC BER performances.

Combining the two aspects, part II starts with the basic process of a FEC performance simulation methodology that takes into consideration of all possible error patterns from DFE burst effect, followed by the introduction of an analytical model for typical multi-stage FEC applications. FEC codewords interleaving, RS symbol distribution and bit-level multiplexing which are typically applied in multi-lane, multi-stage Ethernet systems are modeled. The algorithmic model is also scalable to handle the cases where correlated errors triggered in very short time periods are averaged out in a longer time window and yields an averaged raw BER, thus the commonly seen FEC simulations could be on the optimistic side.

II. Simulation Model for Multi-stage FEC Link

A. Analytical FEC performance simulation method

High speed links are typically mixed error systems: randomly occurred errors triggered by impairments such as inter symbol interference (ISI), random noise, jitter, etc., and burst errors that have dependencies on the previously occurred errors, both exist. In the scenarios where decision feedback equalizer (DFE) is used, once a wrong decision has occurred at the slicer, the error affects the voltage level of the next NRZ bit/PAM4 symbol with certain probability through the feedback loop and resulting in error propagation. For NRZ the maximum error propagation probability is 0.5 and for PAM4, it is 0.75.

For one-tap DFE architecture, there are only two error patterns in the DFE register. Probabilities of single burst error with different lengths can be calculated using error propagation probability Pb, through equation (1), where Pr represents the raw BER, and i is burst length:

$$P_{burst\ error}(i) = Pr \cdot Pb^{i-1} \cdot (1 - Pb) \quad (1)$$

In general, for an M–tap DFE ($M \geq 1$), single burst error propagation probabilities of the $2^{\wedge}M$ error patterns can be calculated using detection SNR or vertical bathtub curves.

For RS FEC, as error correction is performed in units of RS symbols, burst errors of different lengths shall translate into the corresponding numbers of RS symbol errors, and the corresponding symbol error ratio (SER) is calculated. Post-FEC Frame Error Ratio (FER) and BER are then calculated according to the maximum correctable number of RS symbols of the RS code. Let i be the number of RS symbol errors in a FEC frame. For RS (n,k,t):

$$FER = \sum_{i=t+1}^{n} SER(i) \quad (2)$$

B. Analytical model for multi-stage and multi-lane systems

A multi-stage FEC system usually consists of multiple cascaded sub-link stages and for one or more of the sub-link stages, there could be multiple lanes in parallel carrying RS symbols. FEC encoding/decoding is only performed at both ends of the system. Figure 1 (a) illustrates a 3-stage, 4-lane system, where the middle stage active components can be retimer(s) or optical modules in practical scenarios; figure 1 (b) shows how the RS symbols are distributed across multiple PMA lanes in 100Gb/s Ethernet application. The number of PMA lanes is 2, 4, 8 and 16 for 50GbE, 100GbE, 200 and 400GbE applications respectively.

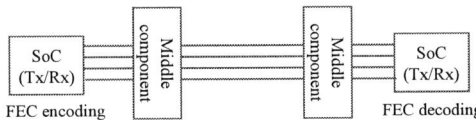

(a). Illustrative topology of a 3-stage, 4-lane FEC system

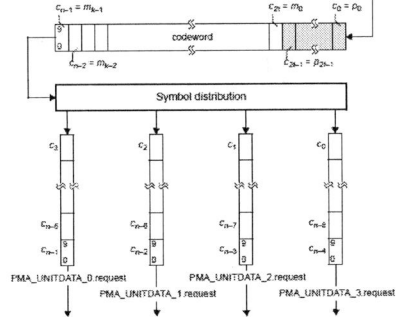

(b). RS symbol distribution across 4-lane in 100GBASE-R [1]

Fig.1. Multi-stage, multi-lane system illustration

For a cascaded stage system, RS symbol errors in a received RS(n,k,t) codeword are introduced independently by each stage. When j RS symbol errors occur in the first stage, there are $n-j$ possible locations for the next stage to introduce RS symbol errors. Therefore, SER of a 2-stage link can be calculated through convolution of the probabilities of different numbers of RS symbol errors in each stage, as in (3). The same method can be extended to 3 or more stages.

$$SER(i) = \sum_{j=0}^{i} SER_{stg1}(j) \cdot \frac{C_{n-j}^{i-j}}{C_n^{i-j}} SER_{stg2}(i-j) \qquad (3)$$

For a multi-lane system where RS symbols are distributed across lanes, there is no overlap of RS symbol errors between lanes in a received RS codeword. The same convolution approach applies. Take 4-lane system for example, SER with i RS symbol errors in a complete RS(n,k,t) codeword is the convolution of SER probabilities of each single lane:

$$SER_{0-1}(i) = \sum_{j=0}^{i} SER0(j) \cdot SER1(i-j) \qquad (4)$$

$$SER_{2-3}(i) = \sum_{j=0}^{i} SER2(j) \cdot SER3(i-j) \qquad (5)$$

$$SER(i) = \sum_{j=0}^{i} SER_{0-1}(j) \cdot SER_{2-3}(i-j) \qquad (6)$$

In (4) to (6), subscripts 0-1, 2-3 represent lane 0 and 1, lane 2 and 3, respectively. The same idea applies to 2, 8 and 16-lane applications. The "aggregated" SER is obtained through SER probability convolution of 2, 8 and 16 lanes.

Note that in some insertion loss dominated links, concentrated ISI patterns in PRBS sequences are more likely to produce concentrated pattern correlated errors. The initial randomly occurred errors (as opposed to burst errors) in such links are no longer evenly distributed. FEC model discussed in this section as in (3) applies for such cases [6], with the exception that that C_{n-j}^{i-j}/C_n^{i-j} is set to 1. SER_{stg1} deals with evenly distributed initial error input, while SER_{stg2} handles the ISI pattern input.

Figure 2 (a) shows the SER simulation result of a cascaded 2-stage, single-lane system, and figure 2 (b) shows the SER simulation result of a single-stage, 4-lane system. In both simulations, DFE error propagation strength of each single lane is denoted by pb. From the analysis, SER of the received RS codeword of a cascaded multi-stage system, or a multi-lane system, has a noticeable degradation compared with any single stage or single lane, and the degradation increases as the number of RS symbol errors increases. This draws attention to post-FEC BER analysis in such multi-stage and multi-lane systems, as the KP4 FEC, i.e., RS(544,514,15) in the 56G and 112G era, is affected by SER with RS symbol error number increasing to more than 15.

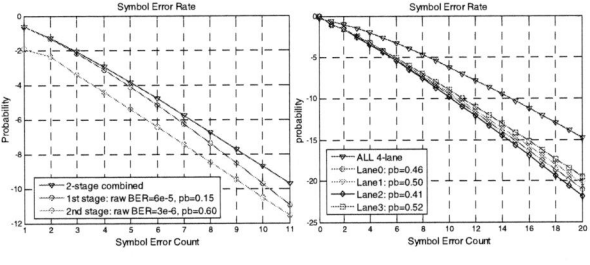

(a). 2-stage, 1-lane system   (b). 1-stage, 4-lane system

Fig.2. SER Simulation for multi-stage and multi-lane systems

C. Analytical model for FEC codewords interleaving effect

FEC codewords interleaving/de-interleaving helps break long burst errors occurred in physical channels and scatter the long bursts induced consecutive RS symbol errors across 2 or more received FEC codewords. For 2:1 FEC codewords interleaving (as defined in 400GBASE-R), pairs of FEC codewords are interleaved/de-interleaved on a 10-bit basis.

With 2:1 interleaving, the probabilities of i consecutive RS symbol errors in a FEC codeword are calculated:

$$P\ with\ interleaving(i)$$
$$= 0.5 \cdot P(2 \cdot i - 1) + P(2 \cdot i) + 0.5 \cdot P(2 \cdot i + 1) \qquad (7)$$

Where, P on the right side of the equation represents the RS symbol error probabilities without interleaving effect.

D. Analytical model for PMA bit-level multiplexing effect

In 50GbE, 100GbE, 200GbE and 400GbE PMA, 2:1 bit-level multiplexing is defined for PAM4 applications. De-multiplexing in the receiving direction converts the PAM4 symbols to pairs of bits in two PMA lanes (MSB and LSB). Thus, one Gray-coded PAM4 symbol error maps into one bit error either in LSB (2/3 chance) or in MSB (1/3 chance). For general analysis, define the burst error length as $(i\text{-}1)\cdot5+j$, where i is an odd integer ($i=1, 3, 5\ldots$), j ranges from 1 to 10.

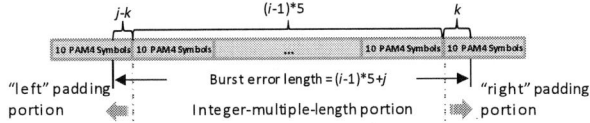

Fig.3. Single burst length definition for 2:1 bit-mux analysis

For the burst error defined in Figure 3, RS symbol error probabilities from the "left" padding portion plus the integer-multiple portion of the burst error are calculated as follows:

$$\forall k \in [0,9], M = \begin{cases} 0, & k \ge 1 \\ 1, & k = 0 \end{cases}$$

978-1-7281-6162-4/20 $31.00 © 2020 IEEE

$$P_{consecutive\ symbol\ errors\ left}(k,\ 1)$$

$$= \sum_{j=k+1}^{10} (11-j)^M \cdot P_{burst}\left((i-1)\cdot 5+j\right)\cdot\left(\left(\frac{1}{3}\right)^{j-k}+\left(\frac{2}{3}\right)^{j-k}\right) \quad (8)$$

$$P_{consecutive\ symbol\ errors\ left}(k,\ 2)$$

$$= \sum_{j=k+1}^{10} (11-j)^M \cdot P_{burst}\left((i-1)\cdot 5+j\right)\cdot\left(1-\left(\frac{1}{3}\right)^{j-k}-\left(\frac{2}{3}\right)^{j-k}\right) \quad (9)$$

RS symbol error probabilities of the "right" part padding:

For $k=0$: $\quad P_{consecutive\ symbol\ errors\ right}(k,0:2)=[1\ 0\ 0]$

For $k \geq 1$: $\quad P_{consecutive\ symbol\ errors\ right}(k,0)=0$

$$P_{consecutive\ symbol\ errors\ right}(k,1)=\left(\frac{1}{3}\right)^k+\left(\frac{2}{3}\right)^k$$

$$P_{consecutive\ symbol\ errors\ right}(k,2)=1-\left(\frac{1}{3}\right)^k-\left(\frac{2}{3}\right)^k \quad (10)$$

The RS symbol error probabilities of the burst error of length $(i-1)\cdot 5+j$ can then be calculated through convolution:

$$P_{consecutive\ symbol\ erros}(i:i+3)$$

$$= \sum_{k=0}^{9}\left(P_{consecutive\ symbol\ errors\ left}(k,1:2)*P_{consecutive\ symbol\ errors\ right}(k,0:2)\right) \quad (11)$$

With bit-level multiplexing, post-FEC BER performance degrades, which can also be found in discussions in [2, 3, 4].

III. PERFORMANCE STUDY FOR MULTI-STAGE FEC LINK

A. Two-stage FEC link with retimer(s)

A two-stage, single-lane backplane link with a retimer is investigated. RS (544,514,15) error correction is applied at the end stage of the link. The transceivers are based on an IBIS-AMI model for a 56Gbps PAM4 ADC-based receiver architecture. Receiver equalization consists of CTLE/AGC, 24-tap FFE, and 1-tap DFE. The end-to-end link insertion loss keeps the same (about 40dB bump to bump); the lengths (and loss) of the sub-links are different, depending on the location adjustment of the retimer under practical engineering constrains. 0.36mVrms crosstalk noise is applied in AMI simulations for sub-link 2.

Table 1 summarizes the simulated raw BER for each sub-link stage and the post-FEC BER of the end-to-end link:

TABLE I. FEC SIMULATION SUMMARY OF A 2-STAGE LINK

Insertion Loss (dB)			Raw BER		End-to-end Post-FEC BER
Sublink1	Sublink2	End-to-end	Sublink1	Sublink2	
24.7	15.4	40.1	4.38E-05	1.01E-05	5.57E-27
25.5	14.7	40.2	2.50E-05	7.12E-06	2.60E-28
26.4	19.9	40.3	2.20E-04	9.48E-06	5.43E-16
28.0	12.1	40.1	1.01E-04	9.78E-07	5.20E-17
30.0	10.1	40.2	2.42E-04	6.29E-06	1.33E-15
31.5	8.7	40.2	3.79E-04	7.02E-06	2.11E-17
33	8.0	41	7.14E-04	5.11E-05	2.14E-15

From the results, post-FEC BER varies about 10 orders while adjusting the location of the retimer in the end-to-end link; the raw BER of the two sub-links hardly changes by 1 order. Case-by-case post-FEC BER analysis helps to evaluate such multi-stage link systems for margin budgeting.

B. Three-stage FEC link of E-O-E System

A three-stage E-O-E system is investigated. In this simulation, optical link SER from a reported 400GBASE-LR8 field data for 27km SMF [5] is used as an example, the measured raw BER is 8.71e-5. One-tap DFE receiver model is used for 400GAUI-8 host chips on both ends of the three-stage link. 2:1 FEC codewords interleaving and 2:1 PMA bit-level multiplexing are enabled.

The DFE error propagation probability factor Pb for the 8 individual lanes in VSR sub-link stages 1 and 3 is swept. From the simulation results in Figure 4, reducing DFE error propagation strength, which means to reduce the DFE tap weight, helps to enhance post-FEC BER performance of the end-to-end link; however, the gain is reducing as Pb reduces incrementally from 0.5 to 0.2 for this system.

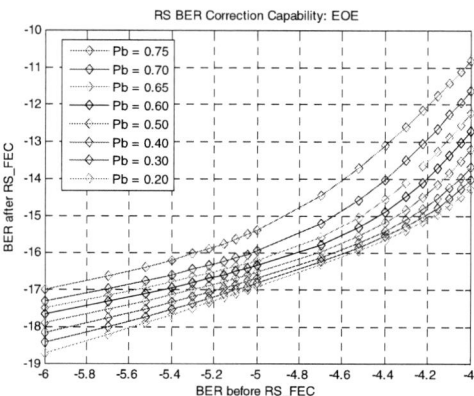

Fig.4. End-to-end FEC performance simulation for E-O-E link

IV. CONCLUSIONS

A simulation method for analyzing end-to-end RS FEC performance for multi-stage link systems is discussed. Examples with a retimer link system and a 400GAUI-8 E-O-E system show that post-FEC BER performance for multi-stage systems largely depends on link configuration, SerDes equalization architecture as well as system link impairments. Accurate post-FEC performance evaluation for such multi-stage systems relies on case-by-case analysis.

REFERENCES

[1] IEEE Standards for Ethernet: IEEE Std 802.3TM-2018.

[2] X. Dong, C. Huang and G. Zhang, "End-to-end FEC Performance Analysis for Multi-stage PAM4 Systems", DesignCon 2020, Santa Clara, CA.

[3] C. Liu, "100+ Gb/s Ethernet Forward Error Correction (FEC) Analysis", DesignCon 2019, Santa Clara, CA.

Also available online: www.signalintegrityjournal.com/articles/1286-gbs-ethernet-forward-error-correction-fec-analysis

[4] P. Anslow, "RS(544,514) FEC Performance", IEEE P802.3cd Task Force, Whistler, Canada, May 2016.

[5] C.Cole, Y.Zhou, K. Smith, M. Gilson, P. Brooks and C.Yu, "400GBASE-LR8 Measurement Data for Reaches >10km", www.ieee802.org/3/B10K/public/18_07/cole_b10k_01_0718.pdf

[6] X. Dong, C. Huang and G. Zhang, "QPRBS31 Correlated Error Analysis in 56G PAM4 FEC Systems", DesignCon 2020, Santa Clara, CA.

Gauss-Newton Method for Performance Evaluation of Decoupling Capacitors on Resonant Parallel Plates

Ihsan Erdin
Engineering Design Services
Celestica Inc.
Ottawa, ON, Canada
ierdin@ieee.org

Ram Achar
Dept. of Electronics
Carleton University
Ottawa, ON, Canada
achar@doe.carleton.ca

Abstract—**A Gauss-Newton (G-N) based method is proposed for optimal placement and performance evaluation of local decoupling capacitors on resonant parallel-plates. Multiple power pins are used as a leverage for simultaneous placement optimization of multiple capacitors by utilizing matrix calculus methods. The algorithm converges in a few iterations, which is a big improvement against the competing evolutionary methods. The proposed method is tested on a sample case and the results are observed to be valid across a practically wide frequency range.**

Index Terms—**decoupling capacitors, gauss-newton, newton-raphson, power delivery network, power integrity.**

I. INTRODUCTION

With design concerns like limited space resources and manufacturing yield issues of printed circuit boards (PCB), the selection and placement of decoupling capacitors have always been among the major priorities of power integrity (PI) analysis. The complexity of a PCB structure makes model generation for PI analysis very challenging especially at high frequencies. While some design parameters like the capacitance value, type, parasitics, etc. could be relatively easy to address, calculations for the optimum placement configuration could be extremely challenging especially for design techniques that rely on numerical EM methods.

Optimal placement and value of decoupling capacitors have long been researched in the literature and industrial circles [1]-[8]. Due to the large number of optimization parameters involved, evolutionary algorithms have become prominent and some of these techniques were implemented in commercial PI analysis tools. Despite their success, the large number of iterations and lack of convergence to a unique solution even for the same initial conditions stand out as the two major drawbacks of evolutionary algorithms. It's possible to address both of these issues by using gradient methods which can converge to a unique solution in a few iterations under favorable initial conditions.

From a practical point, the implementation of evolutionary algorithms is fairly straightforward, provided that a suitable objective function can be found. On the other hand, considerable preliminary work is needed for gradient algorithms. Especially, the calculation of derivatives could be quite involved in the case of multiple capacitors and power pins.

In the past, some research efforts have been devoted with the application of Newton-Raphson (N-R) method to basic cases involving a single capacitor and a power pin port [9]. Although the complexity of the problem impeded the immediate application of the technique to practical cases, the theoretical advancements laid out the groundwork to improve the gradient-based algorithms.

In this paper, N-R method is advanced to the case involving multiple power pins and decoupling capacitors, which represent a more realistic and practical design scenario. The previously developed scalar expressions for the impedance of a power pin and its partial derivative for N-R iterations are extended to multiple pins and capacitors using matrix calculus techniques. The proposed G-N based method allows the *simultaneous* optimization and performance evaluation of *multiple* decoupling capacitors in a few iteration steps.

This document is organized as follows. In Section II, the theoretical analysis is presented. This is followed by Section III, which the proposed model is validated on a numerical example and conclusions are provided in Section IV.

II. THEORY OF THE PROPOSED MODEL

In a system involving P number of power pins and M number of capacitors, the pin impedance matrix $\boldsymbol{Z}_{in}$ can be expressed as follows:

$$\boldsymbol{Z}_{in} = \boldsymbol{Z}_{PP} - \boldsymbol{Z}_{PM}\left[\boldsymbol{Z}_{MM} + \boldsymbol{Z}_c\right]^{-1}\boldsymbol{Z}_{MP} \qquad (1)$$

where the impedance matrices, $\boldsymbol{Z}_{PP}$ and $\boldsymbol{Z}_{MM}$ with entries $Z_{ij}(s)$ ($i,\ j \in \{1,\ ...,\ P+M\}$) represent distributed circuit parameters of power pins and capacitor ports, respectively. The parameter s is the complex angular frequency. The matrix $\boldsymbol{Z}_{PM}$ and its transpose relate the pin and capacitor ports. $\boldsymbol{Z}_c$ is a diagonal matrix of size M, whose elements $Z_c(s)$ are defined by the impedance of decoupling capacitors at the corresponding port locations. The variable $Z_c(s)$ can be represented by its the lead-length parasitic resistance, R, and inductance, L as follows.

$$Z_c(s) = R + sL + \frac{1}{sC} \qquad (2)$$

with C being the capacitance.

978-1-7281-6162-4/20 $31.00 © 2020 IEEE

For rectangular parallel plates of size a and b which are separated by a gap of d, $Z_{ij}(s)$ is given by the following compact relation [10]:

$$Z_{ij}(s) = \frac{\mu d}{ab} \sum_{n=0}^{\infty} \sum_{m=0}^{\infty} \frac{\chi_n \chi_m \sigma_{nmij} \psi_{nmij}}{\beta_n^2 + \beta_m^2 - \beta^2} \quad (3)$$

where μ is permittivity of the medium; $\beta_n = \frac{n\pi}{a}$ and $\beta_m = \frac{m\pi}{b}$ are the eigenvalues of standing waves inside the parallel plates. β is the wavenumber and $\chi_{nm} = \left\{ \begin{array}{l} 1,\ n,\ m = 0 \\ \sqrt{2},\ otherwise \end{array} \right\}$
The remaining variables in (3) are expressed as follows:

$$\psi_{nmij} = \cos(\beta_m x_i) \cos(\beta_n y_i) \times \\ \cos(\beta_m(x_i + r\cos\phi)) \cos(\beta_n(y_i + r\sin\phi)) \quad (4)$$

where x_i and y_i represent the associated port coordinates, r and ϕ are the distance and angle between the ports, respectively. Here,

$$\sigma_{nmij} = \text{sinc}(\beta_m w_{xi}) \, \text{sinc}(\beta_n w_{yi}) \times \\ \text{sinc}(\beta_m w_{xj}) \, \text{sinc}(\beta_n w_{yj}) \quad (5)$$

where w_{xi} and w_{yi} represent the associated port sizes.

Let the Ψ represent the set of coordinates on the plane pair for decoupling capacitors:

$$\Psi = \{r, \phi | r, \phi \in \mathbb{R}^{2M}; r_1, \phi_1, ..., r_M, \phi_M\} \quad (6)$$

Assuming there's a set of coordinates Ψ_0 for which $Z_{in}(s, \Psi_0)$ becomes minimum, one can define the impedance compromise parameter, q as follows:

$$q = \left| \frac{Z_{in}(s, \Psi_q)}{Z_{in}(s, \Psi_0)} \right| \quad (7)$$

It has been established that the elements of Ψ_0 are those closest to the observed pin location [9]. In order to evaluate the performance of decoupling capacitors with respect to the specified pin location, the vector function $\boldsymbol{f}(s, \Psi)$ is defined as follows:

$$\boldsymbol{f}(s, \Psi) = |\boldsymbol{Z}_{in}(s, \Psi)| - q |\boldsymbol{Z}_{in}(s, \Psi_0)| \quad (8)$$

The analytical nature of (3) makes the gradient-based algorithms a prime candidate for the solution of Ψ_q. The Newton-Raphson (N-R) method is a suitable choice for the iterative solution to the vector function $\boldsymbol{f}(s, \Psi)$ as follows:

$$\Psi_{k+1} = \Psi_k - \nabla \boldsymbol{f}_k^{-1}(s, \Psi) \boldsymbol{f}_k(s, \Psi) \quad (9)$$

From (9), it is clear that $P \geq 2M$ is required. If $P > 2M$ $\nabla \boldsymbol{f}_k(s, \Psi)$ becomes overdetermined and (9) needs to be modified as follows:

$$\Psi_{k+1} = \Psi_k - \nabla \boldsymbol{f}_k^+(s, \Psi) \boldsymbol{f}_k(s, \Psi) \quad (10)$$

where $\nabla \boldsymbol{f}_k^+(s, \Psi)$ is the pseudo-inverse of $\nabla \boldsymbol{f}_k(s, \Psi)$ which is defined as

$$\nabla \boldsymbol{f}_k^+(s, \Psi) = [\nabla \boldsymbol{f}_k^T(s, \Psi) \nabla \boldsymbol{f}_k(s, \Psi)]^{-1} \nabla \boldsymbol{f}_k^T(s, \Psi) \quad (11)$$

The relation (10) is also referred to as Gauss-Newton iteration [11]. Regardless of the deterministic nature of $\nabla \boldsymbol{f}_k(s, \Psi)$, calculation of its entries requires the partial derivatives of $\boldsymbol{Z}_{in}(s, \Psi)$ which can be expressed as follows by using the chain rule in matrix calculus [12]:

$$\boldsymbol{Z}'_{in}(s, \Psi) = 2 \boldsymbol{Z}_{MP} [\boldsymbol{Z}_{MM} + \boldsymbol{Z}_c]^{-1} \boldsymbol{Z}'_{PM} - \boldsymbol{Z}_{MP} \\ \times [\boldsymbol{Z}_{MM} + \boldsymbol{Z}_c]^{-1} \boldsymbol{Z}'_{MM} [\boldsymbol{Z}_{MM} + \boldsymbol{Z}_c]^{-1} \\ \times \boldsymbol{Z}_{PM} \quad (12)$$

The parameters Z_{ij} constitute the building blocks of (9), (10) and (12) as the key elements of Newton's algorithm whose convergence depends on the precise evaluation of $\nabla \boldsymbol{f}_k(s, \Psi)$. In the evaluation of the derivatives, the analytical expression of the parameters Z'_{ij} were derived in [9].

III. NUMERICAL EXAMPLE

The validity of the proposed method will be demonstrated on a pair of rectangular parallel-plates which represent power and ground plane conductors. The two planes are sized on x-y coordinates as 125 mm by 75 mm with a dielectric thickness of 0.127 mm which separates them in the z-dimension. The effective permittivity and the loss tangent of the dielectric slab are 4.5 and 0.02, respectively. Sixteen power pins with a 1 mm pitch spacing are shown with the cluster of circular pegs in Fig. 1. Three decoupling capacitors are initially placed within 0.8 mm distance of a power pin which is selected as the observation port for the proposed calculation method. The placement of capacitors is marked with the three grey square pegs in Fig. 1. All capacitors are modeled as 100 nF with lead length parasitics of R_p = 58 mΩ and L_p = 200 pH.

The pin impedance for this initial configuration is calculated as $30 m\Omega$ and taken as a reference for values pertaining to other placement configurations. As the 3 capacitors are moved away from the observation pin, its impedance tends to increase. In order to find the placement configuration for which the pin impedance increases by a specified amount q, the capacitors can be gradually moved away from the observation location as the impedance is checked at incremental distances. While this could be regarded as a viable method for the assessment of a single capacitor, the number of radial and angular possibilities for the simultaneous displacement of all 3 capacitors are overwhelmingly large even for the simple case discussed in this example.

A practical solution comes with the proposed G-N iterations. In this example $q = 2$ is assumed. The required placement of capacitors is then calculated with the proposed method after only three iterations at the selected 60 MHz and shown in Fig. 1 with 3 black square pegs. The analysis takes into account the mutual interaction among the capacitors as well as with the power pins while evaluating their placement. The placement configuration can be interpreted as the region within which

978-1-7281-6162-4/20 $31.00 © 2020 IEEE

the effectiveness of capacitors will not be reduced more than half with respect to the power pin cluster.

The resulting pin impedance curve is plotted in Fig. 2 as a function of frequency in comparison to the reference case. The impedance values at 60 MHz are indicated with circular marks ($30m\Omega$ for the reference and $60m\Omega$ for the final case). Although the frequency is set to 60 MHz in the G-N iterations, a similar $q = 2$ ratio can be observed for the pin impedance magnitude up to 250 MHz between the two configurations. Besides that, calculations are also conducted at other frequencies and similarly fast convergence is observed for the G-N iterations as summarized in Table I.

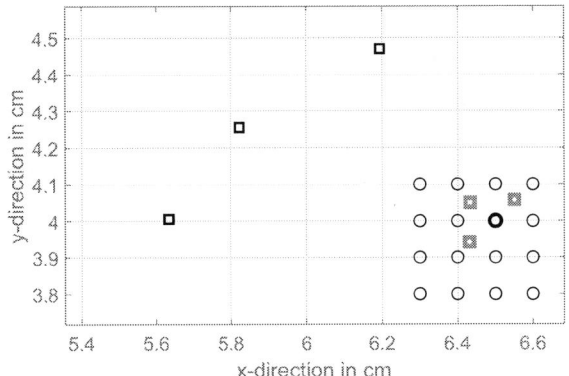

Fig. 1. The assumed configuration for the numerical example with the power pin cluster of circular pegs and decoupling capacitors (square pegs). The initial placement of the capacitors are marked with thick grey square pegs inside the BGA area close to the reference pin (marked with the bold circle) among the others. The calculated final positions of the capacitors are marked with black square pegs outside the BGA area.

The simulations are conducted in an *Intel i7* platform using *MATLAB R2019b*. With the selected initial placement of the 3 capacitors, the proposed algorithm takes a few iterations as summarized in Table I. It should be noted that the proposed G-N method for the optimal placement of multipin/multicapacitor problem can also be applied to other impedance criteria like the specification of a target impedance instead of the impedance increasing to a certain value.

TABLE I
NUMERICAL DATA FOR G-N ITERATIONS AT SELECTED FREQUENCIES

Frequency (MHz)	$\|Z_{in}(r_q)\|$ ($m\Omega$)	$\|Z_{in}(r_{q0})\|$ ($m\Omega$)	# G-N iterations	time (sec)
60	60.0	30.0	3	32.9
120	131.0	65.0	6	69.6
180	240.0	120.0	5	59.9
240	585.0	262.0	2	22.6

IV. CONCLUSION

In this work, the G-N method is developed for the analysis of PDNs involving multiple power pins and decoupling capacitors. By utilizing matrix calculus techniques, G-N method is

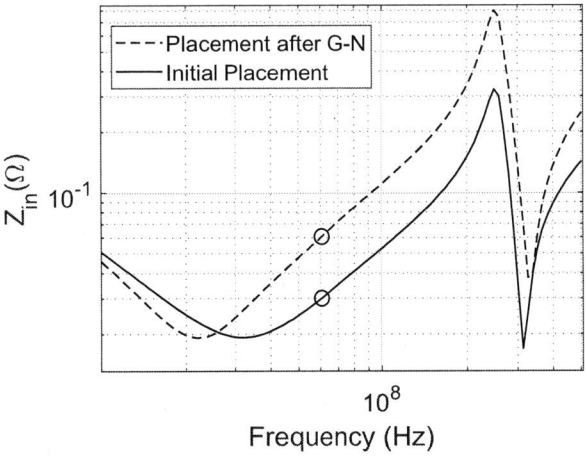

Fig. 2. Magnitude of the input impedance of the highlighted power pin of the configuration shown in Fig. 1. The calculation impedance is marked with circles on the two curves.

tailored for PI optimization and shown as a viable alternative to the evolutionary algorithms. The proposed method allows the simultaneous optimization and performance evaluation of multiple decoupling capacitors in a few iteration steps.

REFERENCES

[1] Y. Chen, Z. Chen, and J. Fang, "Optimum placement of decoupling capacitors on packages and printed circuit boards under the guidance of electromagnetic field simulation," *Proc. 46th Electron. Components and Technol. Conf.*, pp. 756–760, May 1996.

[2] J. Choi *et al.*, "A methodology for the placement and optimization of decoupling capacitors for gigahertz systems," in *Proc. 13th Int. Conf. VLSI Des.*, pp. 156–161, 2000.

[3] S.Kahng, "GA-optimized decoupling capacitors damping the rectangular power-bus' cavity mode resonances," *IEEE Microwave and Wireless Components Lett.*, vol. 16, no. 6, pp. 375–377, Jun. 2006.

[4] K. Wu *et al.*, "Optimization for the locations of decoupling capacitors in suppressing the ground bounce by genetic algorithm," *Progress Electromagn. Res. Symp.*, Aug. 2005.

[5] J. N. Tripathi, P. Damle, and R. Malik, "Minimizing core supply noise in a power delivery network by optimization of decoupling capacitors using simulated annealing," *IEEE 21st Workshop on Signal and Power Integrity (SPI)*, pp. 1–3, May 2017.

[6] K. Bharath, E. Engin, and M. Swaminathan,"Automatic package and board decoupling capacitor placement using genetic algorithms and M-FDM," *45th ACM/IEEE Des. Automation Conf.*, pp. 560–565, Jun. 2008.

[7] A. E. Engin, "Efficient sensitivity calculations for cptimization of power delivery network impedance," *IEEE Trans. Electromagn. Compat.*, vol. 52, no. 2, pp. 332–339, Dec. 2010.

[8] I. Erdin and Ram Achar, "Multi-pin optimization method for placement of decoupling capacitors using genetic algorithm," *IEEE Trans. Electromagn. Compat.*, vol. 60, no. 6, pp. 1662–1669, Dec. 2018.

[9] I. Erdin and R. Achar, "Decoupling capacitor placement on resonant parallel-plates via driving point impedance," *IEEE Trans. Microw. Theory Tech.*, vol. 67, no. 6, pp. 2162–2171, Jun. 2019.

[10] G. Lei, *et al.*, "Wave model solution to the ground-power plane noise problem," *IEEE Trans. Instrument. Measurement*, vol. 44, no. 2, pp. 300–303, Apr. 1995.

[11] J. Nocedal and S. J. Wright, *Numerical Optimization* Springer, 2006, ch. 10.

[12] I. Erdin and R. Achar, "On the effective range of decoupling capacitors including mutual coupling," *2020 IEEE 24th Workshop on Signal and Power Integrity (SPI)*, May. 2020.

Measurement Uncertainty Propagation in the Validation of High-Speed Interconnects

Cemil S. Geyik[1,2], Michael J. Hill[1], Zhichao Zhang[1], Kemal Aygün[1] and James T. Aberle[2]

Assembly and Test Technology Development, Intel Corporation, Chandler, Arizona, USA[1]
School of Electrical, Computer and Energy Engineering, Arizona State University, Tempe, Arizona, USA[2]
cemil.s.geyik@intel.com

Abstract—**Validating the performance of high-speed interconnect modeling against measurements of fabricated test structures requires an understanding of the robustness of the measurement methods as well as the physical variations present in an imperfectly fabricated test structure. This paper presents a methodology for evaluating the performance of interconnect modeling considering the actual metrology variation and the real-world manufacturing tolerances used to fabricate the test vehicle. By ensuring that measurement results, inclusive of operator and equipment variations, overlap the modeling inclusive of expected manufacturing variations, confidence in the high-speed interconnect modeling is established.**

Index Terms—**measurement uncertainty, manufacturing variations, reproducibility, correlation.**

I. INTRODUCTION

Measurement-to-modeling correlation is a critical step in validating the electrical performance of high-speed interconnects [1]–[3]. A good correlation ensures that interconnect behavior can be reliably predicted for any new technology, material, or process. However, achieving a good correlation for multiple metrics is not a simple task considering the high number of factors in the correlation flow which influence the final performance [1], [4]. Furthermore, as network connectivity roadmaps target 100 Gbps and beyond [5], correlation is becoming increasingly challenging for high-speed interconnects due to increased performance sensitivity to any variation.

A measurement result is incomplete unless accompanied with an estimate of the uncertainty associated with the measurement [6]. There are many possible sources of uncertainty, not necessarily independent, including the impact of environmental conditions, personal bias in reading instruments, finite discrimination threshold, approximations and assumptions incorporated in the measurement method [7]. Considering all the challenges in high-speed interconnect validation, it is not surprising that poor correlation occurs more often than is desirable. To ascertain whether a correlation is *good* or *poor*, one needs to understand how the uncertainty propagates to the outcome, and not just focus on the outcome itself.

II. UNCERTAINTY QUANTIFICATION

Measurement uncertainty can be quantitatively determined by metrology capability analysis (MCA) [8], which comprises three parts: accuracy, repeatability, and reproducibility. In this paper, the focus is on reproducibility, which is defined as the closeness of the agreement between the results of measurements of the same measurand carried out under changed conditions of measurement [9]. The changed conditions include repeated device under test (DUT) insertion and measurement instrument calibrations by multiple test equipment operators at different times. This process provides information on the measurement variability introduced by all temporal and spatial variations of any influence quantity. Environmental conditions, e.g., temperature and relative humidity, can have a profound adverse impact on the material properties, and loss [1], [10]. These factors should be maintained at the same intended use condition throughout the course of the experiment.

Reproducibility bounds the usefulness of any measurement and can be expressed quantitatively in terms of the dispersion characteristics of the results. Fig. 1 illustrates the factors affecting correlation quality, among which S-parameters, dielectric permittivity, and cross-section dimensional measurements are prioritized and elaborated in this section.

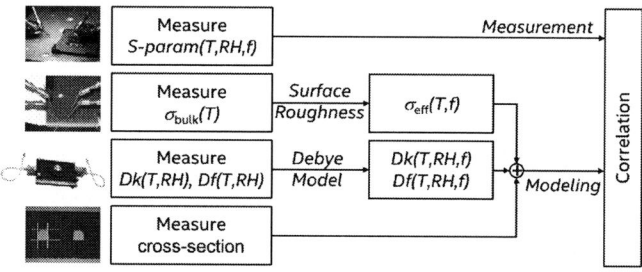

Fig. 1. Measurement-to-modeling correlation flow. T, RH and f indicate dependence to temperature, relative humidity and frequency, whereas σ_{bulk}, σ_{eff}, Dk, and Df refer to bulk and effective conductivity, dielectric constant and dissipation factor, respectively.

A. S-parameter Measurements

A state-of-the-art four-port performance network analyzer (PNA) was utilized to measure the S-parameters of a differential stripline (DSL) package test structure up to 67 GHz. To understand the impact of measurement variation, 3 different operators collected data on 3 different days, calibrating the PNA before each measurement. This yielded 9 measurements of the DUT that could be used to calculate the mean (μ) and control limits ($\mu \pm 3\sigma$), where σ is the standard deviation. Fig. 2 shows the reproducibility results for several differential performance metrics obtained by post-processing single-ended measurement results: return loss (RL), insertion loss (IL), time domain reflectometry (TDR), and phase delay (PD).

978-1-7281-6162-4/20 $31.00 © 2020 IEEE

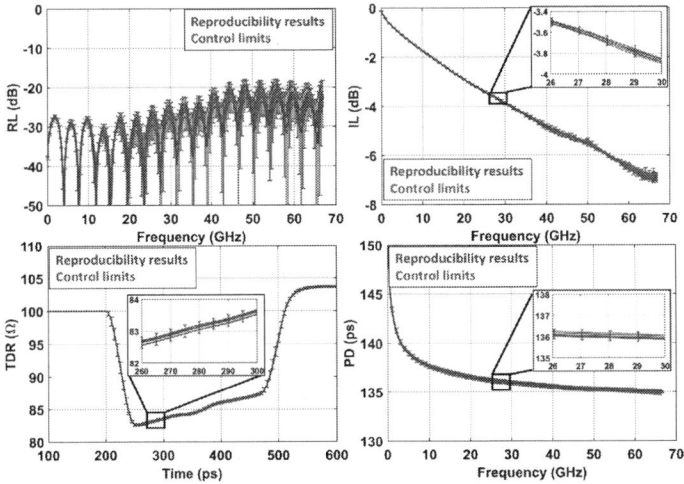

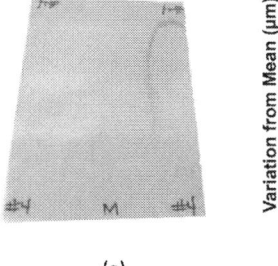

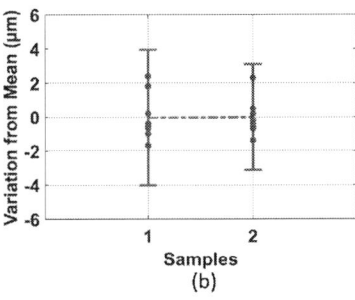

Fig. 2. Four-port PNA reproducibility results for differential metrics referenced to 85 Ω. Error bars indicate $\mu \pm 3\sigma$ control limits.

Fig. 3. (a) A typical dielectric sample received from vendors, and (b) thickness measurement reproducibility results. Error bars indicate $\pm 3\sigma$ control limits.

RL is observed to be very sensitive. This could be due to RL being calculated relative to a reference impedance. Therefore, for well-matched lines, any small dispersion from an already small reflection leads to high relative uncertainty. Variation in IL increases with frequency but relative standard deviation $(3\sigma/\mu)$ remains under 4%. Variations in TDR and PD are very small and practically constant over time and frequency with $3\sigma/\mu$ of 0.25% and 0.15%, respectively.

B. Dielectric Permittivity Measurements

In [10], an MCA was performed on the dielectric measurement metrology, utilizing a split post dielectric resonator (SPDR). From that study, it was found that operator variation in the measurement of sample thickness (required to extract dielectric constant (Dk) from SPDR [11]) was a key limiter to reproducibility. This is because the relative errors in thickness result in an almost one-to-one relative error in Dk.

In this study, a separate MCA was performed on dielectric sample thickness measurement utilizing a micrometer. A typical dielectric sample provided by vendors is shown in Fig. 3(a). Three different operators measured two samples with different thicknesses on three different days. The variations from mean value are shown in Fig. 3(b). It appears that the thickness variation shows small dependence on the mean considering one sample is more than twice as thick. As a result, reproducibility is expressed in absolute terms, i.e., $3\sigma \approx 4$ μm. This result also means that thicker samples would yield smaller relative variation in thickness, and hence smaller relative variation in extracted Dk.

Measurement dynamics and sample thickness variation lead to a combined uncertainty of $3\sigma/\mu \approx 3\%$ in Dk. This result implies that any simulation should be performed at both the upper and lower bounds from the dielectric characterization to bound the expected impact of the measurement variability.

C. Cross-section Dimensional Measurement

High fidelity geometrical representation of a transmission line can be achieved by cross-sectioning and is essential

for a good correlation. Cross-section dimensional features become more critical considering today's on-package high-speed interconnect loss is largely dominated by conductors due to thinner substrate and low loss dielectric materials [2]. Cross-section pictures of typical package traces along with dimensional features are shown in Fig. 4(a).

An MCA was performed on a cross-section dimensional measurement utilizing a visualization software. Three different operators measured four separate dimensional features on three different days. The variation of each dimensional feature from its mean value is shown in Fig. 4(b). The main source for uncertainty is the lack of clarity on where the features start and end due to manufacturing process variations and surface roughness. For larger design rules, this ambiguity might cause a relatively small uncertainty; however, for today's high-density package design rules, the resulting uncertainty is not negligible. Reproducibility results show that 3σ control limits for each dimensional feature can be as large as 0.7 μm.

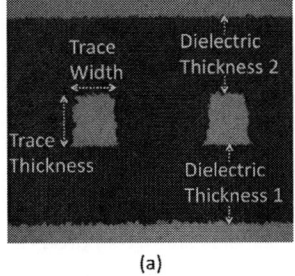

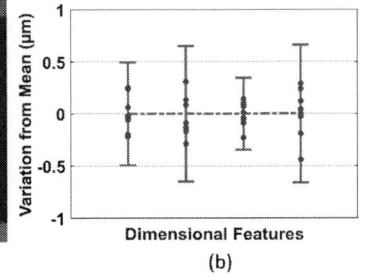

Fig. 4. (a) Cross-section pictures of typical package traces with dimensional features illustrated, and (b) cross-section dimensional measurement reproducibility results. Error bars indicate $\pm 3\sigma$ control limits.

III. MEASUREMENT-TO-MODELING CORRELATION

A package test vehicle was designed and manufactured including a DSL routed on the layer below the surface with a length of 20 mm. Measurement-to-modeling correlation is shown in Fig. 5 at typical use condition for packages, i.e., 90°C and 0% RH [1]. PNA measurement was performed after prebaking to ensure no moisture was left, and on a temperature chuck to achieve this use condition. Modeling results were generated using dielectric and conductor material properties and surface roughness characterized at the same use condition along with cross-section dimensions.

978-1-7281-6162-4/20 $31.00 © 2020 IEEE

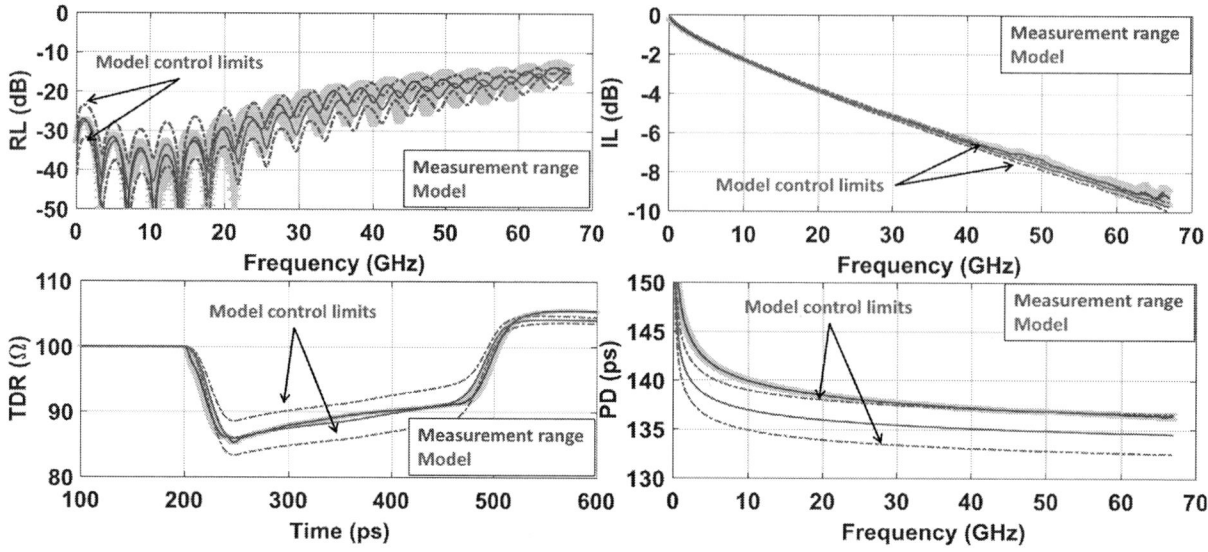

Fig. 5. Correlation at typical use condition for packages. Uncertainty incorporated into measurement (shaded) and propagated to modeling outcome (dash).

Simple visual assessment of measurement and modeling (illustrated by blue and orange solid lines, respectively) may indicate a good correlation for all performance metrics except for the phase delay. A comparison of only single measurement and model lines without any sensitivity analysis is insufficient to evaluate the correlation quality. Propagation of measurement uncertainty is required for a better judgment.

Measurement uncertainty from Section II-A was incorporated into measured S-parameter data as a shaded area, i.e., measurement range. Measurement uncertainty of Dk from Section II-B and measurement uncertainty of cross-section dimensions from Section II-C were incorporated into modeling data as model control limits. Combined uncertainty in standard deviation for each performance metric was calculated using response surface methodology and statistical design of experiments (DOE) [12]. Subsequently, Monte Carlo method was performed to understand the impact of the variabilities on the modeling results. As can be seen, when the dielectric measurement and cross-section dimensional variations are incorporated into the modeling data and compared to the measurement results including the measurement variation, most of the correlation gap in phase delay is accounted for. This result implies that the phase delay can also be considered to have a good correlation for $f > 30$ GHz. It is worth noting that the underlying surface roughness model has a nature of enveloping full spectrum of interest through ensuring better high frequency correlation.

IV. CONCLUSION

This paper presents a systematic methodology for measurement uncertainty quantification and propagation in high-speed interconnect validation. Measurement uncertainty in S-parameters, dielectric permittivity, and cross-section dimensional measurements is examined. Variability in each measurement step of the use condition-dependent correlation flow is quantitatively determined through rigorous MCAs. Combined uncertainty propagated to the performance metrics increases the confidence in correlations by identifying control limits, and helps to better interpret the correlation quality. The next steps for this work include the investigation of uncertainties introduced by various de-embedding methods.

ACKNOWLEDGMENT

The authors would like to thank J. Sauer, S. Christ, R. Marquez, D. Erickson and L. Wojewoda for providing measurements and useful discussions.

REFERENCES

[1] C. S. Geyik, Y. S. Mekonnen, Z. Zhang, and K. Aygün, "Impact of use conditions on dielectric and conductor material models for high-speed package interconnects," *IEEE Trans. Compon., Packag. and Manuf. Technol.*, vol. 9, no. 10, pp. 1942–1951, 2019.

[2] C. S. Geyik, Z. Zhang, S. R. Christ, L. E. Wojewoda, and K. Aygün, "Temperature impact on surface roughness modeling for on-package high speed interconnects," in *Proc. EPEPS*, 2018, pp. 271–273.

[3] C. S. Geyik, Z. Zhang, and K. Aygün, "Improved package modeling and correlation methodology for high speed IO design," in *Proc. ECTC*, 2016, pp. 985–991.

[4] M. Marin and Y. Shlepnev, "Systematic approach to pcb interconnects analysis to measurement validation," in *Proc. IEEE Symp. on EMC, SI & PI*, 2018, pp. 228–233.

[5] "The 2020 ethernet roadmap." [Online]. Available: https://ethernetalliance.org/technology/2020-roadmap/

[6] B. Taylor and C. Kuyatt, "Guidelines for evaluating and expressing the uncertainty of NIST measurement results," NIST, Tech. Rep., 1994.

[7] "Evaluation of measurement data - Guide to the expression of uncertainty in measurement (GUM)," JCGM, Tech. Rep., 2008.

[8] M. J. Hill and L. E. Wojewoda, "A study of permittivity measurement reproducibility utilizing the Agilent 4291B," *IEEE Trans. Adv. Packag.*, vol. 29, no. 4, pp. 714–718, 2006.

[9] "International vocabulary of metrology - Basic and general concepts and associated terms (VIM 3rd edition)," JCGM, Tech. Rep., 2012.

[10] Y. S. Mekonnen, M. J. Hill, L. Wojewoda, and K. Aygün, "Robust temperature and humidity dependent electrical package material characterization," in *Proc. EPEPS*, 2018, pp. 195–197.

[11] J. Krupka *et al.*, "Uncertainty of complex permittivity measurements by split-post dielectric resonator technique," *Journal of the European Ceramic Society*, vol. 21, no. 15, pp. 2673–2676, 2001.

[12] D. C. Montgomery, *Design and analysis of experiments*. John Wiley & Sons, 2019.

978-1-7281-6162-4/20 $31.00 © 2020 IEEE

A Non-Random Exploration based Method for the Optimization of Capacitors in Power Delivery Networks

Seunghyup Han* and Madhavan Swaminathan*

School of Electrical and Computer Engineering, Georgia Institute of Technology, Atlanta, GA, 30332-0250
3D Systems Packaging Research Center (PRC)
Email: *seunghyup@gatech.edu, *madhavan.swaminathan@ece.gatech.edu

Abstract—This paper proposes a non-random exploration based method to optimize the response of power delivery network (PDN) using the minimum number of capacitors. Unlike previous optimization methods which are based on either full search or random exploration (machine learning etc), the present method requires few simulations to converge to the minimum decoupling capacitor solution. The results show that the proposed method is more robust based on comparisons.

Keywords—*Decoupling capacitor, target impedance, power delivery network (PDN).*

I. INTRODUCTION

Designing a robust power delivery network (PDN) has become challenging with the increase in operating frequency and current load in ICs that have low voltage requirements. A typical technique to minimize impedance of PDNs is by using decoupling capacitors (decaps). The level of voltage fluctuation below the threshold level at the IC port can be guaranteed by assigning proper decaps that reduce the self-impedance below the target impedance in the frequency range of interest. However, as system sizes continue to shrink, using a minimum number of decoupling capacitors in the PDN to meet the target impedance is becoming critical due to space constraints. We therefore discuss a method in this paper to meet this objective.

Numerous methods have been proposed to optimize decap design in PDNs. Stochastic optimization methods such as genetic algorithm and particle swarm optimization have been utilized for the selection and placement optimization of decaps [1], [2]. In [3], [4], a reinforcement machine learning technique is proposed to obtain optimal decap designs. However, these methods are based on random exploration; therefore, a large number of PDN simulations are required to find the optimal design especially when the target impedance is difficult to achieve.

In this paper, we propose a method to optimize PDNs with the minimum number of capacitors. During the iterative process, the decaps are chosen through several steps, thereby optimizing the PDN to meet the target impedance in the frequency range of interest. Compared to full search, machine learning (ML), and commercial tools available, the method discussed in this paper provides for fast convergence with significantly fewer PDN simulations.

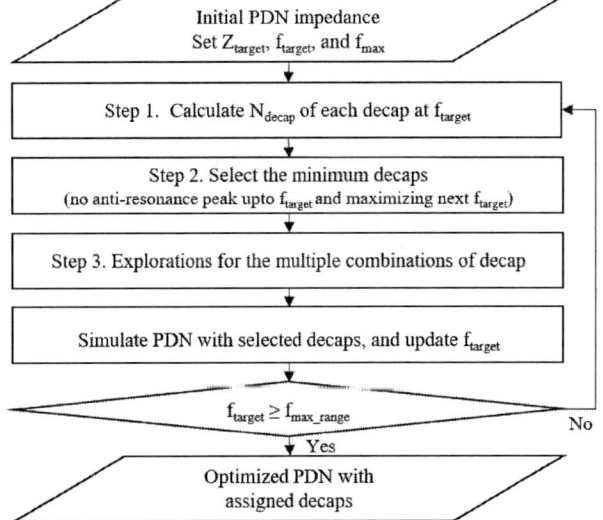

Fig. 1. Flowchart of the proposed decoupling capacitor optimization method

II. PROPOSED METHOD

A. Details of the Technique

The proposed method is based on an iterative process with three steps per cycle, as shown in Fig. 1. In each cycle, decaps are chosen from a decap library and assigned to the PDN. The lowest frequency at which the self-impedance at the IC port is greater than the target impedance is set as the target frequency, f_{target} where this parameter is updated after a decap is assigned in each cycle. For PDN analysis in this section, we use the PDN equivalent circuit shown in Fig. 2. Here, the decap is represented using a capacitor with equivalent series resistance (ESR), inductance (ESL), and surface mount inductance.

In step 1, the number of each decap required to decrease self-impedance below the target impedance at f_{target} is determined using the equation:

$$N_{decap} = \left\lceil \frac{|Z_{decap}(f_{target})|}{Z_{target}} \right\rceil \quad (1)$$

where $Z_{decap}(f_{target})$ is the impedance of each decap at

978-1-7281-6162-4/20 $31.00 © 2020 IEEE

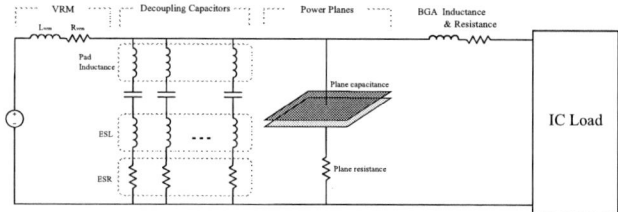

Fig. 2.   Equivalent circuit model of the PDN

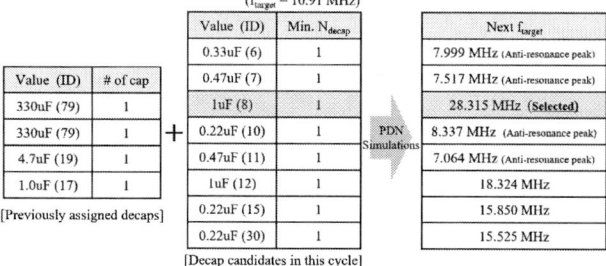

Fig. 3.   Example of how to select the decap at f_{target} in step 2

f_{target} given by:

$$Z_{decap}(f_{target}) = R_{ESR} + j\omega_{target}(L_{ESL} + L_{mount})$$
$$+ \frac{1}{j\omega_{target}C} \quad (2)$$

where R_{ESR}, L_{ESL}, L_{mount}, and C are the ESR, ESL, surface mount inductance, and capacitance of decaps, respectively. The decaps with minimum N_{decap} value are considered the preferred ones to be selected for f_{target}.

In step 2, among the decaps with the minimum N_{decap} value, we select the decap that can maximize the next f_{target} without causing an anti-resonance peak exceeding the target impedance when assigned to the PDN, as shown in Fig. 3. If none of the decaps with the minimum N_{decap} satisfy these conditions, the decaps with the next greater N_{decap} become candidates. This process continues until decaps satisfying the required conditions are found.

The PDN can be optimized with selected decaps using these two steps. However, to minimize the number of decaps in the final decap solution, we need exploration steps to search for combinations of different types of decaps that lead to the minimum. This is implemented in step 3. The value of N_{decap} of the selected decap in step 2 can be 2 or greater. In this case, if f_{target} can be increased by using a smaller number of the selected decaps than N_{decap}, a smaller number of decaps needs to be assigned in this step. Then, other types of decaps are searched for the increased f_{target} in the next cycle. This provides the opportunity to search for combinations of different types of decaps that can achieve a higher f_{target}. Another exploration step is considering decaps that cannot be selected in step 2 as they cause anti-resonances. In rare cases, the anti-resonance peak in the frequency range up to f_{target} can be decreased below the target impedance using the assigned

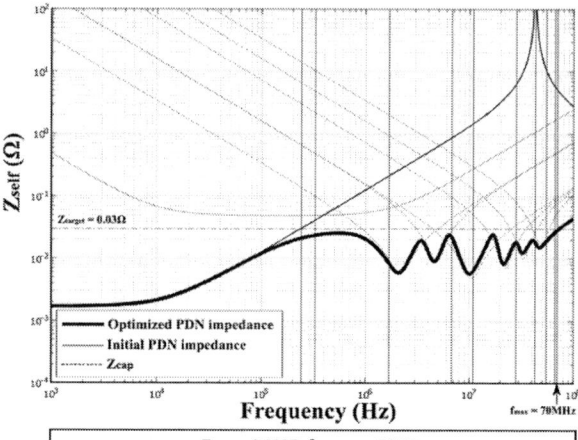

Fig. 4.   Optimized PDN impedance below the target impedance, 0.03 Ω by selected decoupling capacitor using the proposed method

Cycle	Value (Decap ID)	Required #	f_{target}
1	330uF (79)	1	0.32 MHz
2	330uF (79)	1	1.69 MHz
3	4.7uF (19)	1	6.77 MHz
4	1uF (17)	1	10.91 MHz
5	1uF (8)	1	28.31 MHz
6	0.22uF (5)	1	42.85 MHz
7	0.1uF (4)	1	54.95 MHz
8	0.047uF (3)	1	66.23 MHz
9	0.047uF (3)	1	73.45 MHz
Total # of capacitors required		9	

TABLE I.   REQUIRED NUMBER OF PDN SIMULATION FOR THE PROPOSED METHOD, FULL SEARCH, AND DQN-BASED METHOD

f_{max}=70MHz Z_{target}	# of capacitors	Number of PDN simulations		
		Proposed method	DQN-based method	Full search method
0.050 Ω	5	38	5.5×10^3	5.7×10^5
0.040 Ω	6	59	7.1×10^3	3.8×10^6
0.035 Ω	7	42	1.2×10^4	2.2×10^7
0.030 Ω	9	41	1.5×10^4	1.1×10^8
0.027 Ω	10	37	1.8×10^4	5.6×10^8
0.0255 Ω	11	59	3.8×10^4	2.4×10^9

decap in the next cycle. Therefore, in the last step of each cycle, we check whether a higher f_{target} can be achieved by switching the selected decap in a previous cycle to the decaps excluded because of anti-resonance. If a higher f_{target} with a decreased anti-resonance peak is obtained, the selected decap in the previous cycle is replaced with the selected decap.

After completing all the steps for selecting the decap in one cycle, if the next f_{target} is greater than the maximum frequency in the frequency range of interest, the iterative process is completed and the total selected decaps becomes the decap solution for PDN optimization. If the next f_{target} is in the frequency range of interest, we return to the beginning of the cycle and select another decap to increase f_{target} until it reaches the maximum frequency range.

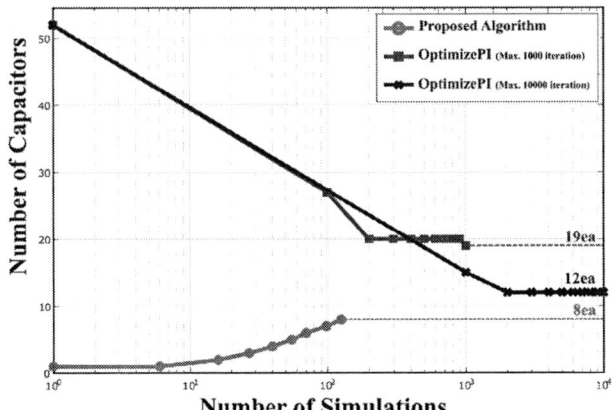

Fig. 5. Convergence comparison for the proposed method and the optimization function when Z_{target} is 0.15 Ω in the frequency range up to 45 MHz

B. Test Results

Using the proposed algorithm, the decap solution that meets the target impedance over a frequency range up to 70 MHz can be obtained, as shown in Fig. 4. Here, the minimum number of decaps are selected in each cycle from the decap library consisting of 83 different decaps to satisfy the target impedance of 0.03 Ω. Note that f_{target} is always increased after each cycle is completed, and the iterative process for the optimization is terminated when f_{target} exceeds 70MHz. To achieve this decap solution, the proposed method only requires 85 PDN simulations.

To verify the performance of the proposed method, the results obtained using the proposed method are compared with those of full search and ML (deep Q-network, DQN) based method for several target impedance cases. The results are summarized in Table I. Here, 35 different decaps in the decap library are considered for the selection. Using the proposed method, the same number of assigned decaps can be achieved as the optimal decap solutions that can be obtained by the full search and DQN based method. To obtain the decap solution using the full search and DQN-based method, a large number of PDN simulations are required, but the proposed algorithm requires less than 100 PDN simulations. This demonstrates the advantages of the proposed algorithm in terms of computing time for PDN optimization.

III. ON-BOARD PDN OPTIMIZATION

We apply the proposed method where data is gathered using a commercial tool (Sigrity OptimizePI) [5] for on-board PDN optimization and compare the result with the decap optimization function provided in the tool. Unlike the case of the PDN equivalent circuit model, Z_{decap} in (2) is unavailable in step 1 because of the parasitic L and R between the cap node and IC port. Therefore, to obtain N_{decap} for each decap in step 1, we use the self-impedance at f_{target} when each decap is assigned to the designated cap node. Here, we assume that the decap placement priority is determined in the order in which it is close to the IC port.

TABLE II. OPTIMIZED DECAP RESULTS WITH THE PROPOSED METHOD AND THE OPTIMIZATION FUNCTION

	Z_{target}= 0.15 Ω, f_{max}= 45 MHz	
Method	**Number of Simulations**	**Number of Decaps**
Proposed method	126	8
Optimization function	1,000 (Max. setting)	19
Optimization function	10,000 (Max. setting)	12

Using the proposed method and the provided optimization function from the tool, we optimize a four-layer PCB board consisting of VRM, power planes, and the target IC. The optimized decap results are provided in Fig. 5 and Table II. Two results from the optimization function in the tool are obtained by setting the different maximum number of PDN simulations, 1,000 and 10,000. As the simulations continue, the optimization function provides fewer number of decaps, and the number of decaps converges to 19 and 12, respectively. Using the proposed method, the decaps keep being assigned until the target impedance is satisfied in the frequency range of interest. The result obtained by the proposed method shows that only 126 PDN simulations are required to obtain the final decap solution, and the number of decaps of the solution is fewer (8 decaps) compared to the optimization function.

IV. CONCLUSION

This paper proposes a non-random exploration based method to minimize the number of decaps for PDN optmization. During the iterative process, the minimum decaps are assigned to the PDN in each cycle to increase the frequency range where the PDN meets the target impedance. For verification, the proposed method is applied to various examples. The results show that the proposed method provides the optimized decap solution with significantly fewer number of PDN simulations compared to full search, ML-based method, and a commercial tool. Our conclusion is that an algorithm based on domain expertise is sufficient for addressing this problem as compared to more sophisticated methods.

REFERENCES

[1] J. Y. Choi and M. Swaminathan, "Decoupling capacitor placement in power delivery networks using mfem," *IEEE Transactions on Components, Packaging and Manufacturing Technology*, vol. 1, no. 10, pp. 1651–1661, 2011.

[2] P. Kadlec, M. Marek, M. Štumpf, and V. Šeděnka, "Pcb decoupling optimization with variable number of capacitors," *IEEE Transactions on Electromagnetic Compatibility*, 2018.

[3] H. Park, J. Park, S. Kim, D. Lho, S. Park, G. Park, K. Cho, and J. Kim, "Reinforcement learning-based optimal on-board decoupling capacitor design method," in *2018 IEEE 27th Conference on Electrical Performance of Electronic Packaging and Systems (EPEPS)*. IEEE, 2018, pp. 213–215.

[4] L. Zhang, Z. Zhang, C. Huang, H. Deng, H. Lin, B.-C. Tseng, J. Drewniak, and C. Hwang, "Decoupling capacitor selection algorithm for pdn based on deep reinforcement learning," in *2019 IEEE International Symposium on Electromagnetic Compatibility, Signal & Power Integrity (EMC+ SIPI)*. IEEE, 2019, pp. 616–620.

[5] OptimizePI. [Online]. Available: https://www.cadence.com/en_US/home/tools/ic-package-design-and-analysis/si-pi-analysis-point-tools/sigrity-optimizepi-technology.html

High-Dimensional Uncertainty Quantification via Active and Rank-Adaptive Tensor Regression

Zichang He and Zheng Zhang

Department of Electrical and Computer Engineering, University of California, Santa Barbara, CA 93106
Emails: zichanghe@ucsb.edu, zhengzhang@ece.ucsb.edu

Abstract—**Uncertainty quantification based on stochastic spectral methods suffers from the curse of dimensionality. This issue was mitigated recently by low-rank tensor methods. However, there exist two fundamental challenges in low-rank tensor-based uncertainty quantification: how to automatically determine the tensor rank and how to pick the simulation samples. This paper proposes a novel tensor regression method to address these two challenges. Our method uses an $\ell_{2,p}$-norm regularization to determine the tensor rank and an estimated Voronoi diagram to pick informative samples for simulation. The proposed framework is verified by a 19-dim phonics bandpass filter and a 57-dim CMOS ring oscillator, capturing the high-dimensional uncertainty well with only 90 and 290 samples respectively.**

I. INTRODUCTION

Fabrication process variations are a major concern in nano-scale chip design. To estimate and quantify the uncertainties caused by process variations, Monte Carlo (MC) is the mainstream uncertainty quantification (UQ) tool used in commercial EDA tools, but it requires a huge amount of simulation samples. Instead, stochastic spectral methods based on generalized polynomial chaos (gPC) [1] offer an efficient alternative by approximating a stochastic circuit performance metric as a linear combination of some basis functions [2–4]. However, stochastic spectral methods suffer from the curse of dimensionality: a huge amount of simulation samples are required when the number of random parameters is large.

Low-rank tensor methods are a promising technique to solve high-dimensional UQ problems [5–8]. In [6], a high-dimensional gPC expansion is obtained via a low-rank tensor recovery, which estimates massive unknown output samples from a few simulation results. However, the method [6] uses a fixed tensor rank, which is hard to estimate *a-priori* in practice. The most recent work [9] uses a greedy rank-1 update until a good accuracy is reached. However, greedy rank-1 tensor update does not provide optimal solutions and can cause over fitting. Besides, it is not clear how to adaptively pick the simulation samples to reduce the computation budget.

Contributions. This paper proposes a novel high-dimensional UQ solver based on tensor regression. In order to automatically determine the tensor rank, we employ a group-sparsity regularization in the training process. We also develop an adaptive sampling strategy to reduce the simulation cost. This method balances exploration and exploitation of our

This work was supported by NSF grants #1763699 and #1846476.

model. Our method is used to quantify the uncertainties of a 19-dim phonic IC and a 57-dim electronic IC with 90 and 290 simulation samples respectively.

II. BACKGROUND

Generalized Polynomial Chaos. Let $\boldsymbol{\xi} = [\xi_1, \ldots, \xi_d] \in \mathbb{R}^d$ be a random vector describing process variations. We aim to estimate the interested performance metric $y(\boldsymbol{\xi})$ (e.g., chip frequency or power) under such uncertainty. A truncated gPC expansion approximates $y(\xi)$ as

$$y(\boldsymbol{\xi}) \approx \hat{y}(\boldsymbol{\xi}) = \sum_{\boldsymbol{\alpha} \in \Theta} c_{\boldsymbol{\alpha}} \Psi_{\boldsymbol{\alpha}}(\boldsymbol{\xi}), \qquad (1)$$

where $\boldsymbol{\alpha}$ is an index vector, and $\Psi_{\boldsymbol{\alpha}}$ is a polynomial basis function of degree $|\boldsymbol{\alpha}| = \alpha_1 + \alpha_2 + \cdots + \alpha_d$. If the joint probability density function of $\boldsymbol{\xi}$ is $\rho(\boldsymbol{\xi})$, then the basis functions satisfy the orthornormal condition:

$$\langle \Psi_{\boldsymbol{\alpha}}(\boldsymbol{\xi}), \Psi_{\boldsymbol{\beta}}(\boldsymbol{\xi}) \rangle = \int_{\mathbb{R}^d} \Psi_{\boldsymbol{\alpha}}(\boldsymbol{\xi}) \Psi_{\boldsymbol{\beta}}(\boldsymbol{\xi}) d\boldsymbol{\xi} = \delta_{\boldsymbol{\alpha}, \boldsymbol{\beta}}. \qquad (2)$$

Once the index set Θ is chosen, we need to determine the unknown coefficient $c_{\boldsymbol{\alpha}}$ for each $\boldsymbol{\alpha} \in \Theta$. The gPC only requires a small number of basis functions and simulation samples when the parameter dimensionality d is small. However, a huge number of basis functions and simulation samples are required when d is large. For instance, in the classical stochastic collocation method [1], the number of simulation samples required to obtain $c_{\boldsymbol{\alpha}}$'s is an exponential or polynomial function of d.

Tensors. A promising tool to overcome the curse of dimensionality is tensors. A d-dim tensor $\mathcal{X} \in \mathbb{R}^{n_1 \times \cdots n_d}$ represents a d-dimensional data array, and it becomes a matrix when $d = 2$. The $(i_1, \cdots, i_d)$-th element of $\mathcal{X}$ can be denoted as $x_{i_1 \cdots i_d}$. Given two tensors $\mathcal{X}$ and $\mathcal{Y}$ of same size, their inner product is defined as:

$$\langle \mathcal{X}, \mathcal{Y} \rangle := \sum_{i_1 \cdots i_d} x_{i_1 \cdots i_d} y_{i_1 \cdots i_d}. \qquad (3)$$

A d-dim rank-R tensor can be written as the sum of R rank-1 tensors, known as a CP decomposition:

$$\mathcal{X} = \sum_{r=1}^{R} \mathbf{u}_r^{(1)} \circ \mathbf{u}_r^{(2)} \cdots \circ \mathbf{u}_r^{(d)} = [\![\mathbf{U}^{(1)}, \mathbf{U}^{(2)}, \ldots, \mathbf{U}^{(d)}]\!], \quad (4)$$

where $\circ$ denotes an outer product. The last term is the Krusal form, where factor matrix $\mathbf{U}^{(k)} = \left[\mathbf{u}_1^{(k)}, \ldots, \mathbf{u}_R^{(k)} \right] \in \mathbb{R}^{n_k \times R}$ includes all vectors associated with the k-th dimension.

978-1-7281-6162-4/20 $31.00 © 2020 IEEE

III. PROPOSED TENSOR REGRESSION METHOD

A. Tensor Regression Formulation

We choose the following index set for the gPC expansion:

$$\Theta = \left\{ \boldsymbol{\alpha} = [\alpha_1, \alpha_2, \cdots, \alpha_d] \mid 0 \leq \{\alpha_i\}_{i=1}^d \leq p \right\}. \quad (5)$$

This specifies a gPC expansion with $(p+1)^d$ basis functions. Let $i_k = \alpha_k + 1$, then we can define two d-dimensional tensors $\mathcal{X}$ and $\mathcal{B}(\boldsymbol{\xi})$ with their $(i_1, i_2, \cdots i_d)$-th element specified as

$$x_{i_1 i_2 \cdots i_d} = c_{\boldsymbol{\alpha}} \text{ and } b_{i_1 i_2 \cdots i_d}(\boldsymbol{\xi}) = \Psi_{\boldsymbol{\alpha}}(\boldsymbol{\xi}). \quad (6)$$

Combining (1), (5) and (6), the truncated gPC expansion can be written as a tensor regression model

$$y(\boldsymbol{\xi}) \approx \hat{y}(\boldsymbol{\xi}) = \langle \mathcal{X}, \mathcal{B}(\boldsymbol{\xi}) \rangle. \quad (7)$$

It is worth noting that tensor $\mathcal{B}(\boldsymbol{\xi})$ depends on $\boldsymbol{\xi}$. When the random parameters $\boldsymbol{\xi}$ are mutually independent, $\Psi_{\boldsymbol{\alpha}}(\boldsymbol{\xi})$ can be written as the product of d uni-variable basis functions for each parameter ξ_k. In this case $\mathcal{B}(\boldsymbol{\xi})$ is a rank-1 tensor.

Our goal is to compute $\mathcal{X}$ given a set of data samples $\{\boldsymbol{\xi}_n, y(\boldsymbol{\xi}_n)\}_{n=1}^N$. Assume that $\mathcal{X}$ has the rank-R decomposition in (4), we can solve the following optimization problem

$$\min_{\{\mathbf{U}^{(k)}\}_{k=1}^d} h(\mathcal{X}) = \frac{1}{2} \sum_{n=1}^N \left(y_n - \langle [\![\mathbf{U}^{(1)}, \mathbf{U}^{(2)}, \ldots, \mathbf{U}^{(d)}]\!], \mathcal{B}_n \rangle \right)^2, \quad (8)$$

where $y_n = y(\boldsymbol{\xi}_n)$ and $\mathcal{B}_n = \mathcal{B}(\boldsymbol{\xi}_n)$.

B. Automatic Rank Determination

The low-rank tensor regression (8) requires the rank of $\mathcal{X}$ to be determined *a-priori*, which is often infeasible in practice. In order to address this issue, we first choose a sufficiently large R such that it is above the actual rank, then we choose a proper rank-shrinking penalty function to regularize (8).

Specifically, we employ a group $\ell_{2,q}$-norm regularization function to shrink the rank of $\mathcal{X}$:

$$g(\mathcal{X}) = \sum_{r=1}^R \left[\left(\sum_{k=1}^d \|\mathbf{u}_r^{(k)}\|_2^2 \right)^{\frac{1}{2}} \right]^{\frac{1}{q}}, \quad q \in [0, 1]. \quad (9)$$

This function puts $\mathbf{u}_r^{(k)}$, the r-th column of each $\mathbf{U}^{(k)}$, in the same group, and measures the $\ell_{2,p}$ norm of all groups. As a result, one can shrink some groups to zero by reducing $g(\mathcal{X})$, leading to an automatical rank reduction. A smaller q leads to a stronger shrinkage, and $q = 1$ corresponds to a group lasso.

By adding the penalty term (9), we have the following improved tensor regression model:

$$\min_{\{\mathbf{U}^{(k)}\}_{k=1}^d} f(\mathcal{X}) = h(\mathcal{X}) + \lambda g(\mathcal{X}). \quad (10)$$

After solving this optimization problem, each obtained factor matrix $\mathbf{U}^{(k)}$ has a few common columns whose values are close to zero. These columns can be deleted and the actual rank of our obtained tensor becomes $\hat{R} \leq R$, where $\hat{R}$ is the number of remaining columns that are not deleted.

It is non-trivial to minimize $f(\mathcal{X})$ since the regularization function $g(\mathcal{X})$ is usually non-differentiable and non-convex

with respect to $\mathbf{U}^{(k)}$'s. Instead, we solve the following optimization problem in practice:

$$\min_{\{\mathbf{U}^{(k)}\}_{k=1}^d} \hat{f}(\mathcal{X}) = h(\mathcal{X}) + \lambda \hat{g}(\mathcal{X}). \quad (11)$$

Here $\hat{g}(\mathcal{X})$ is an upper bound of $g(\mathcal{X})$ obtained via the variational inequality [10]:

$$g(\mathcal{X}) \leq \hat{g}(\mathcal{X}) = \min_{\boldsymbol{\eta} \in \mathbb{R}^R} \frac{\lambda}{2} \sum_{r=1}^R \frac{\sum_{k=1}^d \|\mathbf{u}_r^{(k)}\|_2^2}{\eta_r} + \frac{\lambda}{2} \|\boldsymbol{\eta}\|_q. \quad (12)$$

Once $\{\mathbf{U}^{(k)}\}_{k=1}^d$ is given, the values of $\hat{g}(\mathcal{X})$ and $\hat{f}(\mathcal{X})$ can be estimated by setting the elements of $\boldsymbol{\eta}$ as

$$\eta_r = (z_r)^{\frac{2}{q+1}} \|\mathbf{z}\|_{q_1}^{q_2}, \quad \forall \, r = 1, \ldots, R, \quad (13)$$

where $z_r = \left(\sum_{k=1}^d \|\mathbf{u}_r^{(k)}\|_2^2 \right)^{\frac{1}{2}}$, $q_1 = \frac{2q}{q+1}$ and $q_2 = \frac{q-1}{q+1}$.

Problem (11) can be solved via an alternating algorithm such as a block coordinate descent solver or alternating direction method of multipliers. Due to the page limitation, we omit the details in this paper and will explain the detailed optimization algorithm in an extended journal paper.

C. Adaptive Sampling Strategy

Another fundamental question is how to select the parameter samples $\{\boldsymbol{\xi}_n\}_{n=1}^N$ for simulation. The method in [7] uses some Monte Carlo random samples. Instead, this paper reduces the simulation cost by selecting only a few informative samples for detailed device- or circuit-level simulations.

We first use the Latin Hybercube (LH) sampling to generate an initial sample set Ω. Then we employ an exploration step via the Voronoi diagram to measure the sample density in Ω. Given two distinct samples $\boldsymbol{\xi}_i, \boldsymbol{\xi}_j \in \Omega$, a Voronoi cell $C_i(\boldsymbol{\xi}_i)$ covers the region that are closest to $\boldsymbol{\xi}_i$. It is defined as the intersection of a set of half-planes (hp):

$$C_i(\boldsymbol{\xi}_i) = \bigcap_{\boldsymbol{\xi}_j \in \Omega \setminus \boldsymbol{\xi}_i} \mathrm{hp}(\boldsymbol{\xi}_i, \boldsymbol{\xi}_j) \quad (14)$$

$$\mathrm{hp}(\boldsymbol{\xi}_i, \boldsymbol{\xi}_j) = \{\boldsymbol{\xi} \in \mathbb{R}^d \mid \|\boldsymbol{\xi} - \boldsymbol{\xi}_i\| \leq \|\boldsymbol{\xi} - \boldsymbol{\xi}_j\|\}.$$

It is intractable to calculate an Voronoi cell exactly in a high-dimensional space. However, we can easily estimate it via Monte Carlo [11]. The sample density of C_i is approximated by counting the number of samples that are closest to $\boldsymbol{\xi}_i$. Each sample in Ω determines one Voronoi cell with itself as the center, and we can select a new sample from the cell region with the lowest density.

If the performance metric $y(\boldsymbol{\xi})$ is known to be highly nonlinear, we can further exploit its non-linearity. Given $\boldsymbol{\xi}$ and a Voronoi cell center $\mathbf{a}$, we measure the nonlineary of $y(\boldsymbol{\xi})$ as

$$\gamma(\boldsymbol{\xi}) = |\hat{y}(\boldsymbol{\xi}) - \hat{y}(\mathbf{a}) - \nabla \hat{y}(\mathbf{a})^T (\boldsymbol{\xi} - \mathbf{a})|. \quad (15)$$

We select a new sample as the one with largest $\gamma(\boldsymbol{\xi})$ in a Voronoi cell with the lowest sample density. This method can be easily extended to a batch version by searching the top-K least-sampled regions.

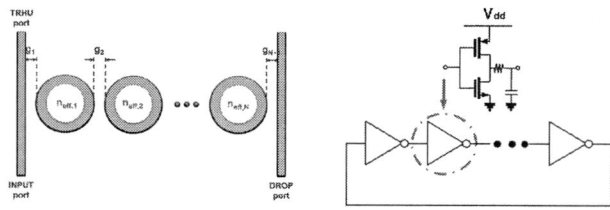

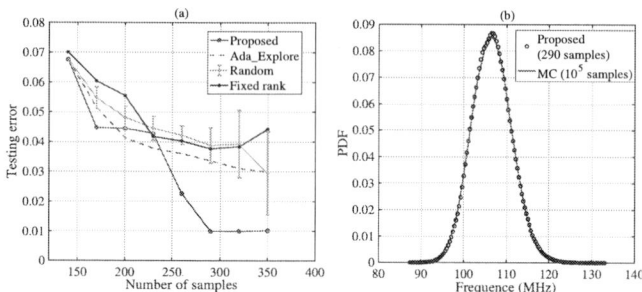

Fig. 1. Left: a photonic band-pass filter with 9 micro-ring resonators. Right: Schematic of a CMOS ring oscillator.

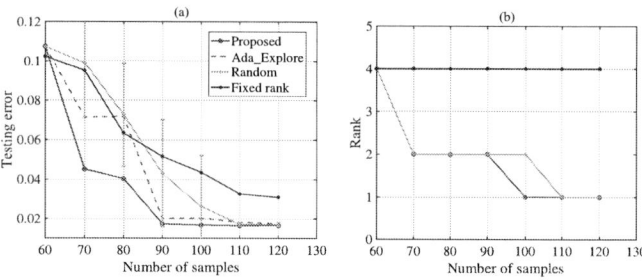

Fig. 2. Result of the photonic filter. (a) Testing error on 10^5 MC samples, achieving 98.3% accuracy. (b) The estimated tensor ranks.

Fig. 3. Results of the CMOS ring oscillator. (a) Testing error on 10^5 MC samples (b) Probability density functions of the oscillator frequency.

TABLE I

COMPARISON WITH THE GPC MODEL WITH A TOTAL DEGREE SCHEME.

	Proposed	Total-degree gPC	MC
# of variables	855	1711	N/A
# of samples	290	1711	10^5
Mean	106.28	106.58	106.53
Stdvar	4.616	6.810	4.641
Error	1%	4.84%	N/A

IV. NUMERICAL EXPERIMENT

We verify our algorithm by a photonic band-pass filter and a CMOS ring oscillator as shown in Fig. 1.

Photonic band-pass filter: This benchmark has 19 Gaussian random parameters describing the variations of the effective phase index (n_{neff}) of each ring, and the gap (g) between adjacent rings and between the end ring and the bus waveguides. We set the highest polynomial order $p = 2$ for each dimension, and initialize $\mathcal{X}$ as rank-4, and use $q = 0.5$ in the regularizer. We initialize 60 samples with a standard LH experimental design, and further select 6 batches with 60 additional samples. The proposed tensor regression framework is compared with a fixed rank method, a random and an adaptive exploration sampling method (shown in Fig. 2).

CMOS ring oscillator: This circuit has 57 Gaussian random parameters describing the variations of threshold voltages, gate-oxide thickness, and effective gate length/width. The simulation results are obtained by calling a periodic steady-state simulator repeatedly. We set basis as order-2 in each dimension, initialize $\mathcal{X}$ as rank-5, and use $q = 0.5$ in the regularizer. We start from 140 standard LH samples, and adaptively select additional 7 batches with 210 samples. The results and comparison are summarized in Fig. 3 and Table I.

V. CONCLUSION

This paper has proposed a tensor regression framework for high-dimensional uncertainty quantification. Our method has addressed two fundamental challenges: automatic tensor rank determination and adaptive sampling. The numerical result has demonstrated the excellent capability of automatic rank determination of our method, and the simulation cost reduction by our adaptive sampling method.

REFERENCES

[1] D. Xiu, *Numerical methods for stochastic computations: a spectral method approach.* Princeton university press, 2010.

[2] Z. Zhang, T. A. El-Moselhy, I. M. Elfadel, and L. Daniel, "Stochastic testing method for transistor-level uncertainty quantification based on generalized polynomial chaos," *IEEE Trans. CAD Integr. Circuits Syst.*, vol. 32, no. 10, pp. 1533–1545, 2013.

[3] Z. He, W. Cui, C. Cui, T. Sherwood, and Z. Zhang, "Efficient uncertainty modeling for system design via mixed integer programming," in *Proc. ICCAD*, 2019, pp. 1–8.

[4] D. V. Ginste, D. De Zutter, D. Deschrijver, T. Dhaene, P. Manfredi, and F. Canavero, "Stochastic modeling-based variability analysis of on-chip interconnects," *IEEE Trans. Compon. Packag. Manuf. Technol.*, vol. 2, no. 7, pp. 1182–1192, 2012.

[5] K. Konakli and B. Sudret, "Polynomial meta-models with canonical low-rank approximations: Numerical insights and comparison to sparse polynomial chaos expansions," *J. Comput. Phys.*, vol. 321, pp. 1144–1169, 2016.

[6] Z. Zhang, X. Yang, I. V. Oseledets, G. E. Karniadakis, and L. Daniel, "Enabling high-dimensional hierarchical uncertainty quantification by anova and tensor-train decomposition," *IEEE Trans. CAD Integr. Circ. Syst.*, vol. 34, no. 1, pp. 63–76, 2015.

[7] Z. Zhang, T.-W. Weng, and L. Daniel, "Big-data tensor recovery for high-dimensional uncertainty quantification of process variations," *IEEE Trans. Compon. Packag. Manuf. Technol.*, vol. 7, no. 5, pp. 687–697, 2017.

[8] M. Chevreuil, R. Lebrun, A. Nouy, and P. Rai, "A least-squares method for sparse low rank approximation of multivariate functions," *SIAM/ASA J. Uncertain. Quantif.*, vol. 3, no. 1, pp. 897–921, 2015.

[9] X. Shi, H. Yan, Q. Huang, J. Zhang, L. Shi, and L. He, "Meta-model based high-dimensional yield analysis using low-rank tensor approximation," in *Proc. DAC*, 2019, pp. 1–6.

[10] R. Jenatton, G. Obozinski, and F. Bach, "Structured sparse principal component analysis," in *Proc. Artif. Intell. Statist.*, 2010, pp. 366–373.

[11] K. Crombecq, I. Couckuyt, D. Gorissen, and T. Dhaene, "Space-filling sequential design strategies for adaptive surrogate modelling," in *Int. Conf. Soft Comput. Techn. Civil, Struct. Environ. Eng.*, 2009, pp. 1–20.

978-1-7281-6162-4/20 $31.00 © 2020 IEEE

A Parallel-in-Time Circuit Simulator for Power Delivery Networks with Nonlinear Load Models

Chung-Kuan Cheng[†‡], Chia-Tung Ho[‡], Chao Jiao[+], Xinyuan Wang[‡], Zhiyu Zeng[+], Xin Zhan[+]

[†]CSE and [‡]ECE Departments, UC San Diego, La Jolla, CA, USA

[+] Cadence Design Systems Inc, 12301 Research Blvd Building V Suite 200, Austin, TX 78759

email: {ckcheng,c2ho}@ucsd.edu, cjiao@cadence.com, xiw193@eng.ucsd.edu, {zzeng, xinzhan}@cadence.com

Abstract—**We apply the parallel-in-time method to the transient simulation of power delivery networks with nonlinear load models. With adaptive Newton-Raphson iterations, the parallel-in-time method achieves considerable speedup compared to the general sequential solver.**

Index Terms—**PDN transient simulation, nonlinear load model, parallel-in-time**

I. INTRODUCTION

In modern VLSI designs, the power integrity becomes a critical issue to ensure the reliability and performance of designs. The challenges of power integrity analysis arise from the tighter noise margin with reducing power supply voltage, higher resistance on metal wires due to scaling, and strong coupling noise between the active devices.

The simulation of power integrity analysis encounters the problems from the increasing size of power delivery networks (PDN) as well as the accuracy of load models. Due to the increasing design complexity, the PDNs could be extremely huge and stiff, which makes the simulation a critical task. To simplify the system-level power integrity analysis, the on-chip macrocells are usually characterized as independent current sources with linear elements. However, the accuracy of power grid analysis is lost and the results could be far from the real cases. An efficient simulation framework is in high demand to handle the issues.

In this paper, we propose a nonlinear macrocell model to capture the dynamic behavior of PDNs and we take advantage of the recent progress in the parallel-in-time approach, such as Parareal (Parallel in Real time) [8] and MGRIT (Multigrid Reduction in Time) [3], and applied the idea to the PDN transient simulations. The main contributions of this paper are listed as follows,

- We adopt a nonlinear voltage-dependent macrocell model in the PDN simulation framework to characterize the dynamic behaviors of whole systems.
- We apply the parallel-in-time method to parallelize the conventional sequential time stepping of the PDN transient simulation.
- We use the adaptive Newton-Raphson (NR) method to solve the nonlinear system efficiently in the iterations of step integrations.

The rest of this paper is organized as follows. In the next section, we introduce the formulation of the PDN transient simulations and nonlinear macromodels. In Sec. IV, we propose the application of the parallel-in-time method to the PDN transient simulations with adaptive NR iterations. Finally, a group of PDNs is used to validate our method. The experimental results are shown in Sec. V.

II. BACKGROUND

Given the circuit netlist and device models, the general formulation is shown as follows,

$$\frac{dq(x)}{dt} + f(x) = Bu(t), \qquad (1)$$

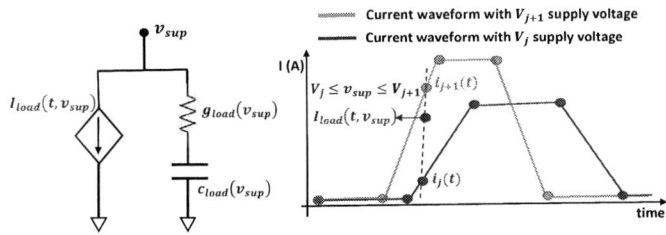

(a) Nonlinear load model (Macrocell) (b) An illustration of calculating $I_{load}(t, v_{sup})$

Figure 1: An example of nonlinear load model in PDN.

where $x \in \mathbb{R}^{n \times 1}$ is the vector of nodal voltages and branch currents. The charge/flux is represented by $q \in \mathbb{R}^{n \times 1}$ and $f \in \mathbb{R}^{n \times 1}$ contains the current/voltage terms. Vector $u(t)$ represents all the external excitations at time t and B inserts the signals to the system. If the element constitutive equations are linearized, we can represent Eq. 1 as the matrix form DAEs,

$$\mathcal{C}(x)\dot{x}(t) + \mathcal{G}(x)x(t) = Bu(x, t), \qquad (2)$$

where matrices $\mathcal{C}(x) \in \mathbb{R}^{n \times n}$ consists capacitance/inductance and $\mathcal{G}(x) \in \mathbb{R}^{n \times n}$ represents the conductance/resistance, respectively. Vector $u(x, t)$ contains the linear and nonlinear input sources. The elements are functions of x.

With given initial state $x(t)$ and assumption that the system is unchanged in the step from t to $t + h$, the linear multi-step integration methods are widely used approximate the solution in Eq. 2, such as Forward Euler (FE), Backward Euler (BE), Trapezoidal (TR) and explicit Matrix Exponential [10]. To guarantee the stability of algorithms [2], [12], we mainly talk about the BE method in this work.

III. NONLINEAR LOAD MODELS IN PDNs

We propose a nonlinear macrocell load model to include the effects of Dynamic Voltage Drop (DvD) in PDNs [6], [7], [11]. A voltage dependent current source $I_{load}(t, v_{sup})$ with series RC, $R_{load}(v_{sup})$ and $C_{load}(v_{sup})$, are used to model the current fluctuation caused by DvD at the power supply node (i.e., v_{sup}), as shown in Fig. 1. Our nonlinear load model provides the fixed pivot points information, which enable us to determine the simulation time points in advance.

The nonlinear load models are generated at different supply voltages (i.e., V_j). During the transient simulation, the values of elements can be interpreted based on v_{sup} at t as

$$I_{load}(t, v_{sup}) = i_j(t) + \frac{(i_{j+1}(t) - i_j(t))}{(V_{j+1} - V_j)}(v_{sup} - V_j)$$

$$g_{load}(v_{sup}) = g_j + \frac{(g_{j+1} - g_j)}{(V_{j+1} - V_j)}(v_{sup} - V_j)$$

$$c_{load}(v_{sup}) = c_j + \frac{(c_{j+1} - c_j)}{(V_{j+1} - V_j)}(v_{sup} - V_j) \qquad (3)$$

where v_{sup} lies between two supply voltages $[V_j, V_{j+1}]$ and the coefficients i, g, c represent the element values at each supply voltage in the macrocell model.

(1) Sequential Method

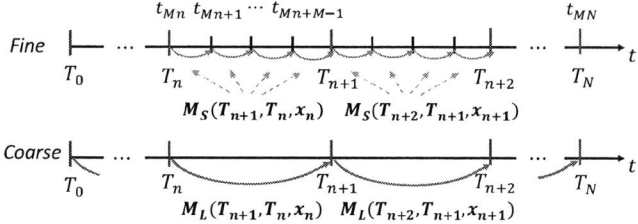

(2) MGRIT Method (Two levels)

Figure 2: In PDN transient simulation, step integrators are applied to (1) general sequential method and (2) MGRIT method with two levels.

IV. A PARALLEL-IN-TIME METHOD FOR PDN TRANSIENT SIMULATION

We propose a parallel-in-time method for nonlinear PDN transient simulations with the MGRIT method [3] and adaptive Newton-Raphson techniques, named as *MGRIT-AdapNR*. Firstly, we discuss the application of the MGRIT method to circuit simulation in Sec IV-A. Then we introduce the adaptive NR method to solve the nonlinear PDNs in Sec IV-B.

A. MGRIT method with Linear Step integrators

Parareal was first presented as a numerical method to solve evolution problems [8] and extended to PDEs with many follow ups. Consider the DAEs in Eq. 2, BE integration starts from

$$x(t + h) = x(t) + h\dot{x}(t + h), \tag{4}$$

which gives

$$\left(\frac{\mathcal{C}(x)}{h} + \mathcal{G}(x)\right)x(t + h) = \frac{\mathcal{C}(x)}{h}x(t) + Bu(x, t + h), \tag{5}$$

where we can define the operator $M = \left(\frac{\mathcal{C}(x)}{h} + \mathcal{G}(x)\right)^{-1}$ on the rhs of equation at t. The DAEs can be solved with linear step integration method with NR iterations.

Fig. 2 demonstrate (1) the general sequential integraton method and (2) the two level MGRIT method. We assume the fine time grids have uniform step size h. Each time interval in the coarse time grid equals $H = Mh$. Let $M_L(T_{n+1}, T_n, x_n)$ denote the long step integration on the coarse time grid from T_n to T_{n+1}, where $T_n = t_{Mn}$. Let $M_S(T_{n+1}, T_n, x_n)$ denote the short step integration on the fine grid which takes M steps from T_n to T_{n+1}. The MGRIT method performs k iterations and approximates the next approximation with the formulation:

$$x_{n+1}^{(k+1)} = M_L(T_{n+1}, T_n, x_n^{(k+1)}) + M_S(T_{n+1}, T_n, x_n^{(k)})$$
$$- M_L(T_{n+1}, T_n, x_n^{(k)}). \tag{6}$$

The first long step integration has to be performed sequentially in order to wait for x_n^{k+1}. The second and the third term only depends on results from the previous iteration, where the integrations between any time interval can be operated in parallel.

The interpretation of Parareal/MGRIT as a time multigrid method is well illustrated in previous work, detailed proof can be found in [4]. The iteration in MGRIT is consistent with the fine grid problem and the algorithm follows the linear convergence of multigrid methods [2], [4], [5].

B. Nonlinear Systems and Adaptive Newton-Raphson Iterations

For a nonlinear system, the implicit formulation Eq. 2 requires NR iterations to achieve a converged solution. We define the residual of the system at t as

$$r(x) \approx Bu(x, t) - \mathcal{C}(x)\dot{x}(t) - \mathcal{G}(x)x(t). \tag{7}$$

Based on the Taylor expansion around the current approximation $x^{(k)}$, the next approximation $x^{(k+1)}$ satisfies

$$0 = r(x^{(k+1)}) \approx r(x^{(k)}) + J(x^{(k)})(x^{(k+1)} - x^{(k)}), \tag{8}$$

where $J(x)$ is the Jacobian matrix with $J_{ij}(x) = \frac{\partial r_i}{\partial x_j}$. In practical circuit simulation, the $J(x)$ is given by the nonlinear elements and choice of multi-step method. The NR iterations follow the relation

$$x^{(k+1)} = x^{(k)} - J(x^{(k)})^{-1}r(x^{(k)}). \tag{9}$$

The corresponding Jacobian is updated at each iteration according to $x^{(k)}$. Either the residual $r(x^{(k+1)})$ is below given tolerance or the change of solution from $x^{(k)}$ to $x^{(k+1)}$ is small enough the iterations are terminated.

Unlike the traditional method where NR iterations are used at each step, adaptive NR (adap. NR) method skips the NR iterations if the change of x at $t + h$ satisfies

$$\|\Delta x^{(0)}\|_\infty \leq \Delta_{th}, \tag{10}$$

where $\Delta x^0 = x^{(0)}(t + h) - x(t)$ and Δ_{th} is the given threshold. Considering the nonlinear macrocell model is less sensitive to its voltage than transistors, we can set larger Δ_{th} to improve the performance.

V. EXPERIMENTAL RESULTS

The *MGRIT-AdapNR* is implemented via the open source software library Xbraid [1] in C++. All experiments are performed on a 1.8GHz Intel Xeon 24-CPUs server.

Table I shows the statistics of PDNs with size ranges from thousands to millions, where the design "genckt30" is created based on the specifications in [9] and used for optimum parameter exploration. For ibmpg1t-nl, ibmpg2t-nl, and ibmpg3t-nl, we extend the original power loads to nonlinear load models with the guidance from industry and use the original PDNs of ibmpg1t, ibmpg2t, and ibmpg3t [9]. The nonlinear load models are updated using Eq. 3 in the transient simulations. We compare *MGRIT-AdapNR* with Sequential solver (*Seq*) using NR iterations at each time step. The maximum absolute error e_{max} and average absolute error e_{avg} are calculated from the probing nodes of each design and reported in the following experiments. The runtime represents walltime.

Table I: Design specifications of PDNs

PDN	#R	#C	#L	#Loads	#Size	#Probing Nodes
genckt30	2.6K	1.4K	0	720	1.6K	90
ibmpg1t-nl	54K	11K	277	11K	40K	24
ibmpg2t-nl	245K	37K	330	37K	165K	20
ibmpg3t-nl	1.6M	201K	955	201K	1M	20

A. Study I: Linear vs Nonlinear Load Model

Fig. 3(a) shows the simulation results of a nodal waveform from ibmpg1t with linear load models and ibmpg1t-nl with nonlinear load models. The simulation time is 3ns with 900 time steps. The maximum IR drop with nonlinear load models is 92% larger, which is underestimated by the linear models. The nonlinear load model is essential for power integrity analysis.

978-1-7281-6162-4/20 $31.00 © 2020 IEEE

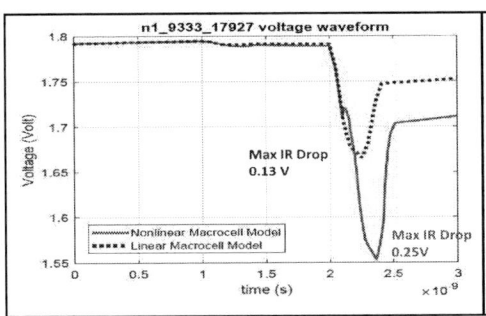
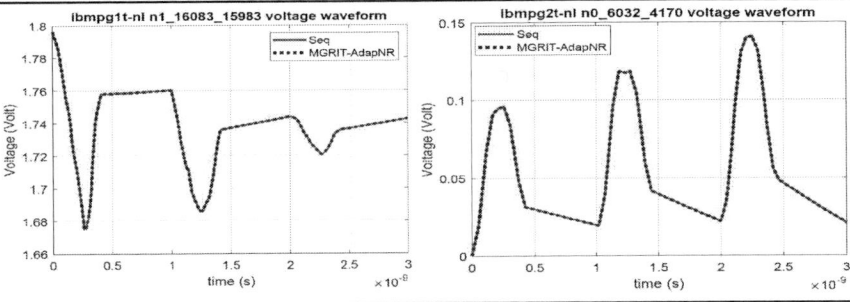

(a) Linear vs Nonlinear Macrocell Model with ibmpg1t PDN

(b) Seq vs MGRIT-AdapNR (Table IV)

Figure 3: (a) Linear vs Nonlinear Macrocell Model; (b) Nodal waveforms of *Seq* and *MGRIT-AdapNR* (Table IV).

B. Study II: Optimum Parameter Exploration

We perform multiple experiments on #Cores, Coarsening Factor (CF), and Maximum Level (ML) of *MGRIT-AdapNR* to find the optimum settings in terms of runtime and accuracy.

#Cores: Table. II shows the *MGRIT-AdapNR* runtime with 4, 8, 16, and 24 cores. Compared to the *Seq*, *MGRIT-AdapNR* with 24 cores achieves 2× speedup and the max error is $3mV$. The sublinear trend of Speedup is mainly caused by the overhead of communication for parallel processing. Once the #Cores exceeds a threshold, the speedup is close to linear [3]. Besides, the difference in max and avg errors of different #Cores is less than 1%. *MGRIT-AdapNR* is robust with various #Cores.

Table II: Experimental results of different #Cores using ibmpg1t-nl with 3ns simulation time and 900 time steps.

	#Cores	e_{max} (mV)	e_{avg} (mV)	Runtime (s)	Speedup (X)
Seq	1	-	-	4790.61	1
MGRIT-AdapNR	4	3.00	3.82E-3	6882.63	0.70
	8	3.00	3.83E-3	4250.47	1.13
	16	3.00	3.84E-3	3092.15	1.55
	24	3.00	3.84E-3	2493.53	1.92

Coarsening Factor (CF) and Maximum Level (ML): The CF defines the fine grid and coarse grid ratio at each level and ML defines the maximum level in multigrid. We use "genckt30" with 410K time steps to explore the optimum CF and ML to fully leverage the parallel-in-time advantage. Table III shows the results of *Seq* and *MGRIT-AdapNR* with various combinations of CF and ML. We select CF=2, 6, and 10. Then, we increase the ML from 2 to 10 with increment 2 until the time grids cannot be coarsened any more. From the results, CF=10 and ML=4 achieves the best performance. The max error is less than 1mV.

Table III: Experimental results of *Seq* and *MGRIT-AdapNR* (24 cores), with multiple combinations of CF and ML using genckt30 test case. Simulation time=6ns. #time steps=410K. Time Grid Ratio=(#Finest Time Grids)/(#Coarsest Time Grids).

	CF	ML	Time Grid Ratio	e_{max} (mV)	e_{avg} (mV)	Runtime (s)	Speedup (X)
Seq	-	-	1	-	-	1289.07	1
MGRIT-AdapNR	2	2	2	0.01	4.11E-4	1967.61	0.66
		4	8	0.06	5.86E-3	1011.81	1.27
		6	32	0.16	1.02E-2	793.43	1.62
		8	128	0.22	1.02E-2	730.05	1.77
		10	512	0.22	1.04E-2	710.14	1.82
	6	2	6	0.04	5.75E-3	938.4	1.37
		4	216	0.36	5.84E-3	445.83	2.89
		6	1296	1.10	2.91E-2	426.2	3.02
	10	2	10	0.07	7.70E-3	730.18	1.77
		4	1000	0.14	1.10E-2	390.74	3.30
		6	100000	5.60	2.86E-1	387.27	3.33

C. Main Results

Table IV shows our main results on PDNs in Table I. The simulation time of ibmpg1t-nl, ibmpg2t-nl, and ibmpg3t-nl are 3ns, 3ns, and 2ns with 900, 960, and 630 time steps, respectively. The *MGRIT-AdapNR* multigrid cycles of all three cases are 3.

The *MGRIT-AdapNR* multigrid cycles and Adap. NR reduce the #NewtonIters up to 30%. Compared to *Seq*, *MGRIT-AdapNR* achieves more than 2× speedup with less than 5mV max error. The *MGRIT-AdapNR* successfully captures the transient waveform of nonlinear PDNs, as shown in Fig. 3(b).

Table IV: Experimental results of *Seq* and *MGRIT-AdapNR* (#Core=24, CF=10 and ML=4). d_{max} is the absolute error of the max voltage fluctuation of the probing nodes.

	d_{max} (mV)	e_{max} (mV)	e_{avg} (mV)	#NewtonIters Seq	#NewtonIters Proposed	Runtime (s) Seq	Runtime (s) Proposed	Speedup (X)
ibmpg1t-nl	5.00E-2	3.00	3.84E-3	1982	1521	4790.61	2493.53	1.92
ibmpg2t-nl	8.00E-2	3.40	8.24E-2	2304	1662	17882.07	7947.37	2.25
ibmpg3t-nl	1.00E-2	2.54	3.12E-2	1824	1256	102683.35	43430.18	2.36

VI. CONCLUSION

We develop the *MGRIT-AdapNR* for the transient analysis of PDNs with nonlinear load models, where the time integration is parallelized. Compared to the *Seq*, *MGRIT-AdapNR* achieves 3× speedup on long simulation time (410K time steps) and 2× speedup on the PDNs from 40K to 1M size. Without the limitation of maximum #Cores on our server, we expect that *MGRIT-AdapNR* can achieve more speedups. The future research directions include (i) exploring the performance improvement of *MGRIT-AdapNR* with more cores and (ii) improving the convergence rate using advance integrators such as Matrix Exponential [10].

VII. ACKNOWLEDGMENTS

We acknowledge the support from NSF CCF-1564302 and the advice from Prof. Albert Chern at UC San Diego and Dr. Robert D. Falgout at Lawrence Livermore National Laboratory.

REFERENCES

[1] Xbraid: Parallel multigrid in time. http://llnl.gov/casc/xbraid.

[2] V. Dobrev et al. Two-level convergence theory for multigrid reduction in time (mgrit). *SIAM Journal on Scientific Computing*, 39(5):S501–S527, 2017.

[3] R. D. Falgout et al. Parallel time integration with multigrid. *SIAM Journal on Scientific Computing*, 36(6):C635–C661, 2014.

[4] M. J. Gander et al. Analysis of the pararaeal time-parallel time-integration method. *SIAM Journal on Scientific Computing*, 29(2):556–578, 2007.

[5] M. J. Gander et al. Nonlinear convergence analysis for the parareal algorithm. In *Domain decomposition methods in science and engineering XVII*, pages 45–56. Springer, 2008.

[6] H. Harizi et al. Efficient modeling techniques for dynamic voltage drop analysis. In *Proc. DAC*, pages 706–711. IEEE, 2007.

[7] P.-Y. Hsu et al. Adaptive sensitivity analysis with nonlinear power load modeling. In *SLIP*, pages 1–6. IEEE, 2018.

[8] J.-L. Lions et al. A pararaeal in time discretization of pde's. In *C.R. Acad. Sci. Paris Ser. I Math*, pages 661–668, 2001.

[9] S. R. Nassif. Power grid analysis benchmarks. In *Proc. DAC, ASPDAC*, pages 376–381. IEEE, 2008.

[10] X. Wang et al. Stability and convergency exploration of matrix exponential integration on power delivery network transient simulation. *TCAD*, 2019.

[11] X. Zhang et al. Power distribution network design optimization with on-die voltage-dependent leakage path. In *EPEPS*, pages 87–90. IEEE, 2013.

[12] H. Zhuang et al. From circuit theory, simulation to *spice^{Diego}*: A matrix exponential approach for time-domain analysis of large-scale circuits. *IEEE Circuits and Systems Magazine*, 16(2):16–34, 2016.

Design, Simulation and Measurement of a Flexible Voltage-controlled Oscillator (VCO) Chip with Bending Radius

Seungtaek Jeong, Seongsoo Lee, Seokwoo Hong, Boogyo Sim, Hyunwook Park, Subin Kim, Youngwoo Kim, Keeyeong Son, and Jeongho Kim
TeraByte Interconnection and Package Laboratory
Korea Advanced Institute of Science and Technology
Daejeon, Republic of Korea
seungtaek@kaist.ac.kr

Jaehak Lee, and Junyeop Song
Department of Ultra-Precision Machines and System
Korea Institute of Machinery & Materials
Daejeon, Republic of Korea
jaehak 76@kimm.re.kr

Abstract—In this paper, we designed, simulated and measured an extremely thin flexible LC voltage-controlled oscillator (VCO) with bending radius for flexible electronics. An LC VCO chip is fabricated using a SK Hynix 0.18 µm process. A silicon substrate of the fabricated VCO chip is grinded to 50 µm to achieve the flexibility. The flexible VCO chip is bent with a bending device to check voltage output and phase with bending radius. The bending radius is applied from infinite to 20 mm. The results showed that the designed flexible VCO chip is operating properly with the bent structure.

Keywords— flexible chip, flexible interconnects, flexible PCB, signal integrity, S-parameter, thin silicon substrate, voltage-controlled oscillator.

I. INTRODUCTION

Flexible electronic devices have been in the spotlight in recent years [1]-[4]. Those electronics are useful when the devices are applied to a human body for monitoring human health conditions. Since there are no rigid parts in the devices, they can be worn without discomfort. The flexible electronics use flexible substrates such as Polyimide, Polyethylene terephthalate (PET), and Urethane. Moreover, the attached silicon (Si) chips must flexible to avoid the failure between the flexible substrate and Si chip.

Firstly, electrical characteristics of flexible PCBs have been studied [5]. Through these studies, the results showed that the electrical characteristics of flexible PCB coils are not significantly affected by the variation of bending radius. Moreover, passive interconnection lines on a Si chip is studied [6], [7]. In these studies, the electrical characteristics of the passive interconnection lines using extremely thin Si substrate are analyzed with bending. However, not only the passive interconnection lines, study on active circuits with the extremely thin Si substrate are important the check the feasibility of the flexible electronics implementation.

In this paper, an active chip, including CMOS is designed, fabricated, and measured to see the flexibility of the flexible system operation with mechanical bending. For an active circuit, a voltage-controlled oscillator (VCO) is designed using

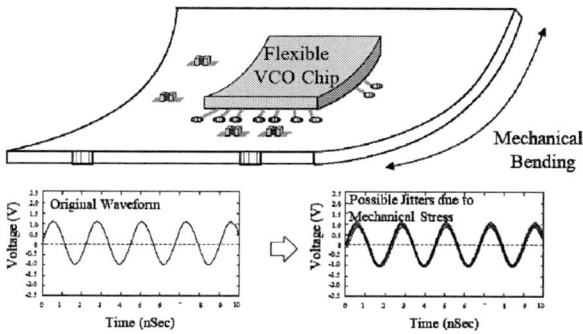

Fig. 1. Flexible voltage-controlled oscillator (VCO) chip with bending radius that may cause time-domain jitter at the output.

the SK Hynix 0.18 µm process. The VCO is the sensitive circuit and output electrical characteristics can be changed with bending which might cause system malfunctions as illustrated is Figure 1. Therefore, we need to prove that the flexible chip operates properly with the bending to realize the flexible electronics.

Firstly, we have designed the LC VCO that operates at 433.42 MHz considering industry science medical (ISM) and telecommunication frequency bands. The ISM bands are radio frequency (RF) bands allocated internationally for the use of RF energy for industrial, scientific and medical applications. Therefore, by using the ISM bands as the communication frequency, lower power wireless devices are frequently used without the permission.

II. DESIGN OF FLEXIBLE VOLTAGE-CONTROLLED OSCILLATOR CHIP ON SI SUBSTRATE

Figure 2 depicts the designed LC VCO using CMOS technology. Two PMOS and two NMOS are used to create negative resistance for an output oscillation. On-chip inductor and varactor circuit are used as a resonant tank for the oscillation. For the LC VCO, the design of the inductor for oscillation is the most important design consideration. For the

978-1-7281-6162-4/20 $31.00 © 2020 IEEE

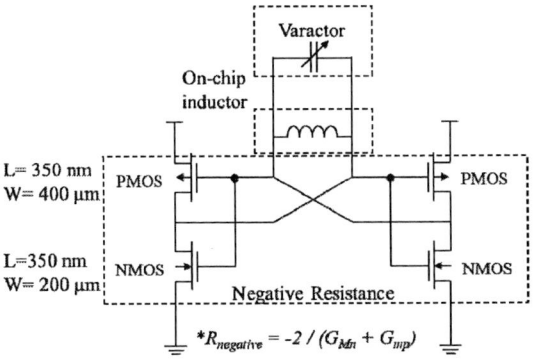

Fig. 2. Designed LC VCO using the SK Hynix 180 nm process. Negative resistance circuit, on-chip inductor, and varactor are designed.

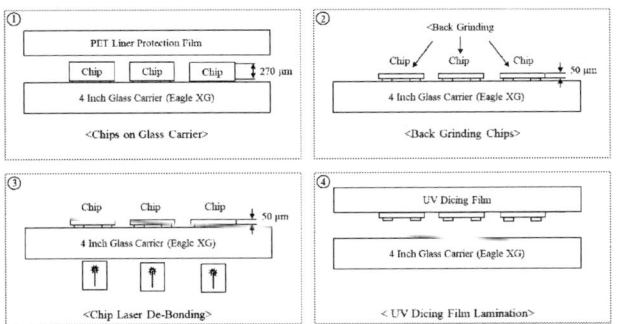

Fig. 3. Chip-level backgrinding process of the fabricated VCO chips to obtained the flexibility.

inductor design, on-chip inductor and off-chip inductor can be considered. In our design, we have designed the on-chip inductor by using a thick metal process.

To achieve a higher Q-factor. The on-chip inductor is optimized using an HFSS 3D EM simulation with the higher Q-factor within the area of 700 μm^2. The line width is designed to be 30 μm and the line to line space is set to 1 μm that is the minimum line to line space set by the manufacturing process. The coil turns are designed to be 4 turns and finally the metal thickness is set to 2 μm using the top thick metal process where the conductor thickness of inner design layers is 0.7 μm.

The designed VCO is fabricated using the SK Hynix 0.18 μm process through multi-wafer project (MPW). In this project, the fabricated chip is only available in chip level. To achieve the flexibility of the Si chip, the Si substrate is grinded to less than 50 μm. However, backgrinding process in the chip level is difficulty in comparison to the back-grinding of the wafer level since the dimension is relatively small and it is hard to handle the flexible chip after the backgrinding process.

TABLE I.   DESIGN PARAMETERS OF THE ON-CHIP INDUCTOR

Dimension	Line Width	Line Space	Metal Thickness	Coil Turn
700 μm^2	30 μm	1 μm	2 μm	4 turns

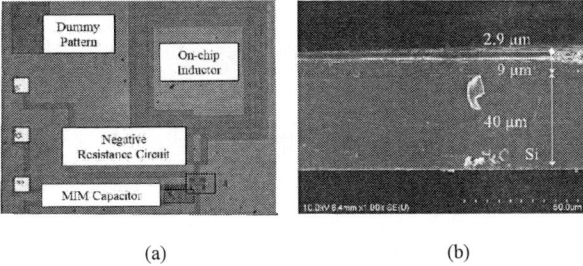

Fig. 4. (a) The top-view of the fabricated flexible VCO chip. (b) The cross-sectional view of the flexible VCO chip after backgrinding.

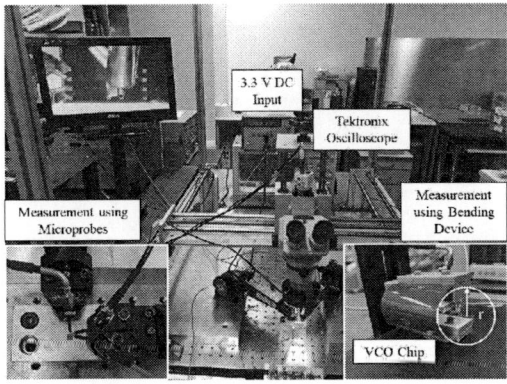

Fig. 5. Measurement setup of the flexible VCO chip using microprobes without bending and with bending using SMA connectors and bending device.

Figure 3 illustrates the backgrinding process of the VCO chip. Firstly, the fabricated VCO chips are attached on the glass carriers because the conventional films for the backgrinding process is not suitable to achieve the equal thickness of the flexible chips after the backgrinding process. Secondly, the VCO chips are grinded to 50 μm to achieve the flexibility. In the wafer level backgrinding process, it is possible to achieve the thickness less than 20 μm [7]. However, since it is the chip level grinding, we decide to have the total thickness of 50 μm. After the grinding process of the chip, the attached flexible VCO chips are detached from the glass wafer using the laser de-bonding process. Finally, the flexible chips are attached to the UV dicing film for the measurement. Figure 4 (a) shows the top view of the fabricated VCO chip and Figure 4 (b) shows the cross-sectional view of the flexible chip after the backgrinding process. The cross-sectional view shows the top metal thickness of 2.9 μm and the total thickness of the design layers as 9 μm. Finally, the Si substrate is grinded to 40 μm to have the flexibility.

III. MEASUREMENT OF THE FLEXIBLE VCO WITH BENDING RADIUS

To measure the flexible chip with the bending radius, the bending device is needed to apply the bending radius the flexible chip. Also, the flexible PCB is designed to measure the

978-1-7281-6162-4/20 $31.00 © 2020 IEEE   35

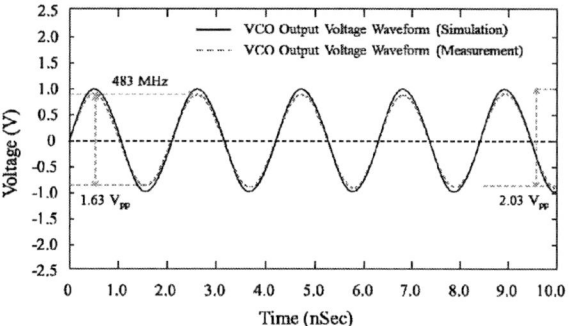

Fig. 6. Comparison of the output voltage waveform simulation and measurement results of the designed VCO without bending.

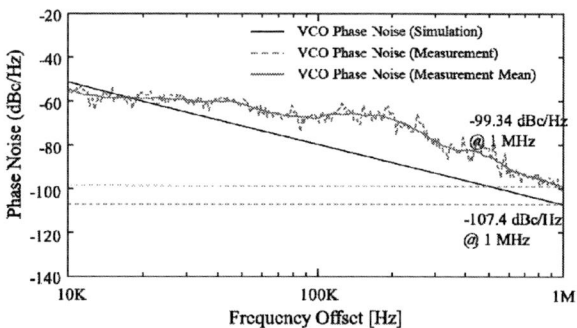

Fig. 7. Comparison of the phase noise simulation and measurement results of the designed VCO without bending.

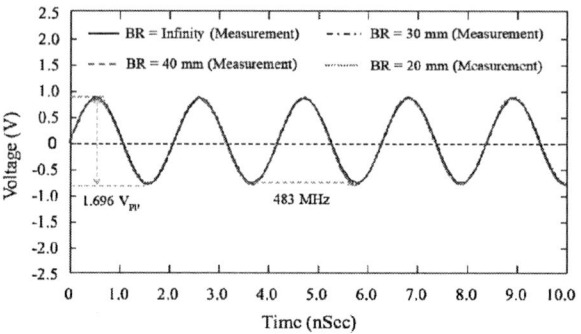

Fig. 8. VCO output voltage waveform measurement results of the flexible VCO chip with bending radius.

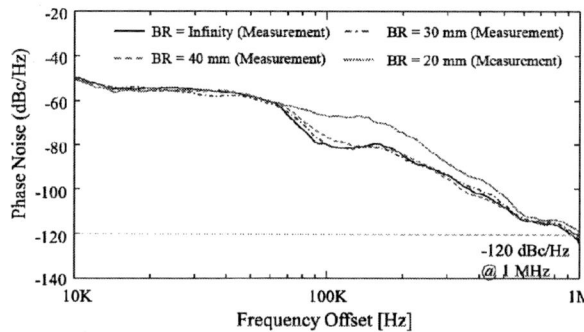

Fig. 9. Phase noise measurement results of the flexible VCO chip with bending radius.

flexible chip as shown in Figure 5. To measure the VCO output with the bending radius, the input voltage of 3.3 V is applied. To measure the output voltage, Tektronix oscilloscope is used.

Firstly, we have tested the flexible chip without bending by using microprobes to compare with the simulation results. Figure 6 shows the comparison between Cadence Virtuoso simulation and measurement results of the flexible chip without bending. The voltage amplitude difference between simulation and measurement is occurred by the higher equivalent series resistance (ESR) value of the on-chip inductor. The oscillation frequency of 483 MHz is occurred with the identical simulation and measurement conditions. Figure 7 shows the comparison of the phase noise simulation and measurement results of the designed VCO from 10 kHz to 1 MHz away from the center frequency. The simulation results showed -99.34 dBc/Hz at 1 MHz and the measurement results showed -107.4 dBc/Hz at 1 MHz. There is a difference in the phase noise results in the middle frequency region, however, the results showed the similar trends.

Figure 8 shows the measurement results of the flexible VCO chip on the PCB with bending radius. The bending radius of the flexible PCB is varied from infinity to 20 mm. The infinity value of the bending radius means the flexible PCB is in the flat structure. The results showed the designed VCO is

properly operating until the bending radius of 20 mm is applied without the voltage waveform distortions. Finally, Figure 9 shows the phase noise measurement results with the bending radius. The results showed that the similar trends until the bending radius is 30 mm and showed the small difference when the bending radius is 20 mm. The difference in the phase noise when the bending radius is 20 mm occurs by the mechanical bending of the cables and SMA connectors during the measurement of the flexible PCB.

IV. FUTURE WORK

For the future work, the flexible chip with the thickness less than 50 μm can be analyzed with the bending radius. The Si substrate has relatively higher Young's modulus than the flexible PCB. Therefore, the flexible chip is not bent perfectly when we apply the bending radius. This problem can be solved by further reducing the Si substrate thickness.

V. COCLUSIONS

In this paper, we have designed the flexible VCO chip to see the feasibility of the flexible electronics. The designed VCO chip is fabricated for the measurement of the flexible chip with the bending. The measured voltage waveforms and phase noise results were compared with the virtuoso simulation results without bending radius. The results showed the correlation between the simulation and the measurement. Finally, the VCO chip is measured with the bending radius on the flexible PCB. The bending radius is applied from infinity to 20 mm. The results showed the proper VCO chip operation without the voltage waveform and the phase noise distortions. Finally, we can conclude that the flexible chip with the bending radius is realizable for the flexible electronics.

ACKNOWLEDGMENT

This research was supported by "Development of Interconnection System and Process for Flexible Three-Dimensional Heterogeneous Devices" funded by MOTIE (Ministry of Trade, Industry and Energy) in Korea. This work was supported in part by the National Research Foundation of Korea Grant funded by the Korean Government (2019M3F3A1A03079612) In addition, we would like to acknowledge the technical support from ANSYS.

REFERENCES

[1] Chin-Chin Tsai, "Recent development in Flexible Electronics," *16th Opto-Electronics and Communications Conference*, Kaohsiung, 2011, pp. 370-371.

[2] J. N. Burghartz et al., "Ultra-thin Si chips for flexible electronics Process technology, characterization, assembly and applications," *28th Symposium on Microelectronics Technology and Devices (SBMicro 2013)*, Curitiba, 2013, pp. 1-3,

[3] A. Afyf, A. Elouerghi, M. Afyf, M. A. Sennouni and L. Bellarbi, "Flexible Wearable Antenna for Body Centric Wireless Communication in S-Band," *2020 International Conference on Electrical and Information Technologies (ICEIT)*, Rabat, Morocco, 2020, pp. 1-4,

[4] J. H. Lee, C. W. Lee, Y. J. Kim, S. M. Kim and J. Song, "Development of PEB Face-Down Interconnect Process for Wearable Device," *2018 7th Electronic System-Integration Technology Conference (ESTC)*, Dresden, 2018, pp. 1-4,

[5] S. Jeong et al., "Smartwatch Strap Wireless Power Transfer System With Flexible PCB Coil and Shielding Material," *in* IEEE Transactions on Industrial Electronics, vol. 66, no. 5, pp. 4054-4064, May 2019.

[6] S. Jeong et al., "Modeling, Simulation and Measurement of On-chip Interconnects with Extremely Thin Si Substrate for Flexible Electronics," *2019 Electrical Design of Advanced Packaging and Systems (EDAPS)*, Kaohsiung, Taiwan, 2019, pp. 1-3.

[7] S. Jeong et al., "Design and Analysis of Flexible Interconnects on an Extremely Thin Silicon Substate for Flexible Wearable Devices," *2019 Joint International Symposium on Electromagnetic Compatibility, Sapporo and Asia-Pacific International Symposium on Electromagnetic Compatibility (EMC Sapporo/APEMC)*, Sapporo, Japan, 2019, pp. 387-390,

978-1-7281-6162-4/20 $31.00 © 2020 IEEE

A Comparison of Finite vs. Infinite Plane Models of Reference Conductors in Electronic Packages

Yi-Ru Jeong and Ali E. Yilmaz

Department of Electrical and Computer Engineering, The University of Texas at Austin, Austin, Texas, USA

yiruta@utexas.edu, ayilmaz@mail.utexas.edu

Abstract—The accuracy of two layered-medium integral-equation approaches are evaluated using a benchmark package-scale microstrip line: (i) In the typical approach, where dielectric layers are extended to infinity using Green's functions, the currents in/on all conductors (here, a signal line and ground plane) are discretized. (ii) In the reduced-domain approach, where large reference conductors are also extended to infinity using Green's functions, the currents in/on a reduced set of conductors (here, only the signal line) are discretized. The approaches are contrasted by computing network parameters of the microstrip line using lumped-port models and diminishing port-truncation effects using a de-embedding technique. It is shown that the S11 parameter found by the typical approach is highly sensitive to the ground-plane mesh density under the signal line and requires finer meshes to reach to the accuracy of the reduced-domain approach.

Keywords—Integral-equation methods; layered media.

I. INTRODUCTION

Reduced-domain layered-medium integral equation (LMIE) approaches can enable accurate, rapid, and scalable electro-magnetic simulations of modern electronic packages [1]. In these approaches, the number of unknowns is reduced beyond that of the typical LMIE methods by extending large reference conductors (e.g., ground/power planes) to infinity and incorporating them into the Green's functions. While zero-thickness perfectly-conducting layers are commonly used in reduced-domain LMIE methods [2],[3], finite-thickness highly-conducting layers can also be used to more accurately model finite-thickness, finite-conductivity, and surface-roughness effects on reference conductors [1]. Because they discretize the current in/on a reduced set of conductors, such LMIE methods can result in smaller and better-conditioned [1] systems of equations.

In this article, a reduced-domain LMIE method is contrasted to a typical LMIE method by using them to compute the network parameters of a package-scale benchmark microstrip line. Significant differences are observed when the S11 parameters found by the two are compared, even after a de-embedding technique is used to reduce the port discontinuity effects. To investigate the origin of the differences, the results are compared as increasingly finer meshes are used. It is observed that as the mesh density increases, the difference between the de-embedded results decreases and that the reduced-domain LMIE shows more accurate results than the typical LMIE for the same mesh density on the signal line. The differences and the less accurate parameters found by the typical LMIE approach are attributed to the numerical solution rather than analytical modeling via the Green's functions of the currents on the ground plane.

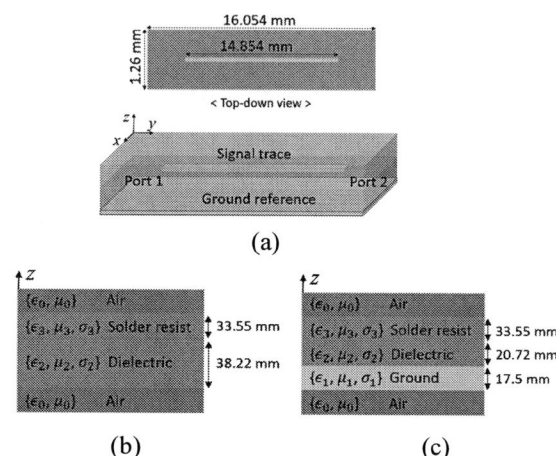

Fig. 1. (a) Top-down and 3-D view of the microstrip line structure. (b) A 4-layer background model for the typical LMIE method [4]. (c) A 5-layer background model for the reduced-domain LMIE method [1].

II. NUMERICAL METHOD

Consider the two different background models of the benchmark microstrip line from [1],[4] shown in Fig. 1. In the typical LMIE approach, the signal line and ground conductors are modeled as residing in a 4-layer planar stratified medium (Fig. 1(b)). In the reduced-domain LMIE approach, only the signal line is modeled as residing in a 5-layer planar stratified medium (Fig. 1(c)) and the reference conductor is modeled as an infinite layer of finite thickness and conductivity. The material parameters of the different layers can be found in [1]. In both LMIE approaches, the surfaces of the conductors that are not modeled as part of the background medium are meshed using triangles (Fig. 2); the surface current density on these conductors is approximated using RWG functions [5]; and impedance-boundary conditions are enforced on these surfaces. Non-radiating lumped-port models are used to extract the network parameters as detailed in [1],[4],[8].

A. Typical LMIE Approach

Layered-medium Green's functions are computed numerically from Sommerfeld integrals with asymptotic-term subtraction scheme and interpolation from sampled results [4],[6],[7]. While dielectric layers are included in the Green's function computation, conductors for the signal traces and the ground reference are meshed as both are modeled as finite-thickness finite-sized conductors. Generally, the unknowns on the reference conductor constitute a significant portion of the total number of unknowns (Fig. 2).

978-1-7281-6162-4/20 $31.00 © 2020 IEEE

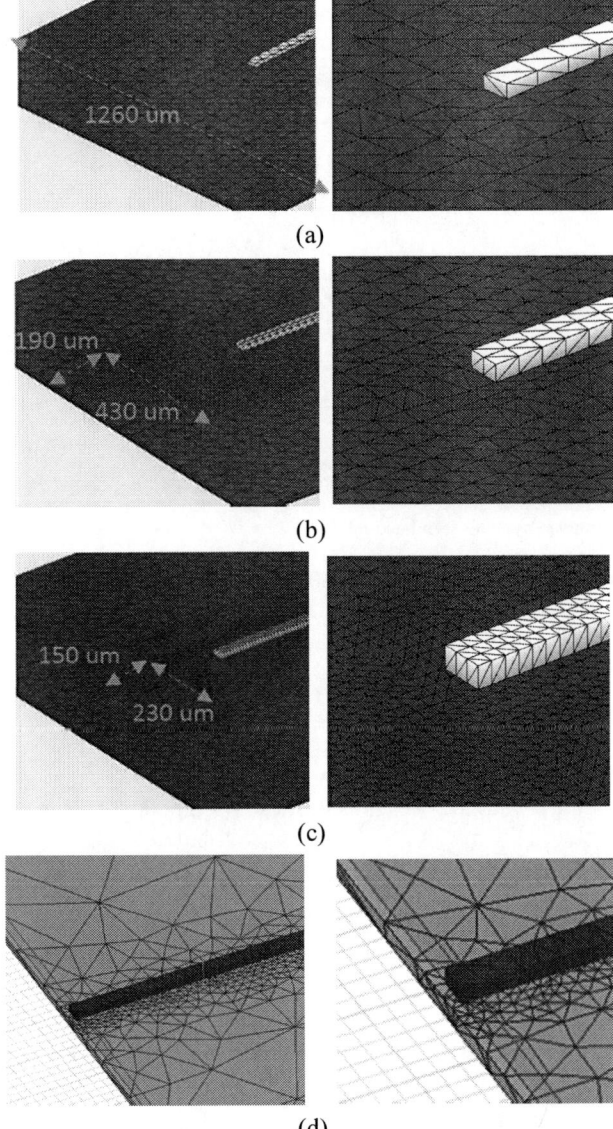

Fig. 2. The various meshes used in this article. The (a) normal mesh, (b) dense mesh, and (c) very-dense mesh used in the LMIE computations (only the signal line portion of the mesh is used in the reduced-domain LMIE approach). Also shown is (d) the surface portion of a typical volumetric mesh generated using a commercial FEM software. The right figures are the same as the left ones, but zoomed in near the end of the signal line.

B. Reduced-Domain LMIE

In this approach, the ground reference is incorporated into layered-medium Green's function as an infinite plane but with finite thickness and conductivity [1]; this removes the need to discretize the current on the reference conductor (simplifying meshing), reduces the number of unknowns to only those assigned to signal traces, and can yield better conditioned systems of equations [1]. On the other hand, introducing a highly-conductive layer to the background medium complicates the accurate computation of layered-medium Green's functions. To ameliorate this issue, log- and linear-scale interpolation [7] are used for near-singular and other interactions, respectively.

III. INFINITE VS. FINITE GROUND MODEL

The microstrip line is analyzed using the typical and reduced-domain LMIE methods to evaluate the differences between finite and infinite ground models. To investigate the accuracy and convergence of the methods, the three kinds of meshes shown in Figs. 2 (a)-(c) are used: (i) In the normal mesh, the edge lengths of the mesh are about the width of the signal trace and all portions of the ground conductor's 6 surfaces are discretized using same edge length as on the signal trace (except on the 4 side surfaces where there are smaller elements). This gives rise to a total of 38 424 triangles of which 1192 are on the signal trace. (ii) In dense mesh, the edge lengths of the signal trace mesh are about half the width of the trace and the mesh density on the ground conductor in the region below the signal trace is also the same. The edge length in other portions of the ground conductor is identical to that in normal mesh. There are a total of 78 432 triangles of which 3570 are on the signal trace. (iii) The very dense mesh is obtained by further refining the dense mesh and reducing the edge length on the signal trace and the portion of the ground conductor below it to about 1/3 of the signal trace width. There are now three regions of increasingly finer meshes on the ground plane for a total of 109 564 triangles of which 17 274 are on the signal trace. For comparison, a typical automatically and adaptively-computed mesh from a commercial software based on the finite-element method (FEM), obtained by setting the meshing frequency to 20 GHz and using default parameters for all mesh settings is shown in Fig. 2(d). Note the density of the mesh on the top and side surfaces of the signal trace and the ground conductor under it.

A. SOC deembedding

In this article, the non-radiating lumped-port model in [8] is used to excite/terminate the microstrip line and extract network parameters; the model uses testing and basis functions with supports over disconnected triangles. While pairs of triangles, one on the signal trace and the other on the ground plane, are used for the typical LMIE approach's (finite ground model) port, a single triangle on the signal trace is used in the reduced-domain LMIE approach's (infinite ground plane model) port.

First, the S-parameters are obtained directly, without any de-embedding techniques, using the normal mesh; the results are shown in Fig. 3 (dashed lines): While the S11 shows good agreement below 5 GHz, it differs significantly in the rest of the frequency range. Next, the effect of the lumped-port models is removed by using the SOC de-embedding technique [9]. During de-embedding, 150-um long feed lines were attached at each port for the typical LMIE approach; this did not yield consistent results for the reduced-domain LMIE approach and the feed lines were extended to 610 um at each port. This decreased the magnitude of S11 from both approaches in Fig. 3 (solid lines). This result is to be expected because the microstrip line is designed to 50 Ω termination and reference impedance for the S-parameters is also 50 Ω. Somewhat surprisingly, while good agreement is still observed below 5 GHz, the difference between the results of the two LMIE approaches *increases* above 5 GHz after de-embedding. As an independent check, S11 is also computed using the commercial FEM-based software with the mesh in Fig. 2(d) and using the wave-port model; this yields yet another result that is different from both LMIE computations and fails to explain the difference.

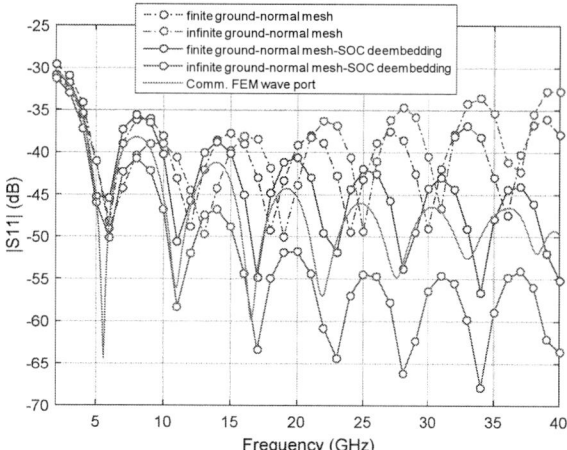

Fig. 3. S11 computed using the two LMIE approaches without any deembedding and S11 computed by SOC deembedding of the results of the two LMIE approaches compared to that from an FEM solver.

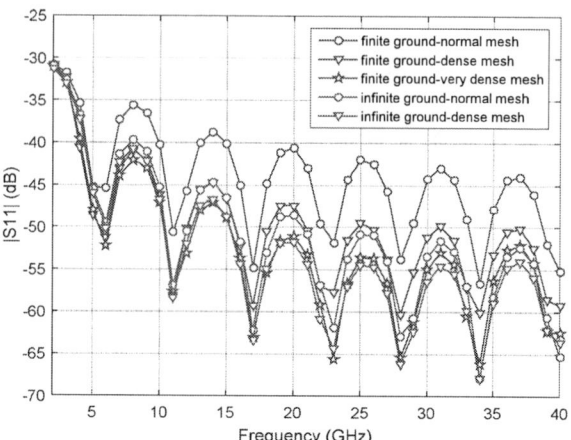

Fig. 4. S11 computed with the typical LMIE approach and the reduced-domain LMIE appraoch using different mesh densities (all of S11 are computed by SOC deembedding).

B. Increasing mesh density

To evaluate the convergence of the results, the denser meshes in Figs. 2(b) and 2(c) are used and the results obtained from LMIE computations are shown in Fig. 4. In both approaches, as the mesh is refined, the magnitude of S11 further decreases and becomes closer to each other; the change is especially significant for the typical LMIE approach using the finite-ground model. Clearly, the normal mesh is not sufficient to obtain an accurate S11 parameter for the typical LMIE approach in this benchmark signal line with a well-matched port. Fig. 4 shows that the typical LMIE approach requires the very dense mesh whereas the reduced-domain LMIE approach requires only the dense mesh to obtain highly accurate S11 results; i.e., they require 171 084 and 11 590 RWG functions, respectively.

IV. CONCLUSION

Two LMIE approaches were compared using a package-scale benchmark microstrip line. While the ground reference is modeled as a finite conductor in the typical LMIE approach, it is modeled as an infinite conductor and incorporated into the Green's function in the reduced-domain LMIE approach. The S11 parameters computed by the two approaches were found to be different even after a de-embedding technique was applied. Using increasingly denser meshes showed that the reduced-domain LMIE method produced more accurate results for the same mesh density on the signal trace.

The results indicate that in order to obtain a highly accurate S11 parameter for this benchmark problem, it is important to also accurately model the currents on the ground plane; i.e., regions on reference planes near the signal trace should also be carefully/densely discretized. Indeed, while the mesh density on signal traces is generally considered significant for accuracy, the mesh density on the ground plane can also affect the accuracy of the results for methods that discretize currents on/in reference planes. Because multi-layer electronic packages generally have multiple ground and power planes, the mesh density on these planes can be a serious impediment for typical LMIE methods, increasing their computational costs significantly. This article shows that reduced-domain LMIE methods can not only reduce the computational costs but also potentially provide more accurate results by modeling these planes using Green's functions.

ACKNOWLEDGMENT

This research was supported by Intel Corporation. The authors thank M. Shattuck from Cadence-AWR for suggesting this study and the Texas Advanced Computing Center for providing HPC resources that have contributed to the research results reported within this article.

REFERENCES

[1] C. Liu and A. E. Yılmaz. "A reduced-domain layered-medium integral-equation method for electronic packages," in *Proc. IEEE 28th Conf. Elect. Perform. of Electron. Packag. Syst. (EPEPS)*, 2019.

[2] B. Wu and L. Tsang,"Fast computation of layered medium Green's functions of multilayers and lossy media using fast all-modes method and numerical modified steepest descent path method," *IEEE Trans. Microw. Theory Tech.*, vol. 56, no. 6, pp. 1446– 1454, June 2008.

[3] Z. Song, H.-X. Zhou, K.-L. Zheng, J. Hu, W.-D. Li and W. Hong, "Accurate evaluation of Green's functions for a lossy layered medium by fast extraction of surface- and leaky-wave modes," *IEEE Antennas. Propgat. Mag.*, vol. 55, no. 1, pp. 92–102, Feb. 2013.

[4] C. Liu, K. Aygun, and A. E. Yılmaz, "A parallel FFT-accelerated layeredmedium integral-equation solver for electronic packages," *Int. J. Num. Model.: Electron. Networks, Dev., Fields*, vol. 33, no. 2, 2020.

[5] S. M. Rao, D. R. Wilton, and A. W. Glisson, "Electromagnetic scattering by surfaces of arbitrary shapes," *IEEE Trans. Antennas Propag.*, vol. 30, no. 3, pp. 409–418, May 1982.

[6] E. Şimşek, Q. H. Liu, and B. Wei, "Singularity subtraction for evaluation of Green's functions for multilayer media," *IEEE Trans. Microw. Theory Tech.*, vol. 54, no. 1, pp. 216–224, Jan. 2006.

[7] H. Yang and A. E. Yilmaz, "A log-scale interpolation method for layered medium Green's function," in *Proc. IEEE Antennas Propag. Soc. Int. Symp.*, July 2018.

[8] C. Liu, J. W. Massey, and A. E. Yılmaz, "A nonradiating finite-gap lumped-port model," *IEEE Antennas Wireless Propag. Lett.*, vol. 17, no. 7, pp. 1339–1343, July 2018.

[9] L. Zhu and K. Wu, "Unified equivalent-circuit model of planar discontinuities suitable for field theory-based CAD and optimization of M(H)MIC's," *IEEE Trans. Microw. Theory Tech.*, vol. 47, no. 9, pp. 1589–1602, Sep. 1999.

Analysis of Power Supply Noise Induced Jitter of I/O Subsystems with Multiple Power Domains

Hyo-Soon Kang
Intel Corporation
San Jose, USA
hyosoon.kang@intel.com

Ashkan Hashemi
Intel Corporation
San Jose, USA
ashkan.hashemi@intel.com

Guang Chen
Intel Corporation
San Jose, USA
guang.chen@intel.com

Xiaoping Liu
Intel Corporation
San Jose, USA
xiaoping.a.liu@intel.com

Wendemagegnehu Beyene
Intel Corporation
San Jose, USA
wendem.beyene@intel.com

Abstract—In high-speed interface designs, multiple power domains are used to improve the performance and minimize noise as well as jitter. Power supply noise induced jitter (PSIJ) is one of major sources of timing uncertainties. To analyze PSIJ, empirical methodologies are often derived in frequency and time domains using the individual DC delay (ps) and sensitivity (ps/mV) of key circuit blocks. In this paper, we examine the validity of PSIJ results from these empirical approaches by comparing those to the results obtained from transistor-based SPICE simulations of complete high-speed data and clock paths such as clock tree, phase interpolator, mux, and buffers under multiple power domains. The accuracy of PSIJ methodology of the overall Input/Output (I/O) subsystem is verified against the combined results of multiple circuit blocks when analyzed separately.

Keywords— DC delay sensitivity, Input/Output (I/O) subsystem, Multiple power domains Power supply noise induced jitter.

I. INTRODUCTION

Many of the critical circuit blocks of clock and data paths in high-speed serial/parallel interface are sensitive to power supply noise. As such, there may exist several power rails to improve the performance, lower-power consumption and to meet predefined voltage levels in the I/O interface specifications. In this study, three main power rails are considered in the I/O PHY circuits as shown in Figure 1. All the clock circuit blocks use VCCA, I/O registers and pre-drivers use VCC, and the output drivers use VCCIO. These power supplies can be delivered from the same VRM with appropriate filtering at board or package level.

The decision to partition the power supply of an interface often involves power noise and timing impact analyses. The off-chip and/or on-chip power integrity analysis evaluates the self-generated and coupled noises. And, timing analysis assesses the jitter impact due the voltage noises. Figure 2 shows an example of various supply noise generation, noise sensitivity of circuits as well as the jitter sensitivity functions of typical circuit blocks in I/O subsystems.

The power supply noise induced jitter (PSIJ) created by the various circuit blocks can take a significant portion of the total link jitter budget. In order to help circuit/system designers, minimize the PSIJ, efficient empirical simulation methodologies of PSIJ analysis have been widely investigated [1, 2]. In these methods, the PSIJ can be analyzed by equation-based numerical calculations, without time-consuming SPICE simulations of the entire circuit blocks. This paper compares the functional block-

based empirical PSIJ analysis and transistor circuit-based SPICE simulation. To verify the jitter transfer function of clock and data paths, the I/O subsystem in an Intel field-programmable gate array (FPGA) is used for transistor-level SPICE simulations. By applying a sinusoidal noise and sweeping the noise frequency, the PSIJ dependency of complete cascaded circuit block on the noise frequency is simulated and compared with the jitter transfer function of the empirical approach of circuit block-based approach.

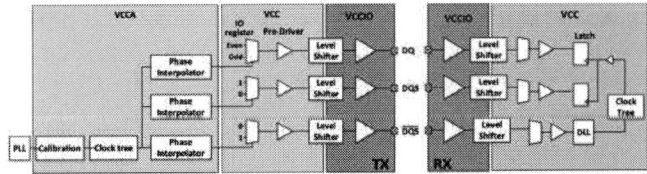

Figure 1: A power supply partition in a typical I/O subsystem for source-synchronous parallel interface.

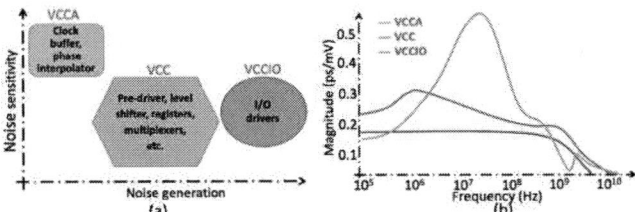

Figure 2: Supply noise partition: (a) noise generation vs. sensitivity, and (b) jitter sensitivity functions for VCCA, VCC, and VCCIO.

The empirical PSIJ analysis consists of deriving a frequency-dependent jitter transfer function (JTF) calculation and modeling of the system-level noise including power distribution network and current noise waveforms as shown in Figure 3. The jitter transfer functions can be calculated with DC delay (ps) and sensitivities (ps/mV) of individual circuit blocks in clock and data paths. As of yet, the jitter transfer function of full-path analysis and comparison have not been accomplished due to the complex design of clock and data paths.

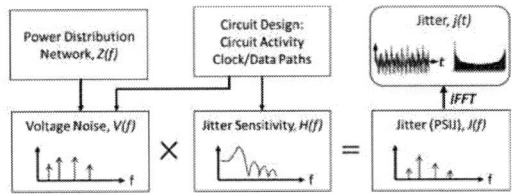

Figure 3: Frequency-domain approach for deriving the power supply noise induced jitter (PSIJ).

II. JITTER TRANSFER FUNCTIONS

1. Open-loop path jitter

The jitter transfer function for open-loop paths can be analytically derived by an equation (2) [1, 3]. The DC delay sensitivity (H_0), is defined as the delay change induced by DC voltage changes as described in equation (1). With this DC delay sensitivity and open-loop delay (τ_d) of the circuit block, the frequency-dependent jitter transfer function can be easily calculated as:

$$H_0 = \frac{\tau_{d\,@Vmin} - \tau_{d\,@Vmax}}{Vmax - Vmin}, \quad (1)$$

$$H(f) = \frac{j}{2\pi f \tau_D} H_0 [1 - e^{-j2\pi f \tau_D}]. \quad (2)$$

The magnitude of the jitter sensitivity can be simplified as:

$$|H(f)| = H_0 \left| \frac{\sin(\pi f \tau_d)}{\pi f \tau_d} \right| = H_0 |\text{sinc}(\pi f \tau_d)|. \quad (3)$$

When the path delay is large, the DC sensitivity (H_0) is high, the path is more sensitive to the low-frequency noise. However, the large path delay induces the lower frequency null point in the frequency-dependent jitter transfer function which can make the path less-sensitive to high-frequency noise. Therefore, large path delay does not always induce the worst-case jitter as in static timing analysis (STA) in which longer delay usually means larger timing loss.

2. Period Jitter of clock signal

For the clock network analysis, period jitter *(PJ)* is a key parameter because data can be generated and recovered by clock rise and/or fall edges. Period jitter is defined as the deviation of the clock period from its ideal period; thus, the jitter transfer function of *PJ* can be calculated by subtracting the jitter of consecutive edges: rise-to-rise or fall-to-fall edge. From the jitter transfer function of open-loop path in equation (1), the jitter transfer of *PJ* can be expressed by equation (4) and (5) with jitter subtracting term of consecutive edges.

$$H_{period}(f) = H(f)(1 - e^{-j2\pi f T_{clk}}) \quad (4)$$

The magnitude of the jitter sensitivity can be simplified as:

$$|H_{period}(f)| = 2H_0 |\text{sinc}(\pi f \tau_d)| \cdot |\sin(\pi f T_{clk})| . \quad (5)$$

Figure 4 shows the time interval error (TIE), which is the time deviation of rise/fall edge from ideal clock edge position from the open-loop sensitivity equation (4), and period jitter sensitivities by equation (5). With the jitter tracking effect of consecutive edges in PJ, the jitter sensitivity has low value in low frequency region, thus the PSIJ becomes smaller.

In high speed parallel interfaces such as memory and chip-to-chip interfaces, source synchronous clocking, which transmits data and clock (strobe) together, is widely used due to the jitter tracking effect of data and clock at the receivers. In such interfaces, the jitter transfer functions *(JTFs)* of data setup and hold time with source synchronous clocking can be expressed by equations (6) and (7), respectively, considering the phase relation between rising and falling edges:

$$H_{setup}(f) = \frac{j}{2\pi f \tau_D} H_0^D [1 - e^{-j2\pi f \tau_D}] - \frac{j}{2\pi f \tau_C} H_0^C [1 - e^{-j2\pi f \tau_C}] \quad (6)$$

$$H_{hold}(f) = \frac{j}{2\pi f \tau_D} H_0^D [1 - e^{-j2\pi f \tau_D}] - \frac{j}{2\pi f \tau_C} H_0^C [1 - e^{-j2\pi f \tau_C}] e^{-j2\pi f T_{Bit}}, \quad (7)$$

where H_0^D and H_0^C is DC sensitivity of data and clock path, τ_D and τ_C are data and clock delays, and T_{Bit} is the unit interval (UI).

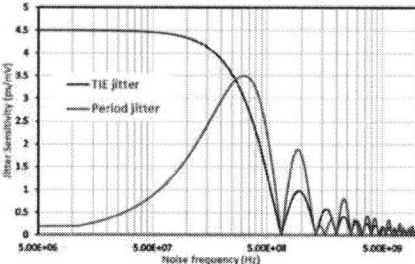

Figure 4: Jitter transfer function of TIE and PJ when the delay (τd) is 1.5ns and clock frequency (1/Tclk) is 1.6GHz.

III. SPICE SIMULATION RESULTS AND CORRELATION

To verify the jitter transfer function of clock and data paths, the I/O subsystem in an Intel FPGA are used for transistor-level SPICE simulations. As shown in Figure 1, the clock is generated from PLL and it goes through calibration circuit, clock tree and phase interpolator for the clock distribution and phase alignment. With the distributed clock, data and strobe signals are generated in IO register blocks. The data and strobe signals are then reshaped in pre-drivers and output buffers to transmit the signals through the channel. Because of the voltage difference between core blocks and output buffers, voltage level shifter is implemented in the design. The data and strobe signals are transmitted together in source synchronous clocking systems with 90-degree phase shift between data and strobe.

1. SPICE simulation setup

The full path circuit blocks from PLL out to output buffers are extracted as SPICE post-layout netlist and it is used for transient simulations. To measure the delay, τ_D, and DC sensitivity, H_0, which are required for the jitter transfer function calculation, the delay values of each block are measured at different DC voltages. In our analysis, we only focused on the PSIJ of clock and data paths in the PHY core blocks, and the output buffer power-supply voltage of VCCIO, which is related to simultaneous switching output (SSO) noise, is connected to ideal voltage of 1.2V. In addition, ideal loading capacitors of 1pF are applied to the output buffers to exclude the channel effect. To simulate the jitter transfer function with the SPICE netlist, a sinusoidal noise signal is applied to power supply nodes while sweeping frequency. The PSIJ impacts to the clock period jitter or data eye-opening are measured at the output buffer.

2. Delay and DC Sensitivity

Figure 5 shows the measured circuit delays of each block at different voltages. It is seen that the delay decreases as supply voltage increases and the DC sensitivity (H_0) can be calculated using the equation (1).

978-1-7281-6162-4/20 $31.00 © 2020 IEEE

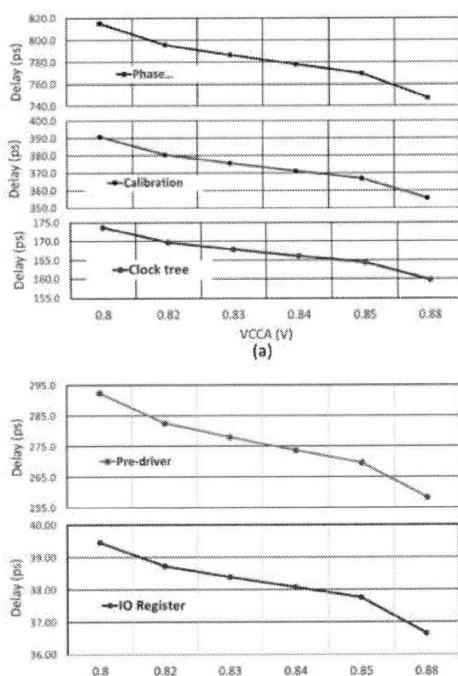

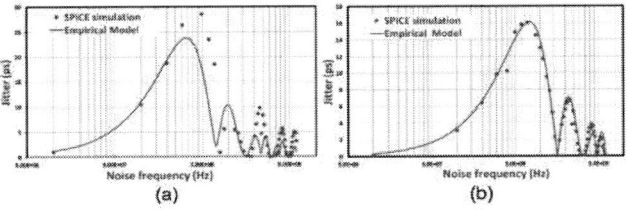

Figure 5: Circuit block delay simulations at different voltages.

3. Clock period jitter correlation

In the SPICE simulation, period jitter (PJ) of the strobe signal is measured at the output buffer with varying VCCA or VCC noise frequency. Figure 6 compares the SPICE results and empirical PSIJ values, where sinusoidal noise is applied to the SPICE simulations and equation (4) is used for the empirical jitter transfer function. The frequency-dependent period jitter characteristics are well correlated between the empirical method and SPICE results. However, there are minor mismatches due to the non-linear delay sensitivity and numerical errors in the eye-opening measurements of the simulations.

Figure 6: Clock (strobe signal) period jitter as a function of (a) VCCA noise frequency and (b) VCC noise frequency.

4. Open-path jitter correlation

With the same SPICE simulations, data signal jitter induced by VCCA or VCC noise is measured and compared with empirical PSIJ analysis results as shown in Figure 7. To analyze open-path jitter characteristics, the data jitter is measured from the eye-diagram with the ideal clock trigger. Jitter transfer function has *sinc* function shape as in equation (3) and frequency-dependent characteristics are well correlated.

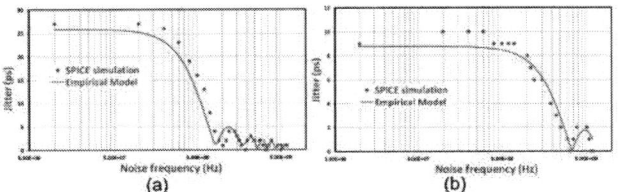

Figure 7: Data jitter without strobe signal triggering as a function of (a) VCCA noise frequency and (b) VCC noise frequency

5. Data jitter in source-synchronous clocking

In source synchronous clocking systems, the data and clock jitter can be tracked out when the noise frequency is relatively low, and the specific phase relationship can be met as described in equations (5) and (6). Figure 8 shows SPICE and empirical PSIJ results in source synchronous clocking systems. To consider the jitter tracking, data eye-diagram is triggered by strobe signals in the SPICE simulations. Because the mismatch between SPICE and empirical analysis in data and strobe signals can be accumulated, the jitter magnitude and null points causing low jitter values have discrepancy, however overall frequency-dependent trends are in a relatively good agreement.

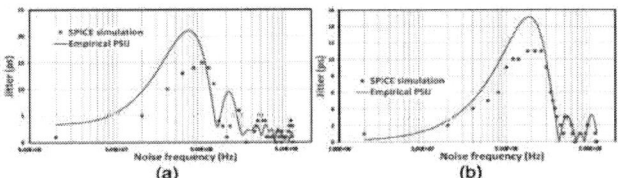

Figure 8: Data jitter with strobe signal triggering as a function of (a) VCCA noise frequency and (b) VCC noise frequency.

IV. CONCLUSIONS

In this paper, the correlations of the SPICE simulation and empirical methodology for power supply noise induced jitter (PSIJ) are presented. Using the I/O subsystem circuit blocks in the FPGA, data and strobe jitters are analyzed by applying sinusoidal noise profiles. By sweeping the noise frequency of each power rail, the jitter transfer function of circuit blocks in each power rail can be measured and correlated with the empirical equations. The strobe signal period jitter, open-loop data path jitter, and data jitter in the source-synchronous clocking are examined through both the SPICE simulations and the empirical PSIJ analysis. The frequency dependent jitter characteristics are well correlated; however, the discrepancies of magnitude and null points are observed, which can be due to the inaccurate data and strobe delay modeling. From the results, the empirical PSIJ analysis methodology mainly focusing on jitter transfer functions is verified.

REFERENCES

[1] J. Jang, O. Franza, W. Burleson, "Compact Expressions for Supply Noise Induced Period Jitter of Global Binary Clock Trees," *IEEE Trans. On Very Large Scale Intergration Systems*, vol. 20, no. 1, pp 66-79, Jan., 2012

[2] R. Schmitt, H. Lan, "Design and Characterization of the Power Supply System for a High Speed 1600 Mbps DDR3 Interface in Wirebond," *DesignCon 2012*, Santa Clara, CA, January 30 – February 2, 2012.

[3] T. Rahal-Arabi, G. Taylor, M. Ma, and C. Webb, "Design and validation of the Pentium® III and Pentium® 4 processors power delivery," *VLSI Symposium*, June 2002, pp. 220-22

Reinforcement Learning-based Auto-router considering Signal Integrity

Minsu Kim, Hyunwook Park, Seongguk Kim, Keeyoung Son, Subin Kim, Kyunjune Son, Seonguk Choi,
Gapyeol Park and Joungho Kim

School of Electrical Engineering
Korea Advanced Institute of Science and Technology
Daejeon, South Korea
E-mail : min-su@kaist.ac.kr

Abstract—In this paper, we propose artificial intelligent (AI)-router, a reinforcement learning (RL)-based auto-router considering signal integrity (SI), for the first time. Our algorithm has two main stages. At first, we design the transformer-based novel neural architecture considering the keep-out region, crosstalk region, and the number of vias for SI optimization. Then, the designed neural network is optimized by the policy gradient, one of the RL algorithms. Compared with the conventional maze routers, the A* algorithm, and the lee algorithm, it is verified that our AI-router outperforms the algorithms in terms of wire-length and crosstalk in a specific test case. Furthermore, it is shown that AI-router successfully performs multi-layer routing which is not feasible with conventional maze routers.

Index Terms—AI-router, reinforcement learning, signal integrity, transformer

I. INTRODUCTION

Recent technology trends require terabytes per second bandwidth. To meet these demands, not only the data rate but also channel density has been significantly increased. Moreover, high integration and miniaturization of modern electronic devices critically increase design complexity in channel routing. Therefore, automatic routing considering wire-length, crosstalk, via, and layer selection became a more challenging problem.

Previous researches on auto-routing mainly consist of two stages that global routing strategies and detailed routing [1]. Global routing strategies provide a feasible routing plan including routing order. After that, each pin-to-pin routing is completed by avoiding obstacles using a maze router. However, existing maze routers have two major limitations. Firstly, existing maze routers are not smart enough to perform detailed routing considering SI [2]-[3]. Because the maze-routers cannot consider crosstalk and vias, several hand-crafted heuristics have to be designed to control the maze router for ensuring SI which leads to increases entire algorithm's complexity. Secondly, existing maze-routers have the disadvantage of modifying the entire algorithm for solving the problem which contains additional constraints and optimization variables.

On the other hands, it has been proven several times that learning-based methods can be used universally in a variety of situations without major modifications to the algorithm. In particular, recent studies have suggested that reinforcement

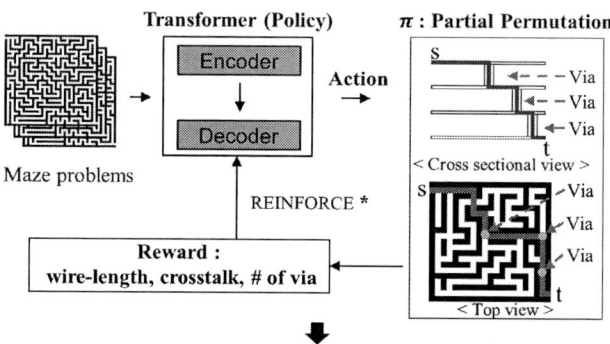

Fig. 1. Overview of training AI-router by RL. Randomly generated maze problems are training data for transformer-based encoder-decoder, which is a policy network. Then the policy network decides action which makes the partial permutation of input coordinate. The value of action is evaluated by the reward function which contains wire-length, crosstalk, and number of vias.

learning shows effective results in combinatorial optimization problems including routing and hardware design [4]-[5].

In this paper, we propose RL-based auto-router considering SI to overcome the limitation of the previous maze routers in terms of SI and versatility. We construct the encoder-decoder model based on the transformer, a state-of-the-art neural network in natural language processing and learning-based combinatorial optimization [5]-[6]. The policy gradient is used for training the encoder-decoder model [7]. For verification, the routing performance of the lee algorithm, the A* algorithm, and the proposed AI-router is compared in terms of wire-length and crosstalk [2]-[3]. Furthermore, we demonstrate that the proposed AI-router performs multi-layer routing which is infeasible to conventional maze routers.

II. PROPOSAL OF REINFORCEMENT LEARNING-BASED AI-ROUTER

As shown in Fig. 1, a randomly generated maze problem is used as training data for the transformer-based encoder-decoder model. Through via, it is possible to move to another layer, then another maze problem of the changed layer is provided. RL is used for training policy networks which determines an optimal routing path. The trained policy network eventually becomes AI-router. Table I contains variables for the equations described in section II.A and II.B.

978-1-7281-6162-4/20 $31.00 © 2020 IEEE

TABLE I
EXPLANATION OF VARIABLES FOR EQUATIONS.

Variables	Representation		Description
X	$\{\mathbf{x} = (x, y, \gamma z, m_o, p_c, p_l)\}$		Input coordinates
m_0	$m_0 = \begin{cases} 1 & (x,y,z) \in O \\ 0 & otherwise \end{cases}$		Obstacle mark
p_c	$p_c = \begin{cases} 2 & (x,y,z) \in C \\ 0 & otherwise \end{cases}$		Penalty of Xtalk
p_l	$p_l = \begin{cases} 0.5 & (x,y,z) \in L \\ 0 & otherwise \end{cases}$		Penalty of loosely Xtalk
γ	$\gamma = 30$		Layer changing penalty
π	$\{\pi_1, \pi_2, ..., \pi_n\}$, $x_{\pi_i} \in X$		Output Permutation
O	$\{(x_i, y_i, z_i)\}$		Obstacle region
C	$\{(x_i, y_i, z_i)\}$		Xtalk region
L	$\{(x_i, y_i, z_i)\}$		Loosely Xtalk region

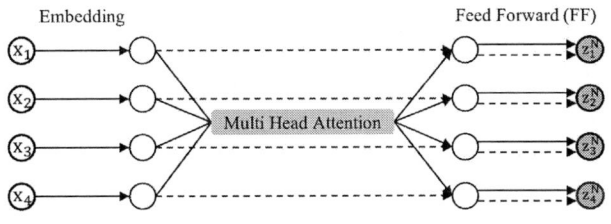

Fig. 2. Conceptual view of encoder architecture. Input nodes **x** are embedded to high dimensional vectors z by multi-head attention, feed forward, and skip connection.

A. Neural Architecture of AI-router

As illustrated in Fig. 2, the encoder architecture mainly consists of multi-head attention (MHA), feed forward (FF), and skip connection (Skip) which embed input node to high dimensional hidden nodes z^N. A detailed equation of the Skip, MHA, and FF is explained in [5]-[6].

First, the input node is embedded through the linear projection as the following equation:

$$z_i^0 = W\mathbf{x}_i + b \tag{1}$$

W and b are learnable parameters. Then, the hidden nodes z^N are made from the embedded node with MHA as the following equation:

$$z_j^{i+1} = Skip_j(FF(Skip_j(MHA_j(z_1^i, z_2^i, ..., z_n^i)))) \tag{2}$$

Then the hidden nodes are propagated forward the decoder as illustrated in Fig.3. The main purpose of the decoder is to output the stochastic policy. To be specific, *context vector* c creates a policy for the next selection, referring to the previous node, the average node of all, the average node of obstacles(keep-out region and pre-routed region), the average node of coupled crosstalk region, the average node of loosely-coupled crosstalk region and the destination(target pin) node:

$$c_k = F_{\theta_1}([z_{k-1}^N, z_n^N]) + F_{\theta_2}([z_{mean}^N, z_o^N, z_c^N, z_l^N]) \tag{3}$$

F_{θ_1} and F_{θ_2} is fully-connected layer with learnable parameter θ_1, θ_2. As shown in (5), the context vector passes MHA and becomes query c'.

$$c_k' = MHA(z_1^N, ..., z_n^N, c_k) \tag{4}$$

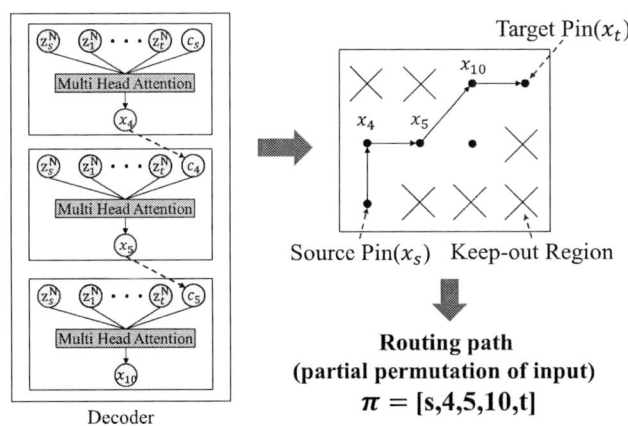

Fig. 3. Conceptual view of proposed decoder processing. The decoder uses the context vector c and MHA to determine the next node. The determined node affects the selection of a next node by changing the context vector c. The decoder eventually creates a routing path of π.

Then the partial policy of selecting the next action can be expressed as follows:

$$u_j = \begin{cases} 10 \tanh\left(\frac{(Qc')^t (Kz_j)}{4}\right) & nodes \in Action \\ -\infty & otherwise \end{cases} \tag{5}$$

$$P_\theta(\pi_i = j | X, \pi_{i-1}, \pi_n) = \frac{e^{u_j}}{\sum_{k=1}^n e^{u_k}} \tag{6}$$

Q and K in (5) are learnable parameters which make query and key, respectively. A detail of the query and key is introduced in [5]-[6]. The *Action* in (5) contains left, right, up, down, 45-degree routing and layer changing through via. Also, obstacles are not possible to be selected by the *Action*.

Therefore the policy of routing permutation π is

$$P_\theta(\pi | X) = \prod_{k=2}^{n-1} P_\theta(\pi_k | X, \pi_{k-1}, \pi_n) \tag{7}$$

B. Training the AI-router with Reinforcement Learning

The policy gradient is used to train the AI-router architecture [7]. When we get permutation of nodes π, we calculate the cost of AI-router as follows:

$$L(\pi) = \sum_{k=1}^{n-1} \left(g(\mathbf{x}_{\pi_{k+1}} - \mathbf{x}_{\pi_k}) + p_c(\mathbf{x}_{\pi_k}) + p_{lc}(\mathbf{x}_{\pi_k}) \right) \tag{8}$$

$$g(\mathbf{x}) = \sqrt{x^2 + y^2 + \gamma^2 z^2} \tag{9}$$

To estimate the crosstalk in the cost function $L(\pi)$, we set the crosstalk region and loosely crosstalk region at $1w$ and $2w$ distances based on the pre-routed channel, respectively. The w is the width of the routing channel. In the end, the total crosstalk is estimated by counting the number of blocks that are invading crosstalk regions on a grid. A block in the grid is a path which is made by node selection from the $\mathbf{x}_{\pi_k}$ to $\mathbf{x}_{\pi_{k+1}}$ in (8).

Also, we assign γ as a penalty for layer change. When changing a layer, it greatly interferes with the routing plan of other layers. Also, the via transition effect adversely affects SI. Therefore, layer change through via is not recommended

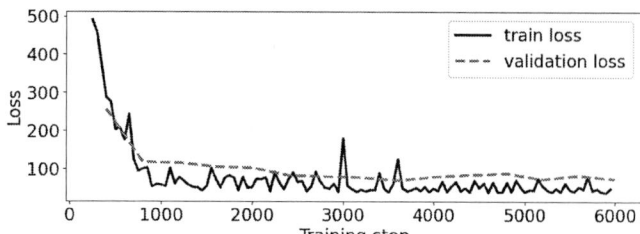

Fig. 4. Loss graph of the AI-router. The validation loss converges just as training loss.

TABLE II
RESULT OF THE TEST CASE IN FIG 5. BOLD MEANS THE BEST PERFORMANCE.

Method	Wire-length	# of Xtalk region ($1w$)
Lee Algorithm	3.8mm	16
A* Algorithm	**3.16mm**	5
AI-router (ours)	**3.16mm**	**0**

except the case that routing is not feasible on a current layer, a high penalty of $\gamma = 30$ is given.

Eventually, the objective function is defined by the expectation of the cost $L(\pi)$ with an output probability of p_θ in (8). θ represents the whole parameters from the encoder-decoder architecture in II.A.

$$J(\theta|X) = \mathbb{E}_{p_\theta}[L(\pi)] \tag{10}$$

We optimize θ using the gradient of $J(\theta|X)$ with Adam optimizer, learning rate 0.00001 [8]. The gradient of the objective function can be derived as follows:

$$\nabla J(\theta|X) = \mathbb{E}_{p_\theta}[(L(\pi) - b(X))\nabla \log p_\theta(\pi|X)] \tag{11}$$

The b in (11) is the rollout baseline for reducing variance [5].

III. VERIFICATION OF THE PROPOSED METHOD

For training AI-router, 12800 randomly generated maze data are used. Each maze problem has a total of 800 nodes with two signal layers. Also, 2560 validation data are used to validate AI-router's learning-convergence and flexibility in various environments. As shown in Fig. 4, the validation loss of AI-router is converged just as it's train loss which means AI-router is stably trained.

To verify the trained AI-router's performance, we set two test cases. Firstly, routing performances of the lee algorithm, A* algorithm, and AI-router are compared in a single-layer test case. Secondly, the AI-router is tested on a multi-layer test case to verify AI-router's capability to perform multi-layer routing.

As illustrated in Fig. 5 and Table II, AI-router makes a better decision that makes shorter and less crosstalk routing than the Lee-router. Also, AI-router makes less crosstalk than the A* algorithm while preserving wire-length. Fig. 7 shows the results of the AI-router in a multi-layer routing test case. It can be observed that the AI-router finds the via location that minimizes crosstalk and wire-length by considering the two layers of keep-out regions and pre-routed regions.

IV. CONCLUSION

In this paper, we proposed an RL-based AI-router considering wire-length, crosstalk, and vias. Our AI-router outperformed conventional maze routers in terms of SI in a

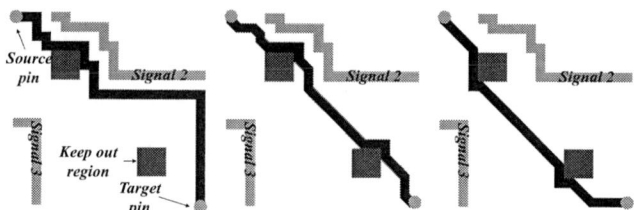

Fig. 5. Test case comparison between lee algorithm, A* algorithm, and AI-router. Because the keep-out region is blocking the middle between the source pin and target pin, the proposed AI-router chooses the routing to bypass downwards considering both crosstalk and wire-length.

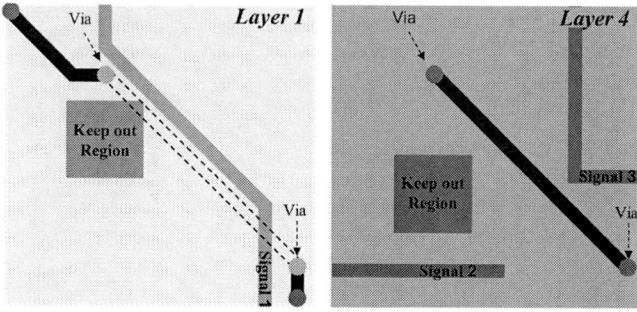

Fig. 6. Result of multi-layer routing by AI-router. Because signal 1 in layer 1 is blocking the path between the source pin and target pin, AI-router decides to change the signal layer through via. The AI-router decides via location considering the routing environment of layer 1 and layer 4.

specific test case. Because of the versatility of the learning-based method, our proposed algorithm can be further applied to various application targets by just fine-tuning the objective function of RL.

ACKNOWLEDGEMENT

We would like to acknowledge the technical support from ANSYS Korea. This work was supported in part by the National Research Foundation of Korea Grant funded by the Korean Government (NRF-2019M3F3A1A03079612)

REFERENCES

[1] J. McDaniel, Z. Zimmerman, D. Grissom, and P. Brisk, "Pcb escape routing and layer minimization for digital microfluidic biochips," *IEEE Transactions on Computer-Aided Design of Integrated Circuits and Systems*, vol. 36, no. 1, pp. 69–82, 2017.

[2] C. Y. Lee, "An algorithm for path connections and its applications," *IRE Transactions on Electronic Computers*, vol. EC-10, no. 3, pp. 346–365, 1961.

[3] P. E. Hart, N. J. Nilsson, and B. Raphael, "A formal basis for the heuristic determination of minimum cost paths," *IEEE Transactions on Systems Science and Cybernetics*, vol. 4, no. 2, pp. 100–107, 1968.

[4] H. Park, J. Park, S. Kim, K. Cho, D. Lho, S. Jeong, S. Park, G. Park, B. Sim, S. Kim, Y. Kim, and J. Kim, "Deep reinforcement learning-based optimal decoupling capacitor design method for silicon interposer-based 2.5-d/3-d ics," *IEEE Transactions on Components, Packaging and Manufacturing Technology*, vol. 10, no. 3, pp. 467–478, 2020.

[5] W. Kool, H. van Hoof, and M. Welling, "Attention, learn to solve routing problems!" in *International Conference on Learning Representations*, 2019.

[6] A. Vaswani, N. Shazeer, N. Parmar, J. Uszkoreit, L. Jones, A. N. Gomez, L. u. Kaiser, and I. Polosukhin, "Attention is all you need," in *Advances in Neural Information Processing Systems 30*, 2017, pp. 5998–6008.

[7] R. J. Williams, "Simple statistical gradient-following algorithms for connectionist reinforcement learning," *Machine Learning*, vol. 8, pp. 229–256, 1992.

[8] D. P. Kingma and J. L. Ba, "Adam: A method for stochastic optimization," in *International Conference on Learning Representations*, 2019.

PCIe Gen-5 Design Challenges of High-Speed Servers

Mallikarjun Vasa, Chun-Lin Liao, Sanjay Kumar, Ching-Huei Chen, Bhyrav Mutnury,

DellEMC Infrastructure Solutions Group.

{Mallikarjun_Vasa, Chun-Lin_Liao, Sanjay_Kumar22, Carol_Chen2, Bhyrav_Mutnury}@dell.com

Abstract— **Supporting PCIe Gen-5 in high-speed servers has become a challenge to designers. Parasitic effects that were benign at PCIe Gen-4 speeds are impacting PCIe Gen-5 operations adversely. Optimizing signal and ground via placement, anti-pad dimensions, AC capacitor placement and trace routing in dense pin area are becoming important as sensitivity to loss, impedance and crosstalk is high at 32 Gbps. In this paper, few approaches to minimize channel impedance discontinuity and near-end/far-end crosstalk (NEXT/FEXT) are discussed. Two channels with optimal design practices and regular design practices are compared and contrasted.**

Keywords—PCIe Gen 5, Via Design, Crosstalk

I. INTRODUCTION

With the increasing demand for bandwidth in server designs, PCI Express (PCIe) has been extended from fourth generation (Gen-4, 16 Gbps) to fifth generation (Gen-5, 32 Gbps). PCIe Gen-5 has 31.25 ps of unit interval (UI) and uses 85-ohms channel impedance as its previous generation. The end-to-end channel loss is 36 dB @ 16 GHz. In order to meet the channel loss requirement, better PCB material(s) and/or cable(s) are needed along with optimal stack-up and layout design practices [1].

To meet the small UI requirement, better impedance control and crosstalk minimization is required on every component such as vias and connectors [2][3]. In the past, sideband termination on PCIe connectors are studied on [4]. Via patterns and optimization methods are presented in [5]-[7]. Most of the work was focused on optimizations on connector or add-in-card (AIC). In this paper, the focus is on the mother board design.

In this study, PCIe Gen-5 design challenges on server mother board channels are introduced in section II. Several channel parameters are analyzed and optimized in section III. Two channels with optimal design and regular design practices are compared and contrasted in section IV. Conclusions are summarized in section V.

II. SERVER DESIGN CHALLENGE

High-speed servers offer feature rich, configurable and customizable PCIe configurations to the customer. As a result of this, high-speed server designs are usually complex. In rack and blade servers, PCIe topologies typically have multiple boards and cables in the path to connect the root-complex to the end-device. Server PCIe topologies are also complicated with many of them having multiple connectors and/or cables in their path. Besides, PCIe channel design in servers needs to reserve channel loss budget for CPU package and AIC loss. On PCIe Gen-4, CPU package loss was typically around 5 dB with only 15 dB channel loss budget reserved for server PCIe channel. On PCIe Gen-5, CPU package loss is at 9 dB and

there is only 17.5 dB channel loss budget allotted for server PCIe channel. From PCIe Gen-4 to Gen-5, the Nyquist frequency got doubled from 8 GHz to 16 GHz, but the loss budget for server design increased from 15 dB to 17.5 dB. This server budget should cover PCB trace loss, via loss, connector loss, cable loss and AC capacitor loss. Channel loss requirement is so restrictive that every channel parameter needs to be optimized to minimize their loss.

III. DESIGN OPTIMIZATION

In this study, several channel parameters are analyzed to optimize PCIe Gen-5 channel. A 20-layer stack-up with 8-signal layers is used. The core and prepreg thickness are 5 and 6 mil, respectively. The total board thickness is 120 mil. Except for a very short trace around breakout area on top layer, all PCIe Gen-5 signals are routed on inner layers and are shielded with two adjacent ground (GND) layers. A via stub of 10 mil is maintained on all channels.

A. Insertion Loss for Different Via Transion Lengths

As per the introduction, the loss budget of the PCIe Gen-5 channel did not double as the data rate. So, designers should intelligently select the routing layer so that overall channel and via loss is within the loss budget. For a 20-layer stack-up with mid-loss material (Dk-3.8 Df-0.015), the loss per via can vary by 0.3 dB or more depending on routing layer. It can be seen from Fig. 1 that inner1 via loss is 0.1 dB and inner8 via loss is ~0.4 dB at 16 GHz.

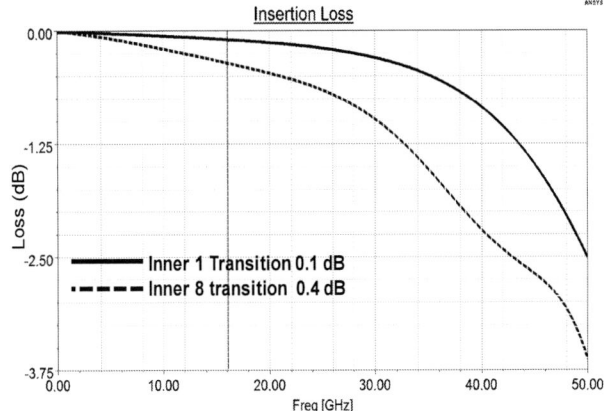

Fig. 1. Simulated Insertion loss comparison for Inner1 to Inner8 transtion

B. Via Anti-Pad Dimensions

Anti-pad refers to the area between signal vias and the ground/power plane, this is necessary to ensure signal via does not short other non-signal layers. The size and shape of the antipad would determine the impedance discontinuity. At high

978-1-7281-6162-4/20 $31.00 © 2020 IEEE

data rates, one common anti-pad cannot be followed for all the via transitions as each layer experiences a different capacitance due to planes, and their impedance are different accordingly, as shown in Fig. 2(B). In Table. 1 it is shown that optimized impedance is the combined effort of anti-pad (Ap), signal to ground distance (Sg), via pitch (Vp) and material property as shown in Fig. 2(A).

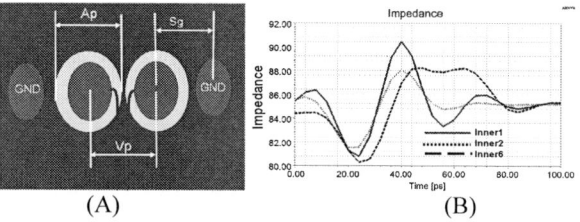

(A)   (B)

Fig. 2. (A) Top view of differential pair via (B) simulated Impedance plot for different layer transitions

TABLE 1. IMPEDANCE VARIATION ON DIFFERENT SIGNAL LAYERS

Layer	Via_Pitch (Vp)	Anti_pad (Ap)	Signal to Ground pitch (Sg)	Impedance
1	38 mil	38 mil	33 mil	(81-91) ohm
2	38 mil	36 mil	33 mil	(82-88) ohm
3	38 mil	34 mil	33 mil	(81-88) ohm
6	38 mil	32 mil	33 mil	(81-91) ohm
8	38 mil	32 mil	33 mil	(80-92) ohm

C. GND Vias

Ground vias are needed to be placed adjacent to PCIe signal transition vias to have good signal return path and crosstalk shielding. On PCIe Gen-4, two symmetric ground vias are typically enough. In this study, two and four adjacent ground vias structures are analyzed to see their crosstalk shielding performance. Three structures are shown in Fig. 3 and their FEXT comparisons are shown in Fig. 4. Structures (A) and (B) have 2 ground vias adjacent to PCIe transition vias, but their ground via locations are different. Since structure (A) has two ground vias between DIFF1 and DIFF3 pairs, it has better crosstalk immunity between DIFF1 and DIFF3. On the other side, structure (B) has better crosstalk performance between DIFF1 and DIFF2. However, four ground vias structure (C) has the best crosstalk performance compared to two ground via structures.

D. GND vias on SMT GND pads

Surface mount (SMT) connector is better for PCIe Gen-5 channel because of its better loss performance. It is recommended to add two ground vias on "heel" and "toe" sides of all ground connector pins as shown in Fig. 5. However, due to routing space limitation, some ground pins could not have two ground vias, and some ground vias could not be placed close to connector ground pins. This section uses simulation data to show the SI risk of several different ground via patterns. Fig. 5 shows the different ground via patterns analyzed in this paper. The patterns include (a) single ground via on different locations (b) two ground vias on different locations and (c) four ground vias structure. With different ground via numbers and locations, the insertion of one PCIe pair and FEXT between

two adjacent PCIe pairs are compared on Fig. 6. From the observation, single ground via the resonant peak on FEXT will happen around 30 GHz. On the two ground via structures, if the ground vias could not be placed close to connector pad, it also could not get good performance. The best pattern is placing 4 GND vias on 10 mil/40 mil away from two sides of connector ground pads.

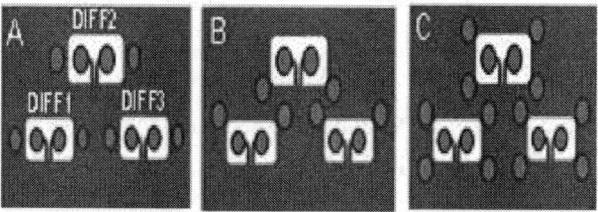

Fig. 3. Comparison of different transition via patterns.

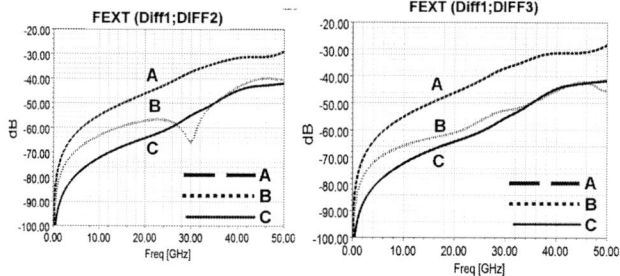

Fig. 4. Comparison of Simulated far end crosstalk (FEXT) between different transition via patterns.

E. Crosstalk in Dense Pin Area

If PCIe pinouts of the chip are not close to the edge of chip but in the inner area, then those PCIe signal needs to penetrate through dense pin area to the outside open area. In the dense pin area, PCIe signal traces should not go through another different pair vias. In Fig. 7, two yellow circles are P/N pins of one PCIe pair (DIFF1) and another PCIe pair (DIFF2 of blue lines) passes through P/N pins of DIFF1. Another pair (DIFF3) do not go through DIFF1 P/N pins. It can be expected that the crosstalk between the two differential pairs (DIFF1 and DIFF2) will be much higher than crosstalk between (DIFF1 and DIFF3). Simulation results are shown in Fig. 8 and it is seen that the crosstalk is ~5 dB higher between DIFF1-DIFF2 compared to DIFF1-DIFF3. In cases when signals penetrate through dense BGA pin area, crosstalk can be reduced by proper layer selection so the via crosstalk can be reduced and selecting a GND pattern around BGA signal vias.

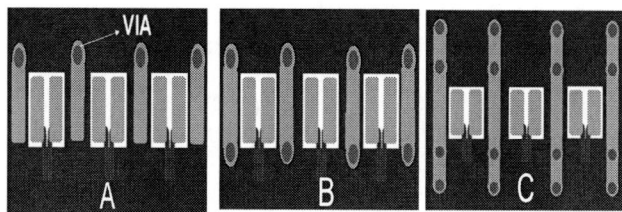

Fig. 5. (A) Single GND via placed Inner Side, (B) Two GND via placed either side, (C) Four GND via placed on either side.

978-1-7281-6162-4/20 $31.00 © 2020 IEEE

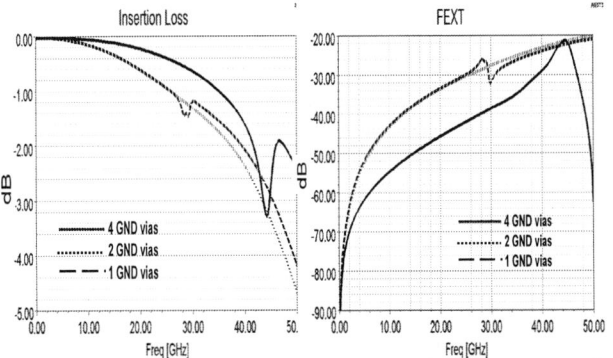

Fig. 6. Simulated Insertion loss and FEXT comparison of three structures of Fig. 5.

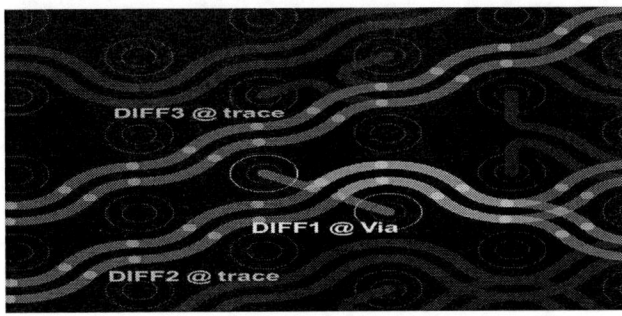

Fig. 7. Differential pairs in pin field area.

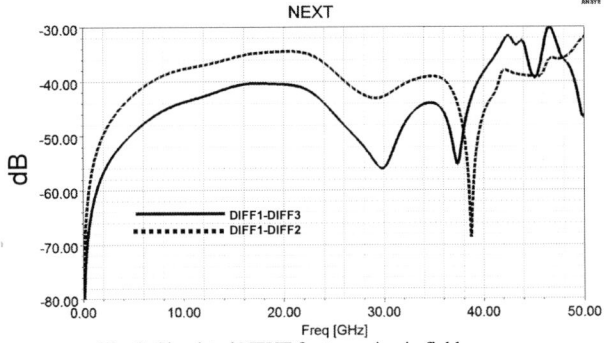

Fig. 8. Simulated NEXT for traces in pin field area.

IV. OPTIMAL VERSUS REGULAR CHANNELS

This section uses two PCIe channels to show how much difference in eye opening can be achieved from optimal and regular channel designs. The optimal channel uses all the best design practices from section III, the regular channel on the other hand uses regular design practices. Individual channel models are cascaded to result in a channel model as shown in Fig. 9. This channel is a PCIe Gen-5 Tx channel, which includes one adapter, one AC cap and two PCIe connectors. Trace length on planar and adapters are 4" and 2", respectively. Optimal channel uses Inner1 transition vias, 5 mil core and 6 mil prepreg. Connector and transition vias have 4 adjacent GND vias. Regular channel uses Inner8 routing, 3 mil core and 4 mil prepreg. Connector and transition vias have only 2 adjacent GND vias. Trace length of both the channels

are same. The cascaded model showed that the optimal channel loss is 3 dB lower and the FEXT and NEXT are 10 dB lower compared to regular channel. Eye diagram of the two channels are shown in Fig. 10, optimal channel has (EH/EW) = (37.0 mV/15.8 ps) and the regular channel (7.0 mV/7.8 ps).

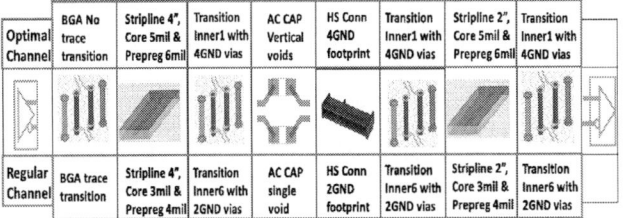

Fig. 9. PCIe Gen-5 channel topology.

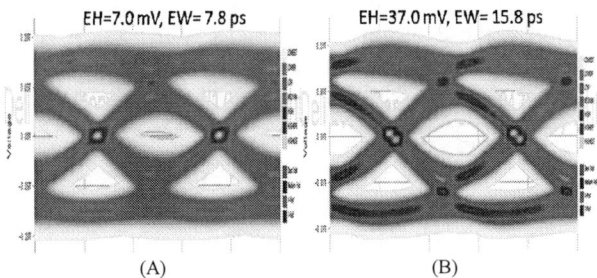

Fig. 10. Comparison between (A) regular channel and (B) optimal channel

V. CONCLUSIONS

In this paper, several channel parameters are optimized to reduce their loss, reflections and crosstalk. At PCIe Gen-5 data rate, any channel imperfection can begin to have a huge impact on SI robustness. Optimizing all aspects of the PCIe channel becomes critical to a successful design. In this paper, the importance of the proposed optimizations is shown using two example channels and it is evident that optimized design practices are needed at PCIe Gen-5 speeds.

REFERENCES

[1] H. F. Xue et al., "PCI Express 5.0 Solution Extending Study Through PCB Stackup and Geometry Optimization," DesignCon 2020.

[2] X. M. Gao, P. Z. Yang, J. X. Zhu, and G. Yang, "Modular Platform Design & Optimization for PCIe 5.0 IPs Validation," DesignCon 2020.

[3] L. Shan, D. Freidman, C. Kennedy, W. Persak and K. Lau, "Backward Compatible Connectors for Next Generation PCIe Electrical I/O," 2018 IEEE 68th Electronic Components and Technology Conference (ECTC), San Diego, CA, 2018, pp. 1798-1804.

[4] Y. Zhou, W. Shi and S. Sudhakaran, "A new approach to mitigate PCI express Gen4 crosstalk from sideband signals in connectors," 2017 IEEE 26th Conference on Electrical Performance of Electronic Packaging and Systems (EPEPS), San Jose, CA, 2017, pp. 1-3.

[5] C. Ye, X. Ye and E. L. Miralrio, "Via pattern design and optimization for differential signaling 25Gbps and above," 2016 IEEE International Symposium on Electromagnetic Compatibility (EMC), Ottawa, ON, 2016, pp. 312-317.

[6] S. Chen, C. Chen, C. Liao, J. Chen, T. Wu and B. Mutnury, "Via optimization for next generation speeds," 2017 IEEE 26th Conference on Electrical Performance of Electronic Packaging and Systems (EPEPS), San Jose, CA, 2017, pp. 1-3.

[7] M. Vasa, A. C. Reddy, B. Mutnury, S. Kumar and Vasanth RD, "High speed interconnect optimization," 2015 Asia-Pacific Symposium on Electromagnetic Compatibility (APEMC), Taipei, 2015, pp. 8-11.

A Shielded-Block Preconditioner for Reduced-Domain Layered-Medium Integral-Equation Methods

Chang Liu, *Student Member, IEEE,* Ali E. Yılmaz, *Senior Member, IEEE*

Department of Electrical and Computer Engineering, The University of Texas at Austin, Austin, Texas, USA

Abstract—A shielded-block preconditioner is adopted to reduce the iteration count of layered-medium integral-equation (LMIE) solvers that include dense metallization layers in the Green's functions. Similar to other sparse block preconditioners, the preconditioner is suitable for integration with parallel LMIE solvers to facilitate the analysis of full-size package models. The performance of the preconditioner is demonstrated by analyzing a via array extracted from a full-size package model.

Key Words—*Integral-equation methods; layered media.*

I. INTRODUCTION

Layered-medium integral-equation (LMIE) methods avoid discretizing and solving for fields in dielectric substrates [1]–[3] and can be enhanced significantly by using parallel iterative solvers, fast matrix-vector multiplication algorithms, and sparse preconditioners [4]. Yet, when used for electromagnetic analysis of full-scale models of modern electronic packages, typical LMIE methods result in large, dense, and poorly conditioned systems of equations; this is in part because they require discretization of currents in/on all conductors.

Reduced-domain LMIE methods, which model contiguous power/ground planes (reference conductors) as highly or perfectly conducting layers in the background medium, reduce the computational domain further and can also yield better conditioned systems of equations; this can be attributed to the fast decay of Green's function components corresponding to horizontal-source to horizontal-observer interactions [3]. When combined with small-aperture modeling methods [5]–[7], reduced-domain LMIE methods can also efficiently analyze complex full-size models by decomposing the package to various subdomains (Fig. 1). For models that include vertical via transitions, however, the Green's function components that correspond to vertical via-to-via interactions (for vias between two highly conductive layers) decay even slower with distance compared to those in the typical LMIE approach [4]. This creates significantly larger off-diagonal matrix entries and can yield rather poorly conditioned matrices that require large number of iterations when iterative solvers are used. This article presents a shielded-block preconditioner [8] that is well-suited for integration into parallel reduced-domain LMIE solvers and demonstrates its effectiveness for a subdomain consisting of a PTH via array at a core dielectric layer.

II. FORMULATION

Consider a full-size electronic package model that is decomposed into smaller subdomains bounded vertically by one or two highly conductive layers (Fig. 1). The fields in these subdomains are coupled through cutouts in reference conductors; using various small-aperture models to accurately

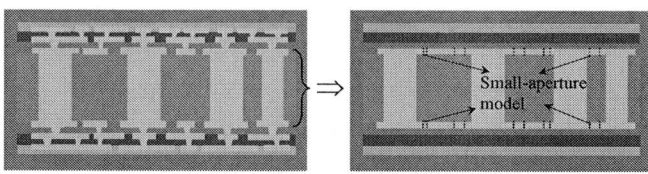

Fig. 1. A subdomain problem consisting of a plated-through-hole (PTH) via array is extracted from a full-size electronic package using small-aperture models to account for coupling through cutouts at the reference conductors.

couple the computations for the different subdomains is an important topic of study discussed elsewhere [5]–[7]. The focus of this article is instead on the full-wave analysis in a certain type of subdomain; specifically, consider a subdomain consisting of two contiguous reference conductors, filled by a core dielectric, and connected by a PTH via array (Fig. 2). The structure is modeled as N^{via} vias residing in a planar layered medium background stratified to $K=5$ layers in the z direction with infinite extent horizontally. Each layer $k=1,...,K$ is assigned an isotropic material with permittivity ε_k, conductivity σ_k, and permeability μ_k, where $k=2,4$ are two highly conductive layers to approximate the reference conductors and $k=3$ is the core dielectric layer embedding the PTH vias.

Consider the extraction of 2-port network parameters for a signal via in the presence of all other structures. Enforcing the standard Leontovich impedance boundary condition on all via surfaces S^{via} to account for finite conductivity and surface roughness effects, yields the LMIE for this subdomain problem:

$$\hat{\mathbf{n}} \times \hat{\mathbf{n}} \times \mathcal{L}^{\text{EJ}}(\mathbf{J},\mathbf{r}) + Z_S(\mathbf{r})\mathbf{J}(\mathbf{r}) = 0 \quad \forall \mathbf{r} \in S^{\text{via}} \tag{1}$$

Here, $\mathbf{J}$ is the surface current density, $\mathcal{L}^{\text{EJ}}$ is the layered-medium electric-field integral-equation operator, and Z_S is the local surface impedance term [3][4]. To extract the network parameters, the delta-gap port model is used: ports 1 and 2 are defined as the circular edges at the top and bottom surface of the signal via and all ports are terminated at 50-Ω loads. The signal via is driven by two different excitations; for the p-th excitation (p=1 or 2), a 1 V time-harmonic voltage source is placed at port p; i.e., $\mathcal{L}^{\text{EJ}}(\mathbf{J},\mathbf{r}) = \delta(\mathbf{r}-\mathbf{r}_p)\hat{\mathbf{h}}_p$ is enforced, where $\mathbf{r}_p$ are points on the circular edges of port p and $\hat{\mathbf{h}}_p$ is the unit normal vector pointing from the port into one of the reference planes.

To solve the LMIE in (1), $\mathbf{J}$ is expanded with a total of N Rao-Wilton-Glisson (RWG) [9] functions on the vias and half-RWG functions on the circular edges touching the highly conductive layers. Galerkin testing of the LMIE in (1) yields the linear system of equations: $\mathbf{Z}_{N\times N}\mathbf{I}_{N\times 1} = \mathbf{V}_{N\times 1}$, where $\mathbf{Z}$, $\mathbf{I}$, and $\mathbf{V}$ are the impedance matrix, unknown coefficient vector, and right-hand-side vector. As there are N^{via} disconnected structures (vias), $\mathbf{Z}$ can be organized into $N^{\text{via}} \times N^{\text{via}}$ blocks

978-1-7281-6162-4/20 $31.00 © 2020 IEEE

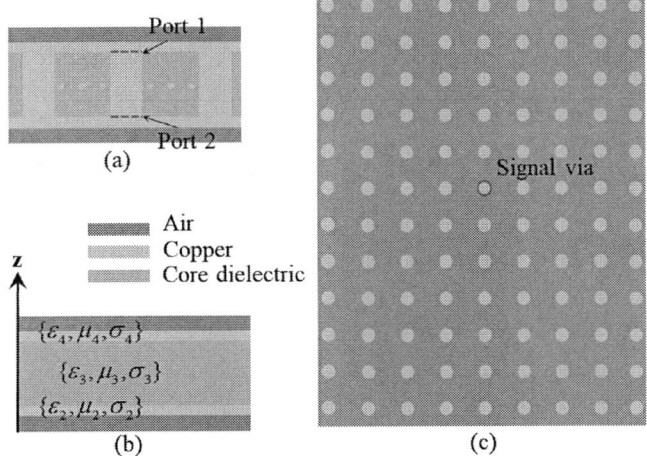

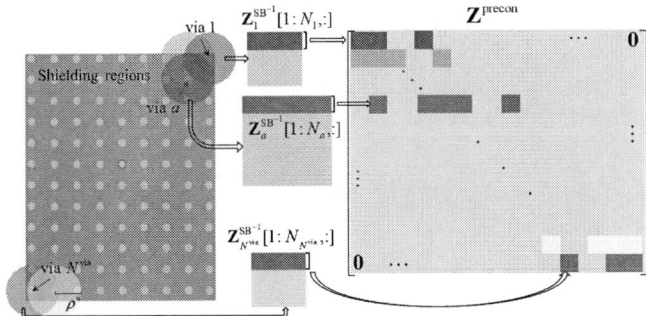

Fig. 3. Shielding regions for the via array and the corresponding matrix pattern for the shielded-block preconditioner.

Fig. 2. A PTH via array with N^{via} vias embedded in a core dielectric layer: (a) cross-section view of the array and the port configuration. (b) The $K=5$-layer background model used by the reduced-domain LMIE method. (c) Top-down view of the via array with the boundary edges of the signal via marked in red.

$$\mathbf{Z} = \begin{bmatrix} \mathbf{Z}_{11} & \cdots & \mathbf{Z}_{1a} & \cdots & \mathbf{Z}_{1N^{\mathrm{via}}} \\ \vdots & \ddots & & & \vdots \\ \mathbf{Z}_{a1} & \cdots & \mathbf{Z}_{ab} & \cdots & \mathbf{Z}_{aN^{\mathrm{via}}} \\ \vdots & & & \ddots & \vdots \\ \mathbf{Z}_{N^{\mathrm{via}}1} & \cdots & \mathbf{Z}_{N^{\mathrm{via}}a} & \cdots & \mathbf{Z}_{N^{\mathrm{via}}N^{\mathrm{via}}} \end{bmatrix} \quad (2)$$

Here, each $N_a \times N_b$ sized block $\mathbf{Z}_{ab}$ stands for the interaction between the observer via b (with N_b testing functions on it) and the source via a (with N_a basis functions on it). To reduce the iteration count and improve solution accuracy, the system of equations can be preconditioned from the left as:

$$\mathbf{Z}^{\mathrm{precon}} \mathbf{Z} \cdot \mathbf{I} = \mathbf{Z}^{\mathrm{precon}} \cdot \mathbf{V} \quad (3)$$

The proposed preconditioner is constructed by assigning each via a "shielding region" by identifying a group of neighboring vias as its "shield" [8]. These regions are dictated by a predefined parameter ρ^s: the axial distance between a via and other vias in its shield is less than ρ^s (Fig. 3). Let N_a^{via} and N_a^{SB} denote the total number of vias and basis functions enclosed in the a-th via's shielding region, and $\mathbf{Z}_a^{\mathrm{SB}}$ denote the $N_a^{\mathrm{SB}} \times N_a^{\mathrm{SB}}$ impedance submatrix that stores all the interactions within the shield; this submatrix can be formulated using the via-to-via interaction blocks in (3):

$$\mathbf{Z}_a^{\mathrm{SB}} = \begin{bmatrix} \mathbf{Z}_{\mathbf{S}_a[1]\mathbf{S}_a[1]} & \cdots & \mathbf{Z}_{\mathbf{S}_a[1]\mathbf{S}_a[N_a^{\mathrm{via}}]} \\ \vdots & \ddots & \vdots \\ \mathbf{Z}_{\mathbf{S}_a[N_a^{\mathrm{via}}]\mathbf{S}_a[1]} & \cdots & \mathbf{Z}_{\mathbf{S}_a[N_a^{\mathrm{via}}]\mathbf{S}_a[N_a^{\mathrm{via}}]} \end{bmatrix}_{N_a^{\mathrm{SB}} \times N_a^{\mathrm{SB}}} \quad (4)$$

Here, the selection array $\mathbf{S}_a$ stores the indexes of the N_a^{via} vias enclosed in the shield of the a-th via; the first via in each shielding region is set to the shielded via itself, i.e., $\mathbf{S}_a[1] = a$. Then, $\mathbf{Z}_a^{\mathrm{SB}}$ for each via a is inverted (capturing the self and mutual via interactions in the shielding region of the via). Next, the first N_a rows of each inverted shielded-block submatrix is

collected and these rows are arranged into rectangular-block-diagonal form (Fig. 3). Specifically, $\mathbf{Z}^{\mathrm{precon}}$ is filled as:

$$\mathbf{Z}^{\mathrm{precon}}\left[\mathbf{C}_a[1:N_a], \mathbf{C}_a[1:N_a^{\mathrm{SB}}] \right] = \mathbf{Z}_a^{\mathrm{SB}^{-1}}[1:N_a, 1:N_a^{\mathrm{SB}}] \quad (5)$$

where $\mathbf{C}_a$ simply maps the (local) indexes of the submatrix $\mathbf{Z}_a^{\mathrm{SB}}$ to the (global) indexes of the impedance matrix $\mathbf{Z}$.

Like other sparse preconditioners, using $\mathbf{Z}^{\mathrm{precon}}$ requires two major computational steps: (i) inverting all shielded-block submatrices to setup the preconditioner, which requires $\mathcal{O}(\sum_a [N_a^{\mathrm{SB}}]^3)$ operations and $\mathcal{O}(\sum_a N_a N_a^{\mathrm{SB}})$ memory space; (ii) multiplying the inverted matrices at each iteration, which requires $\mathcal{O}(\sum_a N_a N_a^{\mathrm{SB}})$ operations per iteration. Therefore, the preconditioner is well-suited for problems with many small disconnected geometries ($N_a \ll N$), e.g., PTH via or stitching via arrays in full-size package models.

III. NUMERICAL RESULTS

To demonstrate the performance of the shielded-block preconditioner, four increasingly larger PTH via arrays are analyzed with a parallel iterative LMIE solver [4] and the resulting number of iterations are compared to those from the element-diagonal [4] and block-diagonal preconditioner. Here, the block-diagonal preconditioner is implemented by setting ρ^s to a very small number such that $\mathbf{Z}_a^{\mathrm{SB}} = \mathbf{Z}_{aa}$. In these examples, the number of vias N^{via} is increased from 25 to 1681, with a uniform spacing of 1 mm in the x and y directions. Every via is a circular cylinder with 750-μm height, 250-μm diameter cylinder, and $\sigma_c = 4.5 \times 10^7$ S/m conductivity. The signal via is always assigned to the center via within these arrays. All vias are embedded in a (core) dielectric layer with relative permittivity $\varepsilon_r = 3.5$, loss tangent $\tan\delta = 0.2$, and relative permeability $\mu_r = 1$. The two reference conductors are modeled as 30μm-thick metal layers and the same conductivity as the vias (Fig. 4). Surface-roughness of the conductors is not modeled. After triangularization, the circular cylinder vias are meshed to 12-edge polygon cylinder vias resulting in 228 unknowns for each via ($N = 228N^{\mathrm{via}}$); thus, the number of unknowns increases from $N = 5700$ to $383\,268$ as the number of vias increases. An FFT-accelerated TFQMR solver combined with different preconditioners is used, and the relative residual error is enforced to be smaller than 10^{-4}. The shielding region parameter ρ^s is set to 1.1 mm for all vias, i.e., $N_a^{\mathrm{via}} = 5$ for all non-boundary vias (Fig. 3).

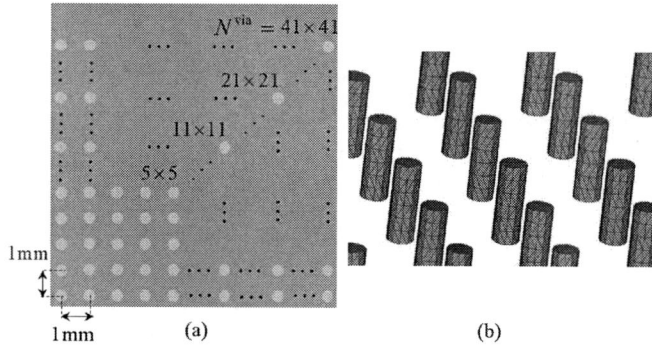

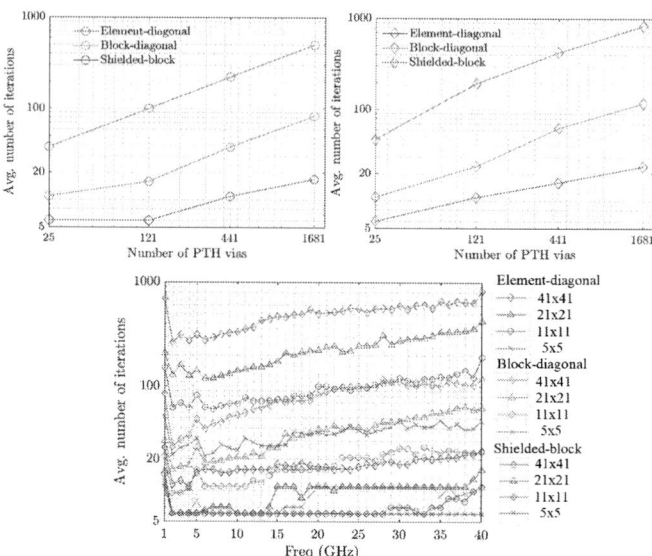

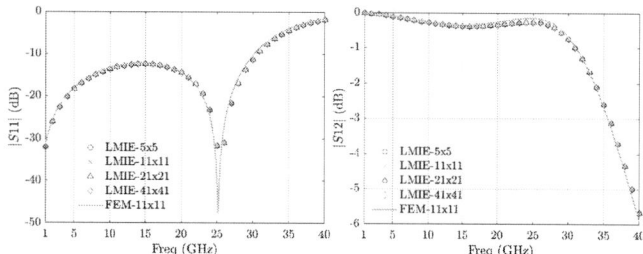

Fig. 4. Four via arrays: (a) top-down view of the 5 × 5, 11 × 11, 21 × 21, and 41 × 41 element arrays with 1 mm uniform spacing, (b) cross-section view of the PTH via arrays embedded in a 5-layer background, and (c) the mesh view of the vias with the reduced-domain LMIE method.

Fig. 5. $|S_{11}|$ and $|S_{12}|$ parameter for the four PTH via array problems obtained from the reduced-domain LMIE method and an FEM solver.

Fig. 6. Comparison of the iteration counts among three preconditioners: Top: average number of iterations vs. problem size at 20GHz (left) and 40 GHz (right); Bottom: average number of iterations for the entire frequency band.

First, the *S*-parameters of the signal via for these four problems are extracted in the 1 to 40 GHz range in Fig. 5. Both $|S_{11}|$ and $|S_{12}|$ parameters are visually identical in all problems, which indicates the effect of the nearby vias on the signal via remains unchanged beyond 25 (5×5) vias. As the frequency increases, less energy can be coupled through the signal via ($|S_{12}|$ drops significantly), while stronger reflection is observed above 30 GHz. Fig. 5 also shows the 11×11 via array's parameters found by a commercial finite-element-method solver, whose results for $|S_{12}|$ and $|S_{11}|$ parameters are within 0.25 dB and 1.5 dB of the LMIE ones at all sampled frequencies, respectively (except at 25 GHz for $|S_{11}|$).

Next, the performance of the three different preconditioners is studied by contrasting the number of iterations they require. As the number of vias increases, the shielded-block preconditioner's iteration count grows the slowest and the method requires much smaller number of iterations (Fig. 6). The iteration counts are also reported at all sampled frequencies in Fig. 6 and the shielded-block preconditioner is found to require < 25 iterations for all cases. The computational costs of the shielded-block preconditioner are sensitive to the shielding parameter ρ^s and mesh density on the vias, i.e., N_a. For the above examples, the shielded-block diagonal preconditioner requires ~10× operations for setting up the preconditioner and ~2× operations to apply it at each iteration compared to the block-diagonal preconditioner. Overall, when the mesh edge lengths on the vias are $\approx \lambda/20$ (λ: the wavelength at the core layer at 40 GHz), the preconditioner setup and multiplication costs are comparable to the impedance-matrix fill and matrix-vector multiplication costs, respectively.

IV. CONCLUSION

A shielded-block preconditioner was adopted and shown to improve the performance of reduced-domain LMIE solvers when applied to subdomains that contain many vertical-source to vertical-observer interactions. The preconditioner can be efficiently parallelized and easily integrated into the parallel reduced-domain LMIE solvers.

REFERENCES

[1] S. Sharma, U. R. Patel, and P. Triverio, "Accelerated electromagnetic analysis of interconnects in layered media using a near-field series expansion of the Green's function," in *Proc. IEEE 27th Conf. Elect. Perform. of Electron. Packag. Syst. (EPEPS)*, Oct. 2018.

[2] G. Bianconi and S. Chakraborty, "Efficient and robust dyadic Green's function evaluation algorithm for the analysis of IC packages and printed circuit board," in *Proc. IEEE 25th Conf. Elect. Perform. of Electron. Packag. Syst. (EPEPS)*, Oct. 2016.

[3] C. Liu, and A. E. Yılmaz, "A reduced-domain layered-medium integral-equation method for electronic packages," in *Proc. IEEE 28th Conf. Elect. Perform. of Electron. Packag. Syst. (EPEPS)*, Oct. 2019.

[4] C. Liu, K. Aygün, and A. E. Yılmaz, "A parallel FFT-accelerated layered-medium integral-equation solver for electronic packages" *Int. J. Num. Model.: Electron. Networks, Dev., Fields*, vol. 33, no. 2, Mar. 2020.

[5] S. Bai *et al.*, "The accuracy of port connections between layers in printed circuit board," in *Proc. IEEE 26th Conf. Elect. Perform. of Electron. Packag. Syst. (EPEPS)*, Oct. 2017.

[6] M. R. Abdul-Gaffoor *et al.*, "Simple and efficient full-wave modeling of electromagnetic coupling in realistic RF multilayer PCB layouts," *IEEE Trans. Microw. Theory Tech.*, vol. 50, no. 6, pp. 1445–1457, 2002.

[7] C-J. Ong *et al.*, "Full-wave solver for microstrip trace and through-hole via in layered media," *IEEE Trans. Adv. Packag.*, vol. 31, no. 2, pp. 292–302, May. 2008.

[8] D. Pissoort *et al.*, "A rank-revealing preconditioner for the fast integral-equation based characterization of EM crystal device," *Microw. Opt. Technol. Lett.*, vol. 48, no. 4, pp. 783–789, Apr. 2006.

[9] S. M. Rao, D. R. Wilton, and A. W. Glisson, "Electromagnetic scattering by surfaces of arbitrary shapes," *IEEE Trans. Antennas Propag.*, vol. 30, no. 3, pp. 409–418, May 1982.

This work was supported by Intel Corporation.

978-1-7281-6162-4/20 $31.00 © 2020 IEEE

On the Accuracy of Cross-Talk Modeling in High-Speed Digital Circuits Using the Accelerated Boundary Element Method

Dongwei Li, Giacomo Bianconi and Swagato Chakraborty

System Design Division, Mentor A Siemens Business

46871 Bayside Parkway, Fremont, CA, USA

dongwei_li@mentor.com, giacomo_bianconi@mentor.com, swagato_chakraborty@mentor.com

Abstract — This paper presents an investigation on how numerical thresholds impact the accurate modeling of cross-talk phenomena for typical interconnect structures in the context of the accelerated boundary element method. Furthermore, a canonical scenario is presented to expose the noise floor of the methodology and its relation to the accuracy controls involved in the solver.

Keywords—Cross-talk, fast iterative solvers, noise floor.

I. INTRODUCTION

Modern trends in electronic design call for 3D full-wave modeling of the interconnects in integrated circuit packages and printed circuit boards. Moreover, the co-existence of high-speed digital interconnects and noise-sensitive RF structures mandate an accurate characterization of cross-talk down to a very low dB level. The boundary element method is especially suited for accurate characterization of this type of phenomenon since it does not require the finite discretization of the space between the aggressor and the victim device. On the other hand, traditional techniques using direct solvers are limited by their prohibitive cost, mandating the need for fast-iterative solvers to analyze complex structures. In this paper, we consider a few typical scenarios in interconnect analysis where we need to accurately model low dB cross-talk. Depending on the coupling mechanism, we demonstrate how different aspects of error control mechanisms in an accelerated boundary element solver play a significant role.

II. METHOD OF MOMENTS

The Boundary Element (BE) or Method of Moments (MoM) is a widely used numerical algorithm for boundary element based electromagnetic analysis of Integrated Circuits (ICs) and Printed Circuit Boards (PCBs) [1]. The MoM, in conjunction with the Dyadic Green's Functions (DGFs), allows numerical modeling of ground planes, dielectric interfaces, and the surrounding environments without explicitly discretizing them.

The spatial domain DGFs can be expressed in the form of a Sommerfeld Integral (SI) as reported in (1):

$$G^{EJ}(\rho, z, z') = \frac{1}{2\pi} \int_0^\infty \tilde{G}^{EJ}(k_\rho, z, z') \cdot J_n(k_\rho \rho) k_\rho^{n+1} dk \quad (1)$$

where ρ, z, z' and J_n are the locations of the source, observation points in a cylindrical coordinate system, and the *n-th* Bessel function of the first kind, respectively. The SI does not have a closed-form and its numerical solution presents some challenges [2]. Over the last decade, several approaches have been proposed to speed-up the SI calculation [3],[4]. It is worthwhile to emphasize that an optimal compromise must be found between accuracy and numerical efficiency.

Once the Dyadic Green's Functions have been evaluated, a matrix equation is solved for the weight coefficients associated with the basis functions used to represent the induced currents. Since the majority of the DGF terms are characterized by a singular behavior when the distance between source and observation point becomes small; the related matrix elements require a special numerical treatment [5]. It is then evident how an additional source of inaccuracy in the Method of Moments can be identified in the finite quadrature orders and singularity extraction techniques required to evaluate the reaction integrals.

The application of the conventional MoM formulation with subsectional basis functions becomes quite inefficient when the structure is electrically large [1], or geometrically complex such as interconnects in IC packages or printed circuit boards. This, in turn, increases both the MoM matrix generation time and *LU* factorization, which present a quadratic and cubic complexity, respectively. A variety of iterative approaches have been deeply investigated over the last decade [6],[7]. These "fast-solver" techniques, able to reduce the setup and solve time complexity to *O(NlogN)* in their multilevel form, are based on an advantageous representation of the far-field interactions, in conjunction with the use of dedicated iterative solvers. However, a numerically efficient error control mechanism is necessary, both for the iterative solution of the associated linear system of equations, as well as for far-field calculations, to guarantee good accuracy and reasonable solution time.

III. BOUNDARY ELEMENTS AND CROSS TALK

When considering MoM in conjunction with DGFs, all the employed unknowns are associated with current density induced over the conducting structure. Accurate characterization of cross-talk between two traces depends on how rigorously the spatial variation of the electric and magnetic field is captured in the space between the two conductors, which is modeled analytically in the Green's function without any explicit discretization of the domain.

With no loss of generality, let us consider the cross-talk between the two differential via transitions shown below (Fig. 1). As we can see, other than the traces themselves, the ground planes are also discretized with a coarser and finer mesh. It is worthwhile noticing how the far (FEXT) and near end cross-talk results (NEXT) are not sensitive to the discretization of the ground planes between the two differential pairs, as long as there is a sufficient amount of mesh elements below the traces to capture the return current accurately.

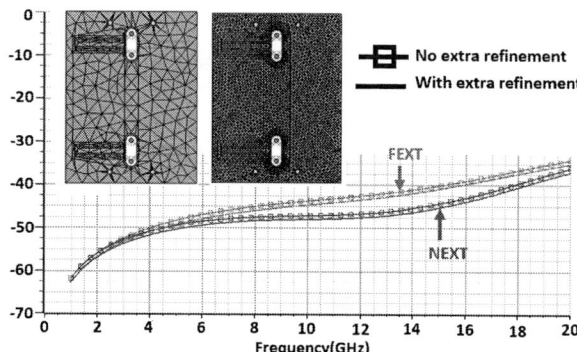

Fig 1. Cross-talk coupling for the differential via transitions.

To further demonstrate the mentioned advantage associated with the boundary element method, we present the field variation between two differential strip-lines (Fig. 2). It is remarkable noticing how even if there is no additional discretization for the ground planes between the differential traces, a smooth variation both for the electric field (b) as well as for the magnetic field (c) is accurately captured.

IV. NUMERICAL RESULTS

A. Different cross-talk creation and mitigation mechanisms

To study the impact of possible numerical tolerances (section-II) on cross-talk calculations, a few accuracy settings have been summarized in Table I with a progressively conservative target. A saturated configuration, '**Ref**', is assumed to be the golden standard for this convergence study.

Three different mechanisms of low-level cross-talk creation are investigated in this paper. Table II summarizes the required settings to achieve convergence at a certain dB level.

Case 1: ***Low far-end cross-talk due to electrically large interconnects***. For this experiment, we have analyzed two pairs of 12'-long differential microstrip lines up to 50 GHz. In this scenario, the FEXT, due to the conduction and roughness loss mechanism, is very low and needs accurate modeling to maintain adequate isolation at the far end. We find that this scenario requires relatively less stringent error thresholds in quadrature, far-field threshold, and iterative solver (Table II), because here the cross-talk is low due to high loss in the conductor, which at high frequency is modeled with surface

impedance model, that accurately does not need to stress the numerical thresholds significantly.

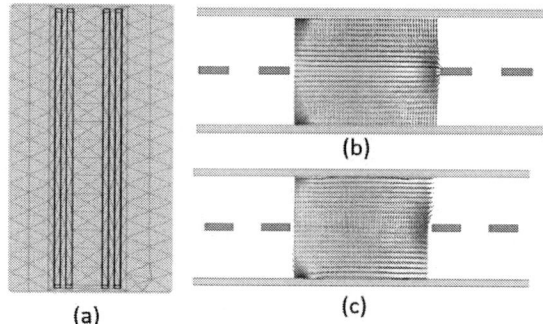

Fig 2. Strip-line differential pair geometry (a), electric field (b), and magnetic field (c) in the space between the differential traces.

Table I. Accuracy settings in Fast Boundary Element Solver

Setting	Quadrature order	Far-field interaction threshold	Iterative solver tolerance
I	1	0.005	$1e^{-3}$
II	2	0.005	$1e^{-3}$
III	2	0.001	$1e^{-4}$
IV	2	0.0005	$1e^{-5}$
V	2	0.0001	$1e^{-6}$
Ref	3	0.00001	$1e^{-8}$

Case 2: ***Cross-talk mitigation by spatial separation***. In particular, we increase the trace separation distance in parallel and broadside coupled traces to achieve low level of cross-talk. As shown in Table II, more stringent error control is required in this case since the low-value cross-talk is associated with far-field interactions between mesh elements or coupling over a small segment of the traces.

Case 3: ***Cross-talk mitigation obtained with isolation planes or stitching vias***. As can be noticed from Table II, the most stringent error control is required in this case. With no loss of generality, let us consider three mesh elements: A, B, C defined over the aggressor, the shielding object, and the victim, respectively. In particular, the mesh element A creates a direct scattered field over C through the Green's function, but at the same time, induces a current on B which, in turn, creates a "nearly" equal and opposite scatterd field on C that has to cancel the previously described contribution from A.

Finally, it is then crucial to quantify the relative cost of the described accuracy settings for more complex scenarios. We have then analyzed three different projects: a 10 layer package with 13 ports discretized with ~100k mesh elements (Project 1), a 10 layer PCB with 20 ports discretized with ~150k mesh elements (Project 2), and a 40 layer PCB with 16 ports and ~150k mesh elements (Project 3). The obtained results are summarized in Table III. It is worthwhile noticing how no more than 3x slow-down is generally required by the more conservative settings.

978-1-7281-6162-4/20 $31.00 © 2020 IEEE

Table II. Summary of the cross-talk study observations.

Case	-40 dB	-60 dB	-80 dB
FEXT for electrically long lines	I	I	III
Broadside coupling with vertical separation	I	II	III
Differential traces isolated by horizontal separation	I	II	III
Differential vias isolated by stitching vias	I	III	IV
Differential vias isolated by shielding planes	I	III	V

Table III. Relative time penalty as a function of the different accuracy settings for three realistic scenarios.

Setting	Relative Time Penalty		
	Project 1	**Project 2**	**Project 3**
I	1.0x	1.0x	1.0x
II	1.4x	1.6x	1.3x
III	1.7x	2.1x	2.1x
IV	2.0x	2.3x	2.8x
V	2.5x	2.8x	6.3x

B. In-pair coupling benchmark

In this section, we will investigate the noise floor associated with a boundary element method by analyzing the differential strip-line shown in Fig. 3. Since the transmission line is a symmetrical structure with homogeneous dielectric, no differential to common mode conversion is expected. The differential to common return/insertion loss can be then considered to be a pure numerical artifact. In particular, four differential strip-line configurations with different coupling levels, namely High Coupling (HC), Medium Coupling (MC), Low Coupling (LC), and Uncoupled (UC) have been analyzed by using HyperLynx Full-Wave Solver [9] and the accuracy setting II (Figs. 3). The authors refer to [10] for all of the physical and electrical dimensions of the different transmission lines. It is remarkable noticing how the noise level is bounded at -60 dB uniformly for all the test cases. Please note that these numerical inaccuracies can be further reduced by increasing the requested level of solution accuracy. To prove that, a lower noise floor is achieved by tuning the accuracy settings to level IV (see Fig. 4). In this case, a -80 dB noise floor level is obtained for the differential to common insertion loss.

V. CONCLUSIONS

The accelerated MoM is a powerful tool to accurately model coupling phenomena down to a very low decibel range. Capturing low-level cross-talk for various scenarios common in modern designs requires different numerical tolerances for some key components of the solver depending on the coupling and isolation mechanism. For a given scenario, it is possible to reduce the noise floor by using more conservative settings in

gradual cost vs. accuracy trade-off. Numerical control required to model low level cross-talk accurately, depends on the mechanism of how that low level cross-talk is created.

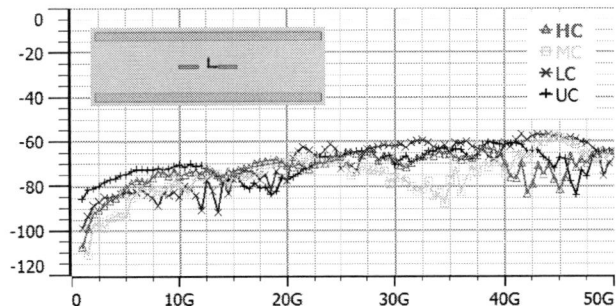

Fig 3. Differential to common return loss conversion.

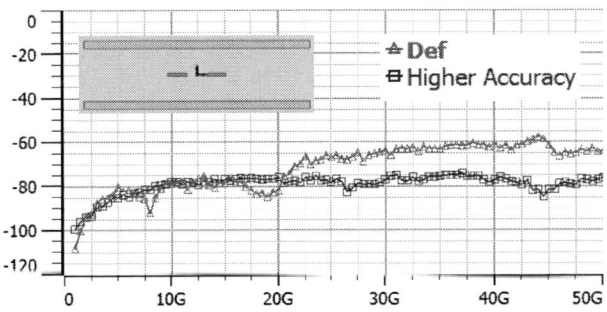

Fig 4. Differential to common insertion loss conversion with higher accuracy level for the HC configuration.

REFERENCES

[1] A. F. Peterson, S. L. Ray, and R. Mittra, Computational Methods for Electromagnetics. New York: IEEE Press, 1998.

[2] W. C. Chew, Waves and Fields in Inhomogeneous Media. New York: IEEE Press, 1995, ser. Electromagnetic Waves.

[3] G. Dural and M. I. Aksun, "Closed-form Green's functions for general sources in stratified media", *IEEE Trans. Microw. Theory Tech.*, vol. 43, no. 7, pp. 1545–1552, Jul. 1995.

[4] G. Bianconi and S. Chakraborty, "Efficient and Accurate Multi-Dimensional Layered Media Green's Function Interpolation Algorithm for the Analysis of Printed Circuit Boards" *in Proc. IEEE Int. Symp. EPEPS*, pp. 115-117., San Jose, CA, USA, 2018

[5] R. D. Graglia, "On the numerical integration of the linear shape functions times the 3D Green's function and its gradient on a plane triangle," *IEEE Trans. Antennas propagat.*, vol. 41, pp 1448-1455, Oct 1993

[6] R. Coifman, V. Rokhlin, and S. Wandzura, "The fast multipole method for the wave equation: A pedestrian prescription," *IEEE Trans. Antennas Propag. Mag.*, vol. 35, no. 6, pp. 7–12, Jun. 1993

[7] D.Gope and V.Jandhyala, "Efficient solution of EFIE via low-rank compression of multilevel predetermined interactions," *IEEE Trans. Antennas Propag.* vol 53(10), pp. 3324-3333, Oct 2005

[8] V. Okhmatovski, M. Yuan, I. Jeffrey and R. Phelps, "A Three-Dimensional Precorrected FFT Algorithm for Fast Method of Moments Solutions of the Mixed-Potential Integral Equation in Layered Media," *IEEE Trans. MTT*, vol. 57(12), pp. 3505-3517, Dec. 2009

[9] https://www.mentor.com/pcb/hyperlynx/full-wave-solver

[10] S. B. Smith, K. Patel, M. Craton, M. Rengarajan, S. A. Blasko, Y. Huang "To Couple or Not to Couple, That is the Question", DesignCon 2017.

Uniformly Accurate Electrostatic Layered Medium Green's Function Approximation via Scattered Field Formulation

Xinbo Li

Dept. of Electrical and Computer Engineering
University of Manitoba
Winnipeg, Canada
lix34545@myumanitoba.ca

Vladimir Okhmatovski

Dept. of Electrical and Computer Engineering
University of Manitoba
Winnipeg, Canada
Vladimir.Okhmatovski@umanitoba.ca

Abstract—**A novel approach to evaluation of electrostatic multilayered media Green's functions is presented. Total field of the point charge in layered substrate is represented as a sum of known closed-form incident field in homogeneous space and scattered field which accounts for the effect of the layers. The former accurately approximates the field near the source while the latter is approximated with cylindrical waves and accurately represents the intermediate and far fields. The cylindrical waves approximation is performed via numerical solution of the differential equation formulated with respect to the scattered field as opposed to the total field as it was done in previous work. The spectral domain scattered field solution is cast into the pole-residue form. It allows for the subsequent analytical evaluation of the Sommerfeld integrals producing closed-form space domain approximation.**

Index Terms—**Green's function, FEM, scattered field formulation**

I. Introduction

Evaluation of the multilayered medium Green's function is vital in many practical applications of electromagnetic analysis including design of microwave circuits and micro strip antennas, modeling of high-speed interconnects. The common practice to obtain the spatial domain Green's function is to solve the 1D spectral domain boundary value problem (BVP) analytically [1], then perform the inverse Fourier-Bessel transform through approximating the Green's function spectrum with known functions allowing for subsequent analytical evaluation of the inverse Fourier-Bessel transform [2], [3]. Depending on the choice of fitting functions, the resultant space domain Green's function loses its accuracy in either near zone [2] or far zone [3]. In the former, the error in the near field is due to spherical waves dominating the solution and being approximated with a counted number of cylindrical waves. In order to develop uniformly accurate approximation, we decompose the spectrum of the total field into the incident and scattered field contributions. The 1D differential equation governing the spectrum of the layered

This work was supported by Collaborative Research and Development Grant from NSERC and Manitoba Hydro International.

medium Green's function is formulated with respect to the scattered field rather than the total field as it was done previously [2], [4]. The boundary conditions at layer interfaces, based on the continuity of Green's function and the normal component of electric flux density are enforced. As a result, the pole-residue approximation of the scattered field spectrum is obtained which leads to accurate cylindrical wave approximation in the space domain in both the intermediate and far zones, since the singularity of Green's function resides in the incident field. The incident field is subsequently added in the analytical form. Hence, the total field accurately describes the field near the source. The proposed method is similar to [5] in that it approximates the scattered field Green's function with a rational function. However, such approximation is achieved through error-controllable solution of the 1D BVP for the scattered field as opposed to the fitting of its analytically determined spectrum with rational function via VECTFIT procedure [6], latter not being an error-controllable process.

II. Numerical Spectral Domain Scattered Potential Evaluation

Consider a parallel plate waveguide bounded by PEC planes situated at elevations $z = 0$ and $z = d$ along the direction of stratification z. The waveguide is filled with planar layered medium. In each layer, the dielectric is assumed to be homogeneous. For a point charge located on the z axis at elevation z' in layer with permittivity ϵ_{src}, the boundary value problem Green's function (i.e. electrostatic potential due to point charge) is governed by the Poisson's equation in cylindrical coordinates

$$\nabla^2 G^{tot}(\rho, z; z') = -\frac{1}{\epsilon_{src}} \frac{\delta(\rho)}{2\pi\rho} \delta(z - z'). \quad (1)$$

Applying forward Fourier-Bessel transform [1]

$$\tilde{G}^{tot}(\lambda, z; z') = \int_0^\infty G^{tot}(\rho, z; z') J_0(\lambda\rho)\rho d\rho, \quad (2)$$

978-1-7281-6162-4/20 $31.00 © 2020 IEEE

to both sides of (1), we reduce it to the following ordinary differential equation (ODE) with respect to the Green's function spectrum $\tilde{G}^{tot}$ [2]

$$\frac{d^2}{dz^2}\tilde{G}^{tot}(\lambda, z; z') - \lambda^2 \tilde{G}^{tot}(\lambda, z; z') = -\frac{1}{2\pi\epsilon_{src}}\delta(z - z'). \tag{3}$$

The spectral domain Green's function $\tilde{G}^{tot}$ is constrained by the boundary conditions at dielectric interfaces and the bounding PEC plates. The boundary conditions at each ith dielectric interface between the layers with elevation $z_{int,i}$

$$\left.\tilde{G}^{tot}\right|_{z_{int,i}^+} = \left.\tilde{G}^{tot}\right|_{z_{int,i}^-}, \quad \epsilon_i^+ \left.\frac{d\tilde{G}^{tot}}{dz}\right|_{z_{int,i}^+} = \epsilon_i^- \left.\frac{d\tilde{G}^{tot}}{dz}\right|_{z_{int,i}^-}, \tag{4}$$

represent the continuity of the electrostatic potential and normal component of the total electric flux, respectively. The PEC boundary conditions for spectral domain Green's function $\tilde{G}^{tot}$ are

$$\tilde{G}^{tot}(\lambda, 0; z') = 0, \qquad \tilde{G}^{tot}(\lambda, d; z') = 0. \tag{5}$$

Both sets of boundary conditions (4) and (5) are resulted from the Fourier-Bessel transform of the corresponding boundary conditions for the Green's function in the spatial domain.

A. Spectral 1D BVP for the Incident Field Green's Function

We define the incident field Green's function as the response to the point charge source in the parallel plate waveguide filled with homogeneous dielectric whose permittivity is ϵ_{src} as in the layered problem. The 1D BVP for the incident field consists of the ODE

$$\frac{d^2}{dz^2}\tilde{G}^{inc}(\lambda, z; z') - \lambda^2 \tilde{G}^{inc}(\lambda, z; z') = -\frac{1}{2\pi\epsilon_{src}}\delta(z - z'), \tag{6}$$

in conjunction with the boundary conditions at the PEC planes

$$\tilde{G}^{inc}(\lambda, 0; z') = 0, \qquad \tilde{G}^{inc}(\lambda, d; z') = 0, \tag{7}$$

and boundary conditions at the ith dielectric interface $z = z_{int,i}$, i.e., both the incident potential and the incident electric flux are continuous:

$$\left.\tilde{G}^{inc}\right|_{z_{int,i}^+} = \left.\tilde{G}^{inc}\right|_{z_{int,i}^-}, \quad \left.\frac{d\tilde{G}^{inc}}{dz}\right|_{z_{int,i}^+} = \left.\frac{d\tilde{G}^{inc}}{dz}\right|_{z_{int,i}^-}. \tag{8}$$

B. Spectral 1D BVP for the Scattered Green's Function

The scattered field Green's function is defined as $\tilde{G}^{sca} = \tilde{G}^{tot} - \tilde{G}^{inc}$. Subtracting the left hand side of equation (6) from the left hand side of (3) and performing the same subtraction for their right hand sides produces the following 1D ODE for the scattered field Green's function

$$\frac{d^2}{dz^2}\tilde{G}^{sca}(\lambda, z; z') - \lambda^2 \tilde{G}^{sca}(\lambda, z; z') = 0. \tag{9}$$

Executing similar subtractions for the left and right hand sides of the boundary condition equations for the total and incident field Green's functions we obtain the boundary conditions for the scattered field

$$\left.\tilde{G}^{sca}\right|_{z_{int,i}^+} = \left.\tilde{G}^{sca}\right|_{z_{int,i}^-}$$

$$\epsilon_i^+ \left.\frac{d(\tilde{G}^{sca} + \tilde{G}^{inc})}{dz}\right|_{z_{int,i}^+} = \epsilon_i^- \left.\frac{d(\tilde{G}^{sca} + \tilde{G}^{inc})}{dz}\right|_{z_{int,i}^-} \tag{10}$$

$$\tilde{G}^{sca}(\lambda, 0; z') = 0, \qquad \tilde{G}^{sca}(\lambda, d; z') = 0. \tag{11}$$

C. Numerical Solution of 1D BVP for Scattered Field Spectrum

A numerical solution of the 1D BVP for the scattered Green's function with Finite Difference (FD) [4] or Finite Element Method (FEM) [7] yields the system of linear equations with respect to the discretized scattered field Green's function $\tilde{\mathbf{G}}^{sca}$

$$([A] + \lambda^2[B])\tilde{\mathbf{G}}^{sca} = \mathbf{b}, \tag{12}$$

where the excitation vector $\mathbf{b}$ has non-zero values at the entries associated with the dielectric interfaces as opposed to the excitation vector in analogous system of algebraic equations occurring in the total field formulation [2] where non-zero values in the excitation vector occur at entries corresponding to the location of the source. Performing eigenvalue decomposition on $[B]^{-1}[A]$ similarly as in [2], we get

$$\tilde{\mathbf{G}}^{sca} = ([A] + \lambda^2[B])^{-1}\mathbf{b} = [E]([D] + \lambda^2[U])^{-1}[T]\mathbf{b}, \tag{13}$$

where $[E][D][E]^{-1} = [B]^{-1}[A]$, $[T] = ([B][E])^{-1}$. Denoting the set of all indices of nodes at the dielectric interfaces as $\{\mathbb{S}_{int}\}$, from (13) we get ith unknown

$$\tilde{G}_i^{sca} = \tilde{G}^{sca}(\lambda, z_i; z') = \sum_{j \in \mathbb{S}_{int}} b_j \sum_{k=1}^{Np} \frac{E_{ik}T_{kj}}{D_k + \lambda^2}. \tag{14}$$

Expression (14) has the same form regardless the numerical method used to solve the 1D BVP for $\tilde{G}^{sca}$. In case second order FEM is used for the solution, for example,

$$b_j = -\frac{2(\epsilon_{i_j+1} - \epsilon_{i_j})}{\epsilon_{i_j+1} + \epsilon_{i_j}}\left.\frac{\tilde{G}^{inc}}{dz}\right|_{z_{int,i_j}}, \tag{15}$$

where i_j is the index of the medium layer associated with the index j that is the index of the node overlapping the dielectric interface. Performing inverse Fourier-Bessel transform [2] on (14), we have

$$G^{sca}(\rho, z_i; z') = \sum_{j \in \mathbb{S}_{int}} C_{i_j}\left(\sum_{k=1}^{N_p} E_{ik}T_{kj}H_0^{(2)}(-i\sqrt{D_k}\rho)\right.$$

$$\left.\sum_{n=1}^{\infty}\frac{g(i_j, n; z')}{D_k - a_n^2} - \sum_{k=1}^{N_p}\sum_{n=1}^{N_t}\frac{g(i_j, n; z')H_0^{(2)}(-ia_n\rho)E_{ik}T_{kj}}{D_k - a_n^2}\right) \tag{16}$$

where

$$C_{i_j} = -\frac{i(\epsilon_{i_j+1} - \epsilon_{i_j})\pi}{(\epsilon_{i_j+1} + \epsilon_{i_j})\epsilon_{src}d^2}, \qquad a_n = \frac{n\pi}{d}, \tag{17}$$

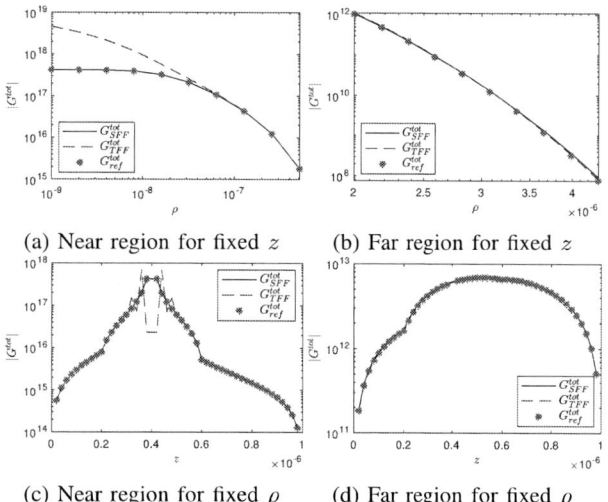

(a) Near region for fixed z (b) Far region for fixed z

(c) Near region for fixed ρ (d) Far region for fixed ρ

Fig. 1: Spatial domain total field Green's function behavior in the near and far zones of the point source.

the square root $\sqrt{D_k}$ is chosen such that the field satisfies the radiation condition, and

$$g(i_j, n; z') = n \cos\left(\frac{n\pi z_{int,i_j}}{d}\right) \sin\left(\frac{n\pi z'}{d}\right). \tag{18}$$

In (16), the following series can be analytically evaluated in closed form by Poisson summation formula. For example, for the case when D_k has positive real value and $z_{int,i_j} > z'$,

$$\sum_{n=1}^{\infty} \frac{g(i_j, n; z')}{D_k - a_n^2} =$$

$$\frac{d^2}{8\pi} \left(\frac{e^{-i\sqrt{D_k}(z_{int,i_j} - z')} + e^{i\sqrt{D_k}(z_{int,i_j} + z' - 2d)}}{1 - e^{-i2\sqrt{D_k}d}} \right.$$

$$- \frac{e^{-i\sqrt{D_k}(z_{int,i_j} + z')} + e^{i\sqrt{D_k}(z_{int,i_j} - z' - 2d)}}{1 - e^{-i2\sqrt{D_k}d}}$$

$$+ \frac{e^{i\sqrt{D_k}(z_{int,i_j} - z')} + e^{-i\sqrt{D_k}(z_{int,i_j} + z' - 2d)}}{1 - e^{i2\sqrt{D_k}d}}$$

$$\left. - \frac{e^{i\sqrt{D_k}(z_{int,i_j} + z')} + e^{-i\sqrt{D_k}(z_{int,i_j} - z' - 2d)}}{1 - e^{i2\sqrt{D_k}d}} \right).$$

Other cases can be treated analogously and are not shown here due to shortage of space.

The second sum over n in (16) in the bracket was an infinite sum. It can be truncated to a given precision since

$$H_0^{(2)}(-ia_n\rho) = H_0^{(2)}\left(-i\frac{n\pi}{d}\rho\right) \sim \sqrt{\frac{2i}{\pi\frac{n\pi}{d}\rho}} e^{-a_n\rho} e^{i\frac{\pi}{4}}, \tag{19}$$

is exponentially decaying in magnitude for large number of n. For given error tolerance δ, we keep the first N_t terms, where

$$N_t = \left\lceil -\frac{d\ln\delta}{\rho\pi} \right\rceil. \tag{20}$$

III. NUMERICAL RESULTS

To validate the proposed scattered field formulation, a three layer substrate with total thickness $d = 1\,\mu\text{m}$ is considered. From bottom to top, the three layers have thicknesses $0.2\,\mu\text{m}$, $0.4\,\mu\text{m}$, and $0.4\,\mu\text{m}$, and relative permittivities $\epsilon_r = 5, 1, 10$, respectively. The source is located at elevation $z' = 0.6\,\mu\text{m}$. The behavior of spatial domain total Green's function G^{tot} obtained using the proposed scattered field formulation (SFF) and the total field formulation (TFF) [2] are shown in Fig. 1. In Fig. 1a and 1b, the observation location is fixed as $z = 0.38\,\mu\text{m}$, whereas in Fig. 1c, ρ is fixed as $0.001d$ and in Fig. 1d, ρ is fixed at $1.6d$.

One can observe that the proposed scattered field formulation maintains high accuracy of the Green's function approximation both in the vicinity of the source, as can be seen from Figs. 1a and 1c, and far from it, as shown in Figs. 1b and 1d. Note that in Figs. 1c and 1d z spans full thickness, i.e. $z \in [0, d]$. In the far region, both methods provide accurate results manifested in Fig. 1b, where $\rho \in [2d, 4.34d]$, and Fig. 1d, in which ρ is fixed at $1.6d$ and observation z covers entire interval between the PEC plates of the waveguide.

IV. CONCLUSION

This paper introduces a new methodology for the closed form evaluation of the electrostatic Green's function in shielded planar layered medium. The proposed scattered field formulation of the Green's function allows to construct uniformly accurate Green's function approximation both in the near and far zone of the point source. The method numerically solves the 1D boundary value problem for the spectrum of the scattered field Green's function. It is followed by the eigenvalue decomposition of the pertinent matrices, which allows to cast the spectrum of the scattered field Green's function into the pole-residue form and enable analytic evaluation of the Sommerfeld integrals. The previous version of the method which performed analogous operations on the total field of the Green's function failed to provide accurate field approximation in the near vicinity of the point source.

REFERENCES

[1] W. C. Chew, *Waves and fields in inhomogeneous media*. IEEE Press, 1995.

[2] A. Cangellaris and L. Yang, "Rapid calculation of electrostatic green's functions in layered dielectrics," *IEEE Transactions on Magnetics*, vol. 37, no. 5, pp. 3133–3136, 2001.

[3] M. Aksun, "A robust approach for the derivation of closed-form green's functions," *IEEE Transactions on Microwave Theory and Techniques*, vol. 44, no. 5, pp. 651–658, 1996.

[4] V. I. Okhmatovski and A. C. Cangellaris, "A new technique for the derivation of closed-form electromagnetic green's functions for unbounded planar layered media," *IEEE Transactions on Antennas and Propagation*, vol. 50, no. 7, pp. 1005–1016, 2002.

[5] V. I. Okhmatovski and A. C. Cangellaris, "Evaluation of layered media green's functions via rational function fitting," *IEEE Microwave and Wireless Components Letters*, vol. 14, no. 1, pp. 22–24, 2004.

[6] B. Gustavsen and A. Semlyen, "Rational approximation of frequency domain responses by vector fitting," *IEEE Transactions on Power Delivery*, vol. 14, no. 3, pp. 1052–1061, 1999.

[7] J. M. Jin, *The finite element method in electromagnetics*. John Wiley & Sons, 2015.

An Efficient and Parallel Electromagnetic Solver for Complex Interconnects in Layered Media

Damian Marek, Shashwat Sharma, Piero Triverio

Edward S. Rogers Sr. Dept. of Electrical and Computer Engineering, University of Toronto, Toronto, ON, Canada

damian.marek@mail.utoronto.ca, shash.sharma@mail.utoronto.ca, piero.triverio@utoronto.ca

Abstract—A novel parallel solver based on the adaptive integral method (AIM) is proposed for the electromagnetic analysis of electrical interconnects in layered media. We show that graph partitioning techniques can be used to optimally distribute, across thousands of processes, the computations related to both matrix filling and system solution. The proposed workload distribution strategy is compared to existing techniques through a scalability study on a large realistic interposer model in layered media.

Index Terms—surface integral equation method, adaptive integral method, parallel algorithms, skin effect modeling

I. INTRODUCTION

Compared to volumetric formulations, surface integral equation (SIE) methods offer an efficient approach for the electromagnetic analysis of interconnects in layered media [1]–[8]. Although efficient, SIE methods are always partnered with a suitable acceleration scheme to increase their scalability. In layered media, the most popular choice is the adaptive integral method (AIM), which accelerates matrix-vector products through fast Fourier transforms (FFTs) [3], [9]. Even with acceleration, existing electromagnetic solvers can hardly handle realistic portions of the intricate interconnect network present in typical integrated circuits, especially those featuring 3D integration. This scenario calls for the development of efficient parallel solvers suitable for large-scale electromagnetic analyses.

To efficiently parallelize an electromagnetic solver one must: i) distribute the workload evenly across processors, and ii) minimize communication between them. In [6], [7], a 3D block decomposition is proposed to efficiently distribute AIM computations. This approach works well for integral equation methods using volumetric meshes, since mesh elements cover the simulation domain in a fairly uniform fashion. However, parallelization efficiency tends to be suboptimal for sparsely-populated problems or for "hollow" meshes, such as those that naturally arise with SIE formulations [10].

In this work, we propose an efficient parallelization strategy for SIE formulations involving layered media, generalizing our previous work in free space [11]. The new method leverages graph partitioning to distribute AIM operations among thousands of computing cores in a balanced fashion that minimizes inter-node communication. A numerical test on a realistic interposer structure, performed with up to 1,600 cores, demonstrates better performance than previously-published solutions.

II. FORMULATION

We consider a structure made by lossy conductors and embedded in a medium stratified along the z axis. The goal is to solve Maxwell equations in order to determine the scattering (S) parameters between some given ports. To model the induced electric current density in conducting objects, we use the augmented electric field integral equation (AEFIE) [5] along with an approximate surface impedance boundary condition (SIBC) [12]. After discretization with Rao-Wilton-Glisson (RWG) [13] and pulse basis functions, the AEFIE can be written as:

$$\begin{bmatrix} jk_0\mathbf{Z}_A + \eta_0^{-1}\mathbf{Z}_s & -\mathbf{D}^T\mathbf{Z}_\Phi\mathbf{B} \\ \mathbf{FD} & \mathbf{C} \end{bmatrix} \begin{bmatrix} \mathbf{J}_s \\ c_0\boldsymbol{\rho}_r \end{bmatrix} = \begin{bmatrix} 0 \\ \mathbf{I}_s \end{bmatrix}. \quad (1)$$

In (1), $\mathbf{Z}_A$ and $\mathbf{Z}_\Phi$ are the discretized vector and scalar potential parts of the single-layer potential operator [14], which involve the Green's function of a layered medium [1]. The SIBC is enforced by $\mathbf{Z}_s$, while $\mathbf{C}$ provides coupling to a Thevenin-equivalent circuit excitation [2], and $\mathbf{I}_s$ is the excitation current. The remaining matrices are identical to [5].

In the AIM, (1) is solved iteratively, and the required matrix-vector products involving $\mathbf{Z}_A$ and $\mathbf{Z}_\Phi$ are accelerated with FFTs. Both $\mathbf{Z}_A$ and $\mathbf{Z}_\Phi$ are expressed as

$$\mathbf{Z} \approx \mathbf{Z}_{NR} + \mathbf{WHP}, \quad (2)$$

where $\mathbf{Z}_{NR}$ models electromagnetic interactions at a short distance, in the so-called near region. Far-region interactions are described by the second term, where $\mathbf{P}$ projects sources onto a regular grid (the AIM grid), $\mathbf{H}$ computes the resulting potentials via FFTs, which are finally interpolated by matrix $\mathbf{W}$ onto the original mesh elements [3].

The efficient parallelization of matrix-vector products involving (2) is difficult, because two different workload decomposition strategies would ideally be required to: 1) distribute the multiplication by $\mathbf{Z}_{NR}$, whose entries are related to the original mesh, and 2) distribute the multiplications and FFTs arising from the $\mathbf{WHP}$ term, which instead involve the AIM grid. Furthermore, undesirable communication will occur in matrix-vector products involving $\mathbf{W}$ and $\mathbf{P}$ if a mesh element and the corresponding AIM grid points are assigned to different processes.

Existing strategies to parallelize AIM-based solvers focus on the efficiency of computations related to the AIM grid, like FFTs, which are typically quite expensive [6], [7], [10].

978-1-7281-6162-4/20 $31.00 © 2020 IEEE

However, for complex multiscale layouts, near-region computations may also be very costly, and need to be carefully balanced across computing nodes to achieve high parallelization efficiency.

III. PROPOSED METHOD

We present a new parallelization strategy for AIM-based solvers which efficiently distributes both near-region and far-region computations across thousands of computing cores, while minimizing the communication required by projection and interpolation steps.

A. 2D Fast Fourier Transforms

In layered media, multiplications by matrix $\mathbf{H}$ in [3] are computed through many 2D FFTs. We parallelize these operations with the pencil decomposition method proposed in [7]. The pencils in [7] decompose the AIM grid into long thin cuboids aligned with the x axis. In this starting configuration, each process computes 1D FFTs along the x axis. Then the result is transposed in a communication step, so that the pencils become aligned to the y axis. Finally, each process computes the remaining 1D FFTs along the y axis.

B. Near-Region Operations

Near-region operations include the computation of the $\mathbf{Z}_{A,NR}$ and $\mathbf{Z}_{\Phi,NR}$ matrices, and the corresponding matrix-vector products. In layered media, the construction of these matrices becomes even more time consuming due to the complexity of the multilayer Green's function.

In [7], a 3D block decomposition method is proposed to distribute near-region computations across processes. In this method, the simulation domain is first partitioned along the x axis into P_x equally sized slabs. Then, each slab is split along the y axis into P_y equally sized pencils. Finally each pencil is split along the z axis into P_z equally sized blocks. The computations associated to each block will be later assigned to $P = P_x \times P_y \times P_z$ processes. Prior to assignment, the workload is balanced with the following procedure [10]:

1) The slab boundaries are shifted one at a time, by small increments, to balance the number of near-region interactions associated to each slab;
2) Then, in each slab, the pencil boundaries are shifted to balance the number of near-region interactions associated to each pencil;
3) Finally, in each pencil the block boundaries are shifted to balance the number of near-region interactions associated to each block.

This method is simple and effective for many cases. However, as will be shown in Section IV, this method may not optimally balance the workload and communication associated with near-region operations for complex and multiscale layouts.

We propose a more efficient solution by generalizing, to structures in layered media, the graph-based approach advocated in [11] for problems in free space. In the proposed approach, near-region computations are distributed in an optimal fashion with the following steps:

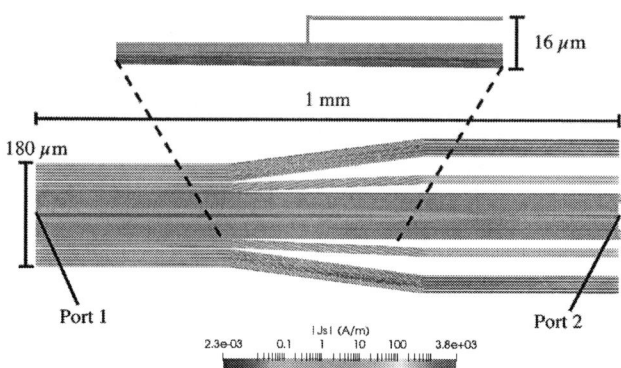

Figure 1. Geometry of the interposer-level interconnect considered in Sec. IV with current density computed at 50 GHz when port 1 is excited.

1) A dual graph of the mesh is constructed, where each triangle is associated with a graph node and graph edges connect adjacent triangles;
2) The number of triangles falling in the near region of a given triangle are assigned as a nodal weight to the corresponding node in the graph;
3) Finally, parMETIS [15], a distributed graph partitioning library, is used to partition the graph into P subgraphs, where P is the desired number of processes. The partition is generated to approximately satisfy two optimality criteria. First, the sum of the nodal weights in each partition should be approximately the same. In this way, the operations required to pre-compute $\mathbf{Z}_{A,NR}$ and $\mathbf{Z}_{\Phi,NR}$ and multiply them against a given vector are evenly distributed among processes. Second, the number of edges cut by the partitioning process is minimized, to minimize inter-process communication.

C. Interpolation and Projection

After the AIM grid pencils have been assigned to processes, there is still freedom for each process to choose one of the mesh partitions generated in Section III-B. To minimize communication in the matrix-vector products involving $\mathbf{W}$ and $\mathbf{P}$, there should be maximal overlap between the mesh partition and the AIM grid pencil assigned to each process. This goal is achieved by the following algorithm:

1) Each process analyzes all mesh partitions and calculates $N_E^{(p)}$, the number of edges of mesh partition p that are within its AIM pencil;
2) Each process ranks all partitions in order of decreasing $N_E^{(p)}$;
3) Finally, starting form the process with lowest rank, each process chooses the partition with highest $N_E^{(p)}$ among those still available.

IV. RESULTS

The scalability of the proposed method was tested by extracting the S parameters of a complex interconnect network from an interposer used for 3D integration (courtesy of Dr.

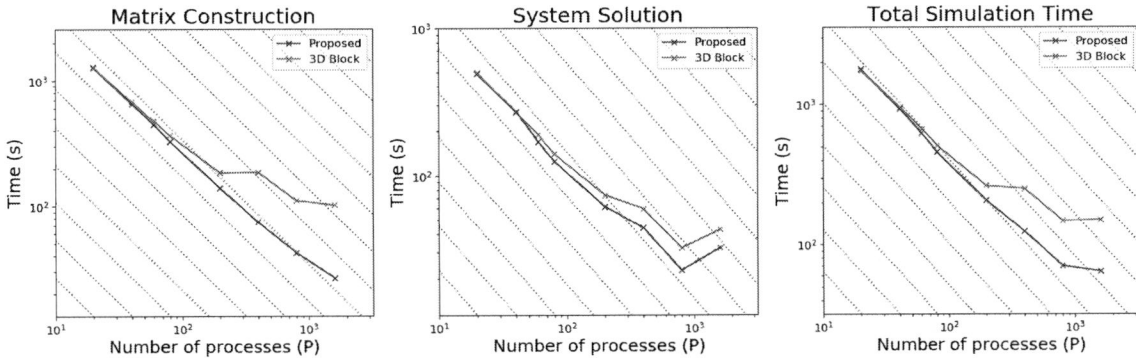

Figure 2. Scalability of the proposed solver and the "3D block" method in [10] for different steps of the solver. Dashed lines represent ideal efficiencies of parallelization and their relative spacing indicates a 2× speedup.

Rubaiyat Islam, AMD). The interposer structure is illustrated in Fig. 1 together with the location of 2 of the 6 ports considered in this test. There are a total of 80 copper wires and one ground plane. The structure is embedded in a two-layer medium bounded above and below by air. The upper layer has $\epsilon_{r1} = 4$, $\sigma_1 = 0$ S/m, and thickness of $h_1 = 27.5$ μm. The lower layer has $\epsilon_{r2} = 11.9$, $\sigma_2 = 10$ S/m, and thickness of $h_2 = 47.5$ μm. The resulting mesh of the structure contained 357,000 triangles and 535,500 edges. The AIM grid dimensions were chosen to be $90 \times 400 \times 8$ ($N_x \times N_y \times N_z$) and the near-region radius was set to 5 grid points.

The proposed method was compared to the 3D block decomposition of [10]. All simulations were run on the Scinet Niagara cluster where each node has 40 Intel Skylake cores running at 2.4 GHz and 202 GB of memory. In this scalability study, the number of processes, P, was varied from 20 to 1,600. The wall time required to precompute the matrices in (1) and solve the system for one frequency point are plotted in Fig. 2. For all P values, the CPU time required by the proposed method to determine an optimal workload distribution is negligible compared to the total simulation time (less than 2%).

The results show that, when the proposed method is adopted, a significant speedup can be realized for both the matrix construction step and the system solution step. For the matrix construction step, when $P = 1600$, the proposed method was 4× faster than the 3D block decomposition method.

V. Conclusion

We presented a new strategy to efficiently parallelize large-scale electromagnetic simulations based on surface integral equations. A novel approach based on graph partitioning is shown to effectively distribute, across thousands of computing cores, the computation of electromagnetic interactions in both the near and the far range. When compared to other state-of-the-art strategies on a complex interconnect network from a 3D integrated circuit, the proposed method showed better scalability.

References

[1] K. A. Michalski and D. Zheng, "Electromagnetic Scattering and Radiation by Surfaces of Arbitrary Shape in Layered Media, Part I: Theory," *IEEE Trans. Antennas Propag.*, vol. 38, no. 3, pp. 335–344, Mar 1990.

[2] Y. Wang, D. Gope, V. Jandhyala, and C. J. R. Shi, "Generalized Kirchoff's current and voltage law formulation for coupled circuit-electromagnetic simulation with surface integral equations," *IEEE Trans. Microw. Theory Techn.*, vol. 52, no. 7, pp. 1673–1682, 2004.

[3] J. Phillips and J. White, "A precorrected-FFT method for electrostatic analysis of complicated 3-D structures," *IEEE Trans. Comput.-Aided Design Integr. Circuits Syst.*, vol. 16, no. 10, pp. 1059–1072, Mar. 1997.

[4] Y. P. Chen, L. Jiang, Z. G. Qian, and W. C. Chew, "An augmented electric field integral equation for layered medium Green's function," *IEEE Trans. Antennas Propag.*, vol. 59, no. 3, pp. 960–968, 2011.

[5] Z. G. Qian and W. C. Chew, "Fast Full-Wave Surface Integral Equation Solver for Multiscale Structure Modeling," *IEEE Trans. Antennas Propag.*, vol. 57, no. 11, pp. 3594–3601, 2009.

[6] C. Liu, K. Aygün, H. Braunisch, V. I. Okhmatovski, and A. E. Yilmaz, "A parallel iterative layered-medium integral-equation solver for electromagnetic analysis of electronic packages," in *Proc. IEEE Conf. on Elect. Perf. of Electron. Packag. and Systems (EPEPS)*, 2017, pp. 1–3.

[7] C. Liu, K. Aygün, and A. E. Yılmaz, "A parallel FFT-accelerated layered-medium integral-equation solver for electronic packages," *International Journal of Numerical Modelling: Electronic Networks, Devices and Fields*, pp. 1–17, Sep. 2019.

[8] S. Sharma and P. Triverio, "SLIM: A Well-Conditioned Single-Source Boundary Element Method for Modeling Lossy Conductors in Layered Media," submitted to IEEE Antennas Wireless Propag. Lett., 2020.

[9] E. Bleszynski, M. Bleszynski, and T. Jaroszewicz, "AIM: Adaptive integral method for solving large-scale electromagnetic scattering and radiation problems," *Radio Science*, vol. 31, no. 5, pp. 1225–1251, Sep. 1996.

[10] F. Wei and A. E. Yilmaz, "A More Scalable and Efficient Parallelization of the Adaptive Integral Method—Part I: Algorithm," *IEEE Trans. Antennas Propag.*, vol. 62, no. 2, pp. 714–726, Feb. 2014.

[11] D. Marek, S. Sharma, and P. Triverio, "An efficient parallelization strategy for the adaptive integral method based on graph partitioning," in *14th Euro. Conf. on Antennas and Propag. (EuCAP)*, 2020, pp. 1–5.

[12] S. Yuferev and N. Ida, *Surface Impedance Boundary Conditions*. CRC Press, 2009.

[13] S. Rao, D. Wilton, and A. Glisson, "Electromagnetic scattering by surfaces of arbitrary shape," *IEEE Trans. Antennas Propag.*, vol. 30, no. 3, pp. 409–418, May. 1982.

[14] W. C. Gibson, *The Method of Moments in Electromagnetics*. Chapman and Hall/CRC, Jul. 2014.

[15] G. Karypis and V. Kumar, "A Parallel Algorithm for Multilevel Graph Partitioning and Sparse Matrix Ordering," *J. Parallel. Distrib. Comput.*, vol. 48, no. 1, pp. 71–95, Jan. 1998.

High-Speed Link Design Optimization Using Machine Learning SVR-AS Method

Hanzhi Ma[1], Er-Ping Li[1], Andreas C. Cangellaris[2], and Xu Chen[2]

[1]College of Information Sci. and Electronic Eng., ZJU-UIUC Institute, Zhejiang University, Hangzhou, China
[2]Department of Electrical and Computer Engineering, University of Illinois at Urbana-Champaign, Urbana, IL

Abstract—**This paper proposes a novel and fast constrained design optimization method based on support vector regression-active subspace method. The proposed optimization method calculates a linear combination of original design parameters named active variable as a low-dimensional representation of high-dimensional design space to transform the non-linear constraint into a reduced linear constraint for optimization problems, which successfully derives a simplified and mathematically solvable equation. A complex high-speed link with 16-dimensional design parameters is utilized to verify this method and results show that the proposed method can efficiently find the optimal design structures compared to interior-point method.**

I. INTRODUCTION

High-dimensional design parameters optimization is one of the most important and challenging problems in the design process of complex high-speed links [1]. Traditional optimization algorithms lead to computational burdens for finding an optimal or relative optimal design due to excessive and repeated link simulation needs. To reduce the computation requirements, surrogate models established by a small amount of simulations or measurements are proposed to replace the link simulations in the optimization routine. Recently, machine learning techniques are also utilized to describes the inverse relationship from desired output to design parameters for efficient parameter optimization. Support vector regression (SVR) can provide the inverse relationship between eye features and geometrical parameters for high-speed links [2], while the combination of deep neural network and symbolic Knowledge Base is used as an intelligent learning architecture for inverse mapping [3].

In this work, we propose a novel and fast constrained design optimization method which uses support vector regression based active subspace (SVR-AS) algorithm [4] to calculate a reduced-dimensional input space of active variables as a linear combination of original design parameters and transforms the non-linear constraint into a linear constraint for optimization simplification. A mathematical formula is provided in this paper from the linear constraint provided by active variable for directly finding the optimal design structure with a specified output and the minimal mean squared distance from the specific design. Numerical results show that compared with interior-point method, SVR-AS based optimization method can successfully and efficiently find the optimal design structures for a complex high-speed link with 16-dimensional design parameters.

II. METHODOLOGY

Let $\boldsymbol{X} = [x_1,...,x_p]^{\mathrm{T}}$ represent the input p-dimensional normalized design parameter space (e.g. board geometry parameters), Y represents the output of interest (e.g. eye opening), and $D = \{(\boldsymbol{X_1}, Y_1), ... (\boldsymbol{X_n}, Y_n)\}$ represents sampling data set. The objective is to find a set of $\boldsymbol{X}$ that is closet to a set of specified numbers (normalized to $x_i = 0$) for desired output Y_0. In other words, we want to find the design closest to a prototype design where the eye opening requirements are satisfied. Considering the mean squared distance between $\boldsymbol{X}$ and the nominal numbers, the optimization problem can be expressed as:

$$\begin{aligned} \min \quad & g\left(\boldsymbol{X}\right) = \tfrac{1}{p} \sum_{1 \leq i \leq p} x_i^2, \\ \text{s.t.} \quad & h\left(\boldsymbol{X}\right) = Y_0. \end{aligned} \tag{1}$$

A. SVR-AS

SVR-AS method can reduce input design space to a low-dimensional representation from the directions that perturbation on design parameters influence more on outputs. The speed and accuracy of SVR compare favorably with artificial neural networks and stochastic collocation for high-speed link models [5], [6]. SVR predictive function with Gaussian Kernel can be expressed as:

$$f\left(\boldsymbol{X}\right) = \sum_{1 \leq j \leq n} (\hat{\alpha}_j - \alpha_j) \exp\left(\frac{-\|\boldsymbol{X} - \boldsymbol{X}_j\|^2}{2\sigma^2}\right) + b, \tag{2}$$

where $\hat{\alpha}_j$ and α_j are Laplace operator, b is the displacement of hyper-plane and σ is the width of Gaussian Kernel.

A symmetric and positively semi-definite matrix is defined using the gradient of Eq. (2) as:

$$\boldsymbol{Z} = \frac{1}{n} \sum_{1 \leq j \leq n} \nabla_{\boldsymbol{X}} f\left(\boldsymbol{X}_j\right) \left(\nabla_{\boldsymbol{X}} f\left(\boldsymbol{X}_j\right)\right)^{\mathrm{T}}, \tag{3}$$

where $\nabla_{\boldsymbol{X}} f\left(\boldsymbol{X}\right) = \left[\frac{\partial f}{\partial x_1}, ..., \frac{\partial f}{\partial x_p}\right]^{\mathrm{T}}$. After eigendecomposition of $\boldsymbol{Z} = \boldsymbol{W}\boldsymbol{\Lambda}\boldsymbol{W}^{-1}$, let $\boldsymbol{W}_1$ be the partial set of eigenvectors corresponding to the largest q eigenvalues and $\boldsymbol{y} = \boldsymbol{W}_1^{\mathrm{T}}\boldsymbol{X}$ denotes active variable as a lower-dimensional representation of the input parameters.

978-1-7281-6162-4/20 $31.00 © 2020 IEEE

B. Design Optimization with Equality Constraints from SVR-AS

SVR-AS provides a sufficient summary plot to illustrate the relationship between active variables and expected output, which can be fitted by polynomial function and then provides the corresponding active variable y_0 for desired output Y_0. Eq. (1) can be derived as:

$$\begin{aligned} \min \quad & g\left(\boldsymbol{X}\right) = \frac{1}{p}\sum_{1 \le i \le p} x_i^2, \\ \text{s.t.} \quad & \phi\left(\boldsymbol{X}\right) = \boldsymbol{W}_1^{\mathrm{T}}\boldsymbol{X} - \boldsymbol{y}_0 = 0. \end{aligned} \quad (4)$$

Lagrange multiplier method can be used for optimal problem with constraints. Lagrange function for Eq. (4) can be expressed as:

$$L\left(\boldsymbol{X}, \boldsymbol{\beta}\right) = g\left(\boldsymbol{X}\right) + \boldsymbol{\beta}\phi\left(\boldsymbol{X}\right), \quad (5)$$

where $\boldsymbol{\beta} = [\beta_1, ..., \beta_q]$ are the Lagrangian multipliers. The number of multipliers is equal to the number of eigenvectors we choose. The solution of the optimal problem is

$$\begin{cases} \frac{\partial L(\boldsymbol{X}, \boldsymbol{\beta})}{\partial \boldsymbol{X}} &= 0 \\ \frac{\partial L(\boldsymbol{X}, \boldsymbol{\beta})}{\partial \boldsymbol{\beta}} &= 0 \end{cases}, \quad (6)$$

which can be further derived as

$$\begin{cases} \frac{2}{p}\boldsymbol{X} + \boldsymbol{W}_1\boldsymbol{\beta}^{\mathrm{T}} &= 0 \\ \boldsymbol{W}_1^{\mathrm{T}}\boldsymbol{X} &= \boldsymbol{y}_0 \end{cases}. \quad (7)$$

Thus, the optimal design parameters can be expressed by

$$\boldsymbol{X} = \boldsymbol{W}_1\left(\boldsymbol{W}_1^{\mathrm{T}}\boldsymbol{W}_1\right)^{-1}\boldsymbol{y}_0. \quad (8)$$

III. Optimization Results and Discussion

A. Optimization Example

A chip-to-chip, realistic high-speed link model is considered as a representative example to verify the proposed optimization method. Fig. 1 illustrates the entire model that consists of transmitter, microstrip line, LGA [7], via model, strip line, via model, LGA, microstrip line and receiver. ANSYS Q3D simulator [8] and Keysight ADS [9] are used to simulate eye opening of the high-speed link. The desired output is eye width of eye opening after receiver. Design parameters of transmission lines of this link shown in the Fig. 2 and Table I are considered as the input design space.

B. Results from SVR-AS based Optimization Method

SVR-AS method uses 300 simulated data samples for forward surrogate model training and generates active variable as a reduced-dimensional design space. Since the first eigenvalue is much larger than the others, active variable in this work is defined as a one-dimensional linear combination of 16 design parameters. Fig. 3 shows the sufficient summary plot between active variable and corresponding eye width. A second-order polynomial function shown in orange color fits this relationship well with 1.49×10^{-24} mean squared error.

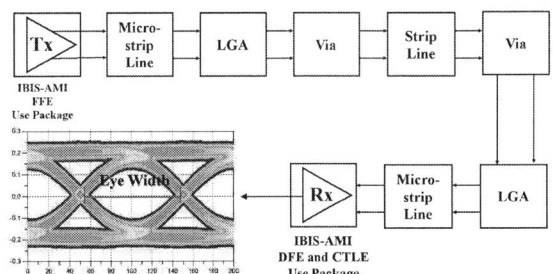

Fig. 1: High-speed link model and eye diagram.

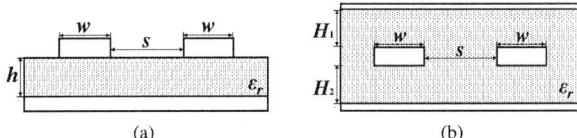

Fig. 2: Geometry of microstrip line and strip line.

Eq. (8) can be used to calculate the optimal design for desired eye width. In practice, it is common that several design parameters are fixed at specific values and thus not in the set of optimization variables. In this work, we fix $\varepsilon_r = 4.4$. Even for these requirements, a variant of Eq. (8) ($\boldsymbol{X}_{\mathrm{rest}} = \boldsymbol{W}_{\mathrm{rest}}\left(\boldsymbol{W}_{\mathrm{rest}}^{\mathrm{T}}\boldsymbol{W}_{\mathrm{rest}}\right)^{-1}\left(\boldsymbol{y}_0 - \boldsymbol{W}_{\mathrm{set}}^{\mathrm{T}}\boldsymbol{X}_{\mathrm{set}}\right)$) still solves the problem directly. Table I shows the optimal results calculated by SVR-AS based optimization method when eye width $Y_0 = 8 \times 10^{-11}$ sec is required. For this optimal design, simulated eye width is 8×10^{-11} sec as required and the mean squared distance from nominal number is 0.003 for normalized design parameters. SVR-AS based optimization method only needs 0.0014 sec to calculate the optimal results on an AMD Ryzen Threadripper 1950X 16-Core Processor without additional simulations or surrogate model predictions.

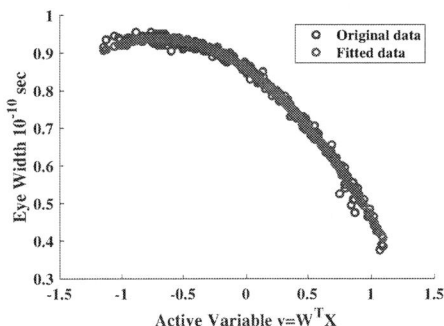

Fig. 3: Sufficient summary plot and its fitted function.

978-1-7281-6162-4/20 $31.00 © 2020 IEEE

TABLE I

DESIGN SPACE AND OPTIMAL RESULTS OF THE HIGH-SPEED LINK

	Parameter	Specified design	Range	Optimal results from SVR-AS based optimization method	Optimal results from interior-point method with SVR model	Unit
Tx microstrip line	ε_r	4.4	3.6-5.2	4.4	4.4	/
	w	5	4.5-5.5	4.9996	4.9995	mil
	s	8	7.2-8.8	8.0011	8.0008	mil
	h	4	3.7-4.3	3.9999	4.0003	mil
	l	11	2-20	11.0920	11.0108	mm
Strip line	ε_r	4.4	3.6-5.2	4.4	4.4	/
	w	5	4.5-5.5	4.9977	4.9980	mil
	s	8	7.2-8.8	7.9975	7.9984	mil
	H_1	8	7.2-8.8	7.9999	8.0000	mil
	H_2	8	7.2-8.8	7.9952	7.9952	mil
	l	275	50-500	323.8759	327.2000	mm
Rx microstrip line	ε_r	4.4	3.6-5.2	4.4	4.4	/
	w	5	4.5-5.5	4.9997	4.9995	mil
	s	8	7.2-8.8	7.9994	8.0000	mil
	h	4	3.7-4.3	3.9996	3.9994	mil
	l	11	2-20	11.0521	11.0630	mm
Simulated eye width				80	80	ps
Eye width evaluations of optimization process				/	693	/
Computation time of optimization process				0.0014	7.69	sec
Mean squared distance from nominal design				0.0030	0.0034	/

C. Comparison and Discussion

Traditionally, interior-point method can solve non-linear constrained minimization problem through a sequence of approximation problems. In this example, interior-point method needs 38 iterations and 693 function evaluations to solve Eq. (1). Eye width is evaluated by SVR predictive model in this paper. Jointly calling ANSYS Q3D Extractor and Keysight ADS can also be used in the optimization stage to replace surrogate model. The optimal results are shown in Table I and the corresponding simulated eye width result is also 8×10^{-11} sec. The mean squared distance from nominal design is 0.0034 for normalized design parameters, which is larger than the result from SVR-AS based optimization method.

Results from Table I illustrate the comparison between SVR-AS method and interior-point method. SVR-AS optimization method can calculate the optimal results accurately with extremely low computation cost after the SVR predictive model is established. However, interior-point algorithm calls SVR predictive model repeatedly during the iterations. Also it is worth mentioning that SVR-AS based optimization method can quickly calculate lots of different settings that satisfies different specific requirements (in this paper, we keep $\varepsilon_r = 4.4$).

IV. CONCLUSION

In this paper, SVR-AS based optimization method is proposed for fast design optimization of complex high-speed link model. SVR-AS based optimization method utilizes the sufficient summary plot calculated by SVR-AS algorithm to successfully transform the complex non-linear constraint optimization into a linear equality constraint minimization problem and provides a directly solvable function for the optimal results calculation. Compared with interior-point method, a traditional non-linear constrained minimization algorithm, the proposed method has an extremely low computation cost

and a better optimal result. Results show that SVR-AS based optimization method is promising for high-speed link predesign.

ACKNOWLEDGMENT

Hanzhi Ma was supported by the Scientific Project Foundation of Zhejiang Lab under Grant No. 2020KC0AB01. Xu Chen was supported by the National Science Foundation under Grant No. CNS 16-24811 - Center for Advanced Electronics through Machine Learning (CAEML).

REFERENCES

[1] H. Kim, C. Sui, K. Cai, B. Sen, and J. Fan, "Fast and precise high-speed channel modeling and optimization technique based on machine learning," *IEEE Transactions on Electromagnetic Compatibility*, vol. 60, no. 6, pp. 2049–2052, Dec. 2018.

[2] R. Trinchero, M. A. Dolatsara, K. Roy, M. Swaminathan, and F. G. Canavero, "Design of high speed links via a machine learning surrogate model for the inverse problem," in *2019 Electrical Design of Advanced Packaging and Systems (EDAPS)*, Dec. 2019, pp. 1–3.

[3] K. Roy, M. A. Dolatsara, H. M. Torun, R. Trinchero, and M. Swaminathan, "Inverse design of transmission lines with deep learning," in *2019 IEEE 28th Conference on Electrical Performance of Electronic Packaging and Systems (EPEPS)*, Oct. 2019, pp. 1–3.

[4] H. Ma, E.-P. Li, A. C. Cangellaris, and X. Chen, "Support vector regression-based active subspace (SVR-AS) modeling of high-speed links for fast and accurate sensitivity analysis," *IEEE Access*, vol. 8, pp. 74 339–74 348, 2020.

[5] T. Lu, J. Sun, K. Wu, and Z. Yang, "High-speed channel modeling with machine learning methods for signal integrity analysis," *IEEE Transactions on Electromagnetic Compatibility*, vol. 60, no. 6, pp. 1957–1964, Dec. 2018.

[6] H. Ma, E.-P. Li, A. C. Cangellaris, and X. Chen, "Comparison of machine learning techniques for predictive modeling of high-speed links," in *2019 IEEE 28th Conference on Electrical Performance of Electronic Packaging and Systems (EPEPS)*, Oct. 2019.

[7] Y. Zhang, J. A. Hejase, J. C. Myers, J. J. Audet, W. D. Becker, and D. M. Dreps, "3D electromagnetic modelling of connector to pcb via transitions," in *2018 IEEE 27th Conference on Electrical Performance of Electronic Packaging and Systems (EPEPS)*, Oct. 2018, pp. 99–101.

[8] *ANSYS® Q3D Extractor, Release 18.0.*

[9] *Advanced Design System, Release 2019.*

A Transmission Line Coupler Component for direct B2B communications

Reiji Miura
The University of Tokyo
Tokyo, Japan
miura@kuroda.t.u-tokyo.ac.jp

Tadahiro Kuroda
The University of Tokyo
Tokyo, Japan
kuroda@kuroda.t.u-tokyo.ac.jp

Mototsugu Hamada
The University of Tokyo
Tokyo, Japan
Hamada@kuroda.t.u-tokyo.ac.jp

Abstract— **A contactless connector by using a transmission line coupler (TLC) in direct board-to-board (B2B) communications is described. We propose a TLC block component, and a new structure of side vias for impedance matching. In this method, error-free communication up to 3 Gbps is possible with a B2B distance of 15 mm.**

Keywords— *contactless connector, transmission line coupler, impedance matching*

I. INTRODUCTION

A practical system consists of multiple module boards, which are connected by a backplane. Signals on a module board are carried to its edge connector for their inter-module communications, which brings power, delay, area penalties.

We develop a transmission line coupler (TLC) technology enabling a direct board-to-board(B2B) high-speed communications, in which transmission lines on two separate printed circuit boards are electromagnetically coupled. In this paper, we present a new TLC applicable for the replacement of the current backplane communication.

The TLC utilizes crosstalk between transmission lines for the communication. This means that the signal is transferred with both electric and magnetic fields distributed along with the transmission lines. With this nature, the impedance matching at the TLC is manageable as we do in the transmission line design. Therefore, a high-speed B2B communication as high as 12Gbps is achieved while the conventional magnetic coupling B2B communication is limited to about 1Gbps. Besides, since it is tolerant to the coupler's misalignment, a conventional backplane connector can be used for the mechanical connection and power supply combined with the TLC to integrate a system with multiple modules. The contactless signal connection is free from the reliability problem of the insertion-extraction of module boards. However, a general backplane has a board pitch of about 15mm, while the conventional TLC was desirable to be used with the gap of up to 5 mm. It is difficult to increase the communication distance.

In this paper, we propose a TLC block component for the direct B2B communication, in which a TLC is elevated in the block component. The component is mounted on a PCB so that the gap of the TLC is decreased. A prototype was fabricated with a TLC capable of communication up to 3.5 Gbps. The height of the prototype component is 5 mm, with which the communication distance is decreased to 5mm (15 mm (board pitch) − 2 x 5 mm (height of two TLC block components)). When a conventional side via is used to lift the signal up, the communication up to 1Gbps is error free. A simulation analysis shows that the impedance mismatch at the

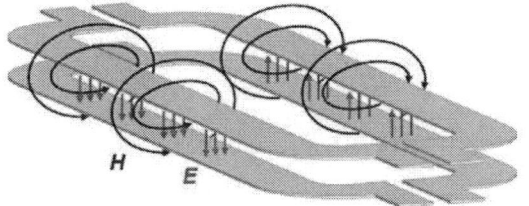

Fig. 1 Basic structure of TLC.

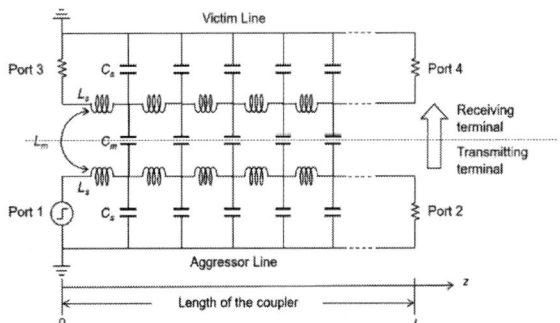

Fig. 2 Equivalent circuit model of two coupled transmission lines.

side via that lifts the signal limits the data rate. Therefore, we propose a new TLC block component with a new structure of side vias, which is error-free up to 3Gbps by improving the signal lifting part while maintaining manufacturing feasibility and achieving impedance matching.

II. TLC FUNDAMENTALS

A. Electrical Characteristics and Design Parameters

Fig. 1 depicts the basic structure of TLC. It consists of two pairs of differential transmission lines, with each pair implemented on the surface layer of a different PCB, and with the two pairs coupled across a gap. Each pair of differential transmission lines is laid out in the shape of a horseshoe. The horseshoe shape enables strong inter-pair coupling across the gap by increasing intra-pair spacing to 1) reduce intra-pair coupling, and 2) accommodate wider traces for better inter-pair coupling.

Based on the circuit model in Fig. 2, the voltage and current of a signal at any particular point z along the transmission line and at any particular time t can be represented by the following set of equations:

$$\frac{\partial}{\partial z}\begin{bmatrix} V_1 \\ V_2 \end{bmatrix} = -\frac{\partial}{\partial t}\begin{bmatrix} L_{11} & L_{21} \\ L_{21} & L_{22} \end{bmatrix}\begin{bmatrix} I_1 \\ I_2 \end{bmatrix}$$
$$\frac{\partial}{\partial z}\begin{bmatrix} I_1 \\ I_2 \end{bmatrix} = -\frac{\partial}{\partial t}\begin{bmatrix} C_{11} & C_{21} \\ C_{21} & C_{22} \end{bmatrix}\begin{bmatrix} V_1 \\ V_2 \end{bmatrix}$$

(1)

978-1-7281-6162-4/20 $31.00 © 2020 IEEE

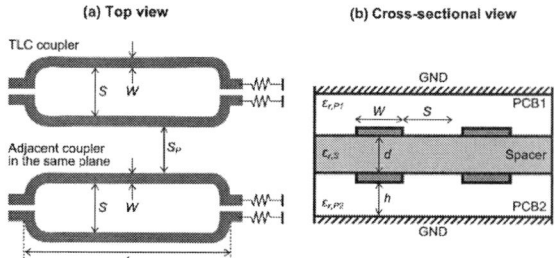

Fig. 3 Physical design parameters of a standard TLC.

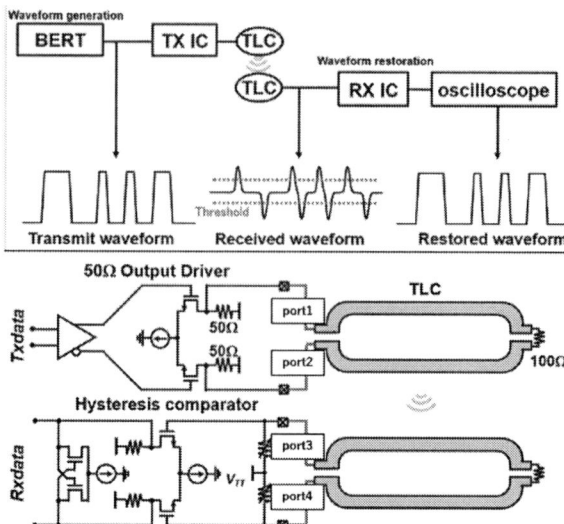

Fig. 4 Communication evaluation method.

where $V_1(z,t)$ and $I_1(z,t)$ are the voltage and current for the transmitting terminal, $V_2(z,t)$ and $I_2(z,t)$ the voltage and current for the receiving terminal, $L_{11}, L_{22}=L_s$ are the self inductances of the individual transmission lines, $C_{11}, C_{22}=C_s+C_m$ are the total capacitances of the individual transmission lines, $L_{12}, L_{21}=L_m$, and $C_{12}, C_{21}=-Cm$. (Note that the definition of the inductance matrix is not completely analogous to that of the capacitance matrix in that $L_{11}, L_{22} \neq L_s+L_m$.) The transmitter voltage at Port 1 is $V_{TX}(t)=V_1(0,t)$, and the received voltage at Port 3 is $V_{RX}(t)=V_2(0,t)$.

Since signal transmission in TLC is achieved through capacitive and inductive coupling, its effectiveness is quantified by the capacitive and inductive coupling coefficients, K_c and K_L, given by

$$K_C = \left| \sqrt{\frac{C_{21}}{C_{11}C_{22}}} \right|, K_L = \left| \sqrt{\frac{L_{21}}{L_{11}L_{22}}} \right| \tag{2}$$

Fig. 3 defines the physical design parameters of the standard TLC. The assumption of homogeneous dielectric medium means that both PCBs and the spacer have the same dielectric constant, in other words, $\varepsilon_{r,P1} = \varepsilon_{r,P2} = \varepsilon_{r,S} = \varepsilon_r$. Since $K_C = K_L = K$, the coupling coefficient K can be computed by considering electric field distribution only, by modeling the coupled traces as parallel-plate capacitors plus fringe effects. Specifically, for identical transmission lines where $C_{11}=C_{22}$, K is given by

$$K = \frac{\varepsilon_0 \varepsilon_r \frac{W}{d} + C_{m,fr}}{\varepsilon_0 \varepsilon_r \frac{W}{d} + C_{m,fr} + 2C_{s,fr} + \varepsilon_0 \varepsilon_r \frac{W}{h}} \tag{3}$$

When the ratio of the line width W to the communication distance d and the ratio of the line width W to the distance h to GND are increased, the coupling force K also increases. However, in order to achieve a communication distance of about 15 mm, we need to set the line width W to 15mm, and the TLC size becomes 75mm×10mm. This is not practical nor acceptable.

B. Communication evaluation method

Figure. 4 shows the communication evaluation method in this study. The transmitter IC is inserted in front of the TLC to work as a buffer and stabilize the amplitude and rise time. After passing through the TLC, the low-frequency component is cut off and the waveform is received as a return-to-zero waveform. A hysteresis comparator is used as a receiver IC to restore this received waveform. The BERT generates PRBS signal, and each waveform is confirmed using an oscilloscope. The transmitter is MAX3842 of MAXIM. The receiver is ADCMP580 of ANALOG DEVICES.

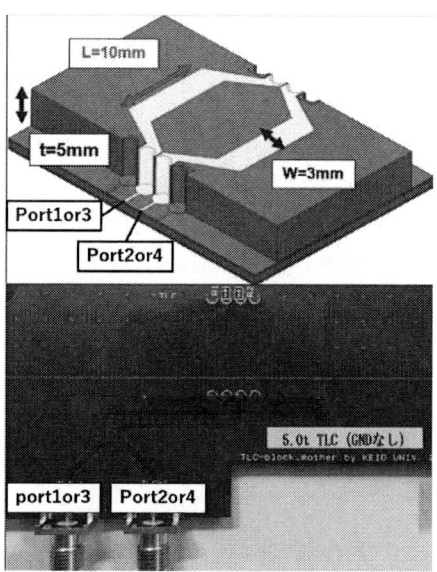

Fig. 5 details of TLC block component.

III. PROPOSED METHOD

This section describes the TLC block component (Fig. 5) proposed in this paper. This TLC is elevated in the block component. The component is mounted on a PCB so that the gap of the TLC is decreased. Further, coupling coefficients K shown in equation (3) can be increased since the distance h to GND becomes large. A prototype was fabricated with line width W=3mm, line length L=10mm and the TLC capable of communication up to 3.5 Gbps.

A. Block height determination

This subsection describes the height t of the block component which lifts the TLC up. We perform simulation by changing the value of the height t, measure the S/N ratio (Signal amplitude / Noise amplitude in Fig.6), and compare them. The results are shown in Fig. 6. We chose t=5mm with the best S/N ratio as the block height.

978-1-7281-6162-4/20 $31.00 © 2020 IEEE

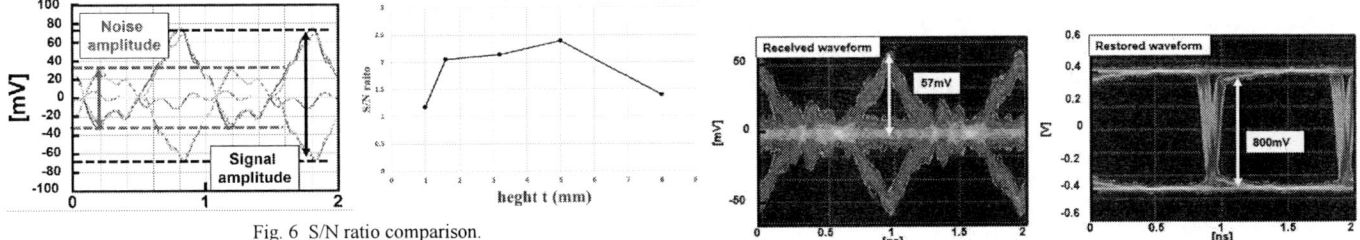

Fig. 6 S/N ratio comparison.

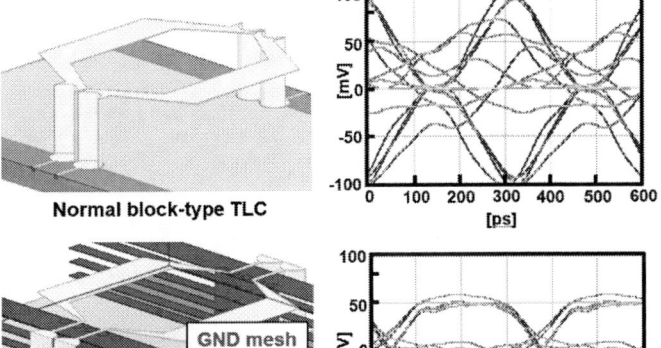

Fig. 7 Waveform obtained in the experiment.

B. Experimental results

We create an experimental system as shown in Fig. 4, and set the input waveform communication speed from BERT to 1 Gbps. The communication distance between TLC couplers is 5 mm, and the distance between boards is 15 mm. Fig.7 shows the received waveform through the TLC, and restored waveform observed with an oscilloscope.

In the received waveform in Fig. 7, the signal amplitude is 57 mV, which is larger than the receiver IC threshold, and the noise amplitude is within the threshold. It is possible to confirm the restoration of the waveform by passing it through the receiver IC.

C. Impedance matching of edge vias

As mentioned above, the TLC used in this paper can communicate at up to 3.5Gbps. The simulation is performed at a board-to-board distance of 1.5 cm and a communication speed of 3 Gbps. However, the noise amplitude becomes too large (Fig.8 normal block-type TLC), and receiver IC can't restore the waveform. The cause of the noise is that the edge via which used to lift the signal up does not have impedance matching.

To match the impedance of the side vias to transmission lines, the insertion of a return path is required along with the side vias. However, having a metal layer perpendicular to the substrate is very difficult from a manufacturing feasibility point of view.

We have propose a new TLC block component in which GND mesh is added along with the edge vias as shown in Fig. 8 for impedance matching. This GND mesh consists of vias and wiring layers commonly used for PCBs. In the received waveform obtained with the new TLC block component, the signal amplitude is about 70 mV, which is larger than the threshold of the receiver IC, and the noise amplitude is within the threshold. Therefore, proposed a new TLC block component with a new structure of the edge via is error-free up to 3Gbps. The cost increase of a new TLC block component is only for the PCB of the block part. Assuming the size of a typical PCB is 20 cm x 20 cm, the cost increase is only a few percent.

IV. CONCLUSION

In this paper, we propose a contactless method by using a transmission line coupler in direct B2B communications. Since it was difficult for TLC to communicate at a distance of 15 mm, we propose the new TLC block component that lifts the TLC from the substrate. Furthermore, error-free communication up to 3 Gbps is possible with a board-to-board distance of 15 mm by improving the impedance of the edge vias that lifts the signal while maintaining manufacturing feasibility.

Fig. 8 Receive waveform comparison (3Gbps).

REFERENCES

[1] S. Kawai, H. Ishikuro, and T. Kuroda, "A 2.5Gb/s/ch Inductive-Coupling Transceiver for Non-Contact Memory Card," IEEE International Solid-State Circuits Conference (ISSCC) Dig. Tech. Papers, pp. 264-265, Feb. 2013.

[2] A. Kosuge, T. Takeya, M. Shioya, M. Taguchi, and T. Kuroda, "A 3Gb/s Non-Contact Inter-Module Link with Duplex Transmission Line Couplers and Low Frequency Compensation Equalizer," JSAP International Conference on Solid-State Devices and Materials (SSDM) Extended Abstracts, pp. 1152-1153, Sep. 2012.

[3] A. Kosuge, T. Takeya, M. Shioya, M. Taguchi, and T. Kuroda, "A 3Gb/s Non-Contact Inter-Module Link with Duplex Transmission Line Couplers and Low Frequency Compensation Equalizer," JSAP International Conference on Solid-State Devices and Materials (SSDM) Extended Abstracts, pp. 1152-1153, Sep. 2012.

[4] W. Mizuhara, T. Shidei, A. Kosuge, T. Takeya, N. Miura, M. Taguchi, H. Ishikuro, and T. Kuroda, "A 0.15mm-Thick Non-Contact Connector for MIPI Using Vertical Directional Coupler," IEEE International Solid-State Circuits Conference (ISSCC) Dig. Tech. Papers, pp. 200-201, Feb. 2013.

[5] W. J. Yun, S. Nakano, W. Mizuhara, A. Kosuge, N. Miura, H. Ishikuro, and T. Kuroda, "A 7Gb/s/Link Non-Contact Memory Module for Multi-Drop Bus System Using Energy-EquipartitionedCoupledTransmissionLine,"IEEEInternationalSolid-StateCircuits Conference (ISSCC) Dig. Tech. Papers, pp. 52-53, Feb. 2012.

[6] A. Kosuge, W. Mizuhara, N. Miura, M. Taguchi, H. Ishikuro, and T. Kuroda, "A 12.5Gb/s/ LinkNon-ContactMulti-Drop Bus System with Impedance-Matched Transmission Line Couplers and Dicode Partial Response Transceivers," in Proc. of IEEE Custom Integrated Circuits Conference (CICC), pp. 7.9.1-7.9.4, Sep. 2012

A Review of 90 Degree Corner Design for High-Speed Digital and mmWave Applications

Heidi Barnes
Design Test Software
Keysight Technologies
Santa Rosa, USA
heidi_barnes@keysight.com

Giovanni Bianchi, Jose Moreira
Hardware R&D
Advantest
Boeblingen, Germany
{giovanni.bianchi; jose.moreira}@advantest.com

Abstract—**The design of 90 degree corners has been an important topic since the start of microwave design several decades ago. With the arrival of standards like 400Gb Ethernet, 5G and WiGig, engineers are now required to create PCB designs with signals in the mmWave frequency range for consumer applications. In this paper we will do a quick review on the design of right angle corners for PCB construction and present some measurement results.**

Keywords— Microstrip routing, Right Angle Corners, mmWave.

I. INTRODUCTION

There was a time where discussing the impact of right angle corners in printed circuit board (PCB) signal traces was restricted to engineers working in niche applications. This is no longer true with some mainstream applications in both RF and digital applications already in the mmWave frequency range. For the benefit of engineers starting to work on these applications without past experience with right angle corners at mmWave frequencies, we present in this paper a review of the key points of designing right angle corners in printed circuit board signal traces. We also present some measured results up to 65 GHz from manufactured test structures using de-embedding techniques to remove the impact of the connector transition.

II. RIGHT ANGLE CORNER DESIGN

To the best of our knowledge, the first studies on microstrip bends are almost 50 years old [1]. The topic was well established in the following 20 years [2-6]. The quintessence, of all that research effort could be condensed in Figure 1. Figure 1a shows the physical structure of a microstrip mitered bend (dash triangle), two short feeding microstrips (gray rectangles) have been added, in order to show how the bend is connected with the rest. The relevant geometrical parameter of the structure is the so-called miter ratio (or miter factor) $M=d/\sqrt{w}$: $M=0$ means a square bend, $M=0.5$ means an isosceles triangle (half of a square). A relatively good approximated equivalent circuit is shown in Figure 1b. The two transmission lines (with identical parameters) TL1,TL3 have the same characteristic impedance as the feeding lines (gray rectangles in Figure 1a, i.e. microstrips with width w); they produce a phase shift on the S-parameters. The two inductors L1, L3 (both with the same

inductance $L_1=L_3$) together with the capacitor C2 (with capacitance C_2) play the main role in the bend's modeling. Intuitively, it can be seen that the higher the M, the higher $L_1=L_3$, the lower C_2, and vice-versa. If M is made such that $Z_0 = \sqrt{2 \cdot L_1/C_2}$, where Z_0 is the characteristic impedance of the feeding lines, then the bend is "seamless" or well matched, that condition usually happens with M close to 0.5. The true optimum value depends on the substrate and on the frequency range. It has to be remarked that the electrical length of a bend could be longer, equal, or even shorter than that of a straight section of microstrip with the same width and length, depending on w, substrate height h and dielectric constant ε_r, and metal thickness t.

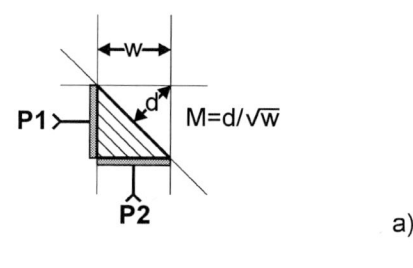

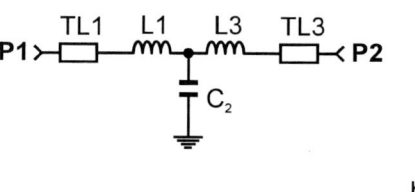

Figure 1: Microstrip bend: a) physical structure, b) equivalent circuit test.

In order to keep the investigation less abstract, one practical example was considered, using a substrate with h=200 μm, ε_r=4, t=30 μm and a corresponding 50 Ω microstrip width of w=397.5 μm. Figure 2 shows three possible bends with different miter ratios, all of them having a 400 μm long 50 Ω microstrip line on both sides. Figure 3 shows the resulting amplitude of reflection (right-y) and transmission (left-y) coefficients. It can be seen that M=0.5 is close to the "optimum", while M=0.4 (closer to a full square bend) is the worst case among the three considered.

978-1-7281-6162-4/20 $31.00 © 2020 IEEE

However, if the frequency range of interest is below 11.8 GHz (where the blue line crosses the green one), the lowest reflection coefficient amplitude is given by M=0.6, not M=0.5. The general conclusion that could be taken from this example is that a true "optimally mitered" bend configuration of general validity simply does not exist, although commercially available RF circuit simulators offer such element. The situation could be even more complicated if real components - instead of perfectly matched 50 Ω S-parameter ports - are connected with the RF ports P1, P2. Real components are never perfectly matched, therefore, the discontinuity due to the bend interacts with the standing wave regime caused by the non-ideal termination. Consequently, the optimum M can further change.

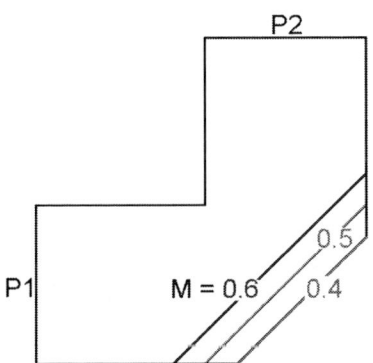

Figure 2: Some microstrip bends with different miter factor.

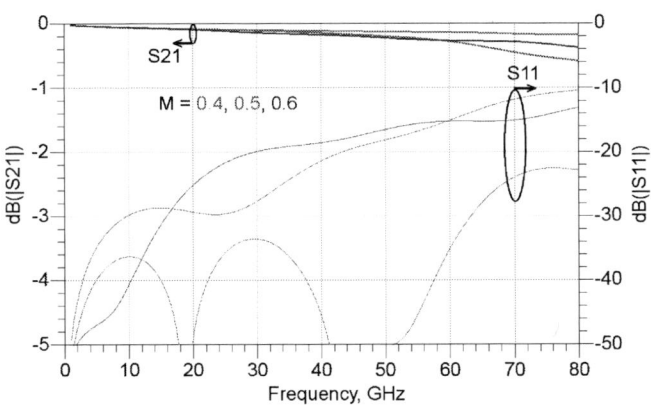

Figure 3: Transmission and reflection coefficients of the bends represented in Figure 2.

One more aspect to consider is the old, and never well answered, question: Is it better to use a smooth curved right angle bend or an abrupt square/miter bend? Again, in order to consider a practical example (still with the same substrate and 50 Ω line as per Figure 2 and Figure 3), let us consider the layouts shown in Figure 4, where the distance between the RF ports P1, P2 is (x=4.4 mm; y=0.9 mm). Three of the many possibilities are shown in Figure 4: 90° bends (blue), 60° round curves (red), 30° curves (green), the corresponding amplitudes of reflection (right-y) and transmission coefficient (left-y) are shown in Figure 5. It can be seen that there is no dramatic difference among the three cases. Moreover, the theoretical

worst case could be the best one in practice, due to inaccuracies of the realization process, limited precision of the simulation models, termination impedances different from the ideal 50 Ω case.

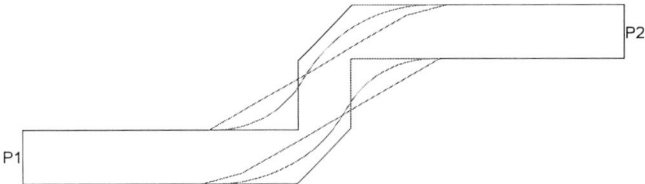

Figure 4: Three possible routings for two offset lines: with 90° bends (blue), 60° round curves (red), 30° bends (green).

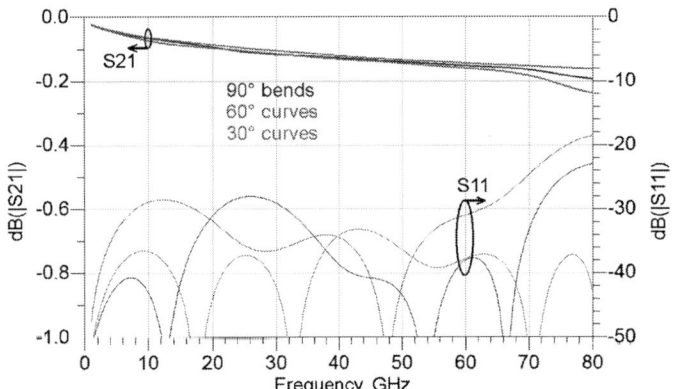

Figure 5: Transmission and reflection coefficients of the RF paths represented in Figure 4.

III. MEASUREMENT EXAMPLE

To demonstrate the impact of the right corner design, the PCB test coupons shown in Figure 6 were manufactured in Rogers 4350B using a 0.5207 mm wide microstrip (20.5 mil) in a 0.254 mm thick dielectric (10 mil). The microstrip traces used a NiAu plating without any soldermask. The connectors are 1.85 mm type from Signal Microwave (ELF67-002) with a custom optimized footprint for the used stackup.

Figure 6: Manufactured PCB test coupons including a straight right angle, a mitered right angle and a curved right angle. Also included in each PCB coupon is a connector de-embedding structure.

To be able to measure and compare the different right angle designs at mmWave frequencies (in this paper up to 65 GHz), it is not only required to have an excellent connector to PCB microstrip transition but also to de-embed the impact of the connector transition. For this the test coupons also include a 28 mm length 2x-thru de-embedding structure so that we can de-embed the connector and its transition plus part of the microstrip trace [7,8]. Figure 7 shows the measured insertion and return loss of the 2x-thru structure where one can observe better than 10 dB return loss up to 65 GHz and also a separation greater than 5 dB between the insertion and return loss. As discussed in [7], this demonstrates that not only that we have an excellent connector transition but also that this 2x-thru structure can be used for de-embedding up to 65 GHz.

The mitered right angle test structure used an M values of 0.5 and the curved right angle used a radius of 2.5 mm. Figure 8 shows pictures of the manufactured right angle structures. As can be observed, PCB manufactured structures are never perfect, a straight angle right corner will have slightly round edges depending on the etching process.

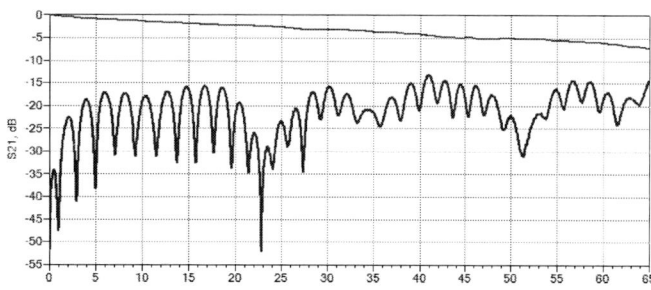

Figure 7: Measured insertion and return loss of the 28 mm 2x-thru de-embedding structure.

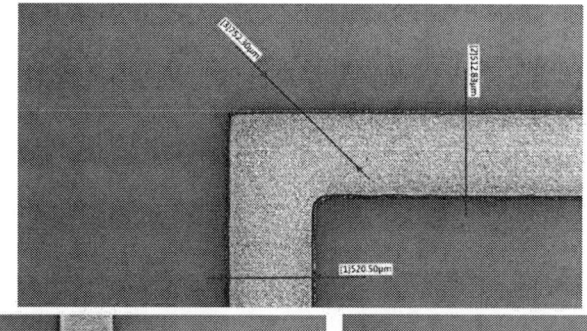

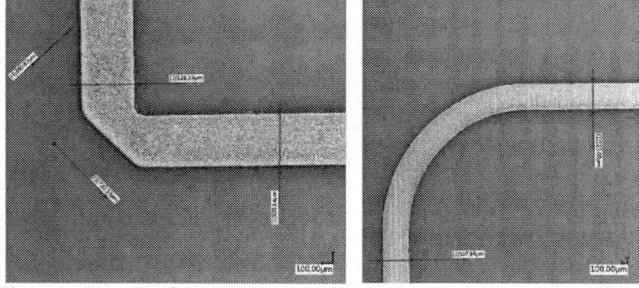

Figure 8: Pictures of the manufactured right angle structures.

Figure 9 presents the measured insertion and return loss results after connector de-embedding for each structure. De-embedding was preformed using the Keysight PLTS 2020 software package and after de-embedding only a 1 mm microstrip trace is present before the right angle structure. The results clearly show the performance advantages of both the mitered right angle and also the curved right angle compared with the straight right angle.

Note that the curved right angle PCB structure is longer than the mitered right angle structured which also implies it is slightly more lossy. The measurements were performed with harmonic sampling from 6.5 MHz to 65 GHz (10.000 points) and an IF bandwidth of 1 kHz.

IV. CONCLUSIONS

This paper has presented a review of the design of right angle corners and showed some measurement examples. In conclusion both mitered and curved right angles showed improvements compared with a straight right angle and similar performances. This paper again stresses the point that square right angles do have bad performance at mmWave frequencies and should be avoided. It is also important to understand that right angle design discontinuities can impact EMI [9] or even DC [10].

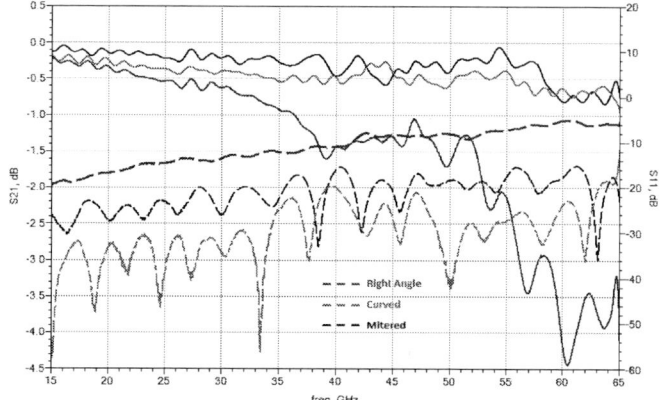

Figure 9: Measured insertion and return loss after connector/signal trace de-embedding.

REFERENCES

[1] R. J. P. Douville and D. S. James, "Experimental Characterization of Microstrip Bends and Their Frequency Dependent Behavior," 1973 IEEE Conference Digest, October 1973, pp. 24-25.

[2] R. J. P. Douville and D. S. James, "Experimental Study of Symmetric Microstrip Bends and Their Compensation," IEEE Transactions on Microwave Theory and Techniques, Vol. MTT-26, March 1978, pp. 175-181.

[3] K. C. Gupta, R. Garg, and R. Chadha, Computer-Aided Design of Microwave Circuits, 1981, p. 195.

[4] M. Kirschning, R. H. Jansen, and N. H. L. Koster. "Measurement and Computer-Aided Modeling of Microstrip Discontinuities by an Improved Resonator Method," 1983 IEEE MTT-S International Microwave Symposium Digest, May 1983, pp. 495-497.

[5] J. Moore, H. Ling, "Characterization of a 90° Microstrip Bend with Arbitrary Miter via the Time-Domain Finite Difference Method", IEEE Transactions on Microwave Theory and Techniques, Vol.MTT-38, April 1990, pp.405-410

[6] B. Wadell, Transmission Line Design Handbook, p. 294,1991.

[7] Heidi Barnes et al., "S-Parameter Measurement and Fixture De-Embedding Variation Across Multiple Teams, Equipment and De-Embedding Tools", Designcon 2019

[8] IEEE 370 Project, standards.ieee.org/develop/project/370.html

[9] Philippe Sochoux et al. "EMI from SerDes Differential Pairs", DesignCon 2009.

[10] Istvan Novak, "The Perils of Right-Angle Turns at DC", Signal Integrity Journal September 2017.

A Tunable Neural Network based Decision Feed-back Equalizer model for High-speed Link Simulation

Thong Nguyen and Jose Schutt-Aine
Department of Electrical and Computer Engineering.
University of Illinois Urbana-Champaign
Urbana, IL 61801, USA
tnnguye3, jesa@illinois.edu

Abstract—This paper presents a model combining a feed-forward neural network (FNN) with a recurrent neural network (RNN) to model Decision Feed-back Equalizer (DFE). By using the FNN as the mapping between the tap values and the dynamic behavior of the DFE, a complete model of the DFE can be constructed for channel simulation. The paper shows a 2-tap DFE example in which excellent agreement between the model generated by the proposed method and transient simulation can be observed.

I. INTRODUCTION

Losses and dispersion caused by passive channels in high-speed serial links are compensated by equalization circuits. Receiver equalization consists of continuous-time linear equalization (CTLE) and decision feed-back equalization (DFE). CTLE is linear and typically modeled as rational transfer functions which can also be incorporated into the channel while DFE is nonlinear, hence, needs separate handling. Leveraging success of deep learning methods in time-series data modeling, [1], [2] uses an Elman RNN [3] to model high-speed link buffers. However, the equalization parameters were fixed when the RNN learns the input - output mapping in [2]. In this work, we extend the work done in [2], allowing modeling a complete model of DFE using an FNN followed by the Elman RNN for sequence mapping.

In the next section, a brief description DFE is presented to recall the problem at hand. Section III is dedicated to explain the details of our proposed architecture allowing a complete DFE *tunable* model. A 2-tap DFE example is shown in Section IV to demonstrate the effectiveness of the proposed method. It also serves as a step-by-step guide on how to train and use the proposed model for simulation. Conclusion and possible further extension of this work is presented in Section V.

II. RECEIVER DECISION FEEDBACK EQUALIZATION (DFE)

A passive, lossy channel acts similarly to a low pass filter in the sense that the higher the frequency component, the more loss happens. As the result, a sharp rising/falling signal after passing through a channel will lose its high frequency contents which makes its rise/fall time slower, the signal spreads out

longer. In most cases, it could spread out to multiple bit periods (also known as unit interval, or UI) causing a phenomenon called intersymbol interference (ISI).

Figure 1 shows how a bit error could happen due to ISI. The figure was exaggerated for illustration purposes. When the output of the prior bit (in this case, a logical 1) spreads out, it interferes with the next bit (in this case, a logical 0), alters the voltage waveform leading to the wrong logical level recognized.

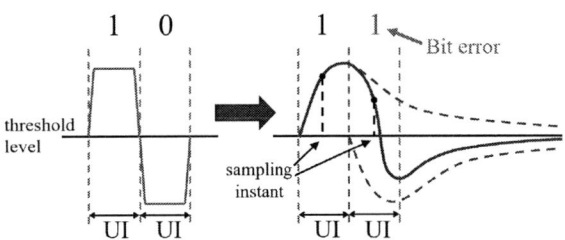

Fig. 1. Bit error caused by ISI.

The basic principle of DFE is using a slicer circuit to sample and quantize the input waveform at locations that are multiple UIs away from each other, normally known as the cursor values (a_i's), then using these cursor values, the DFE circuit subtracts ISI from the incoming signal via a feedback FIR filter

$$v_{out}[n] = v_{in}[n] - \sum_{i=1}^{N} a_i d_{RX}[n-i] \qquad (1)$$

Figure 2 shows an example of different number of tap DFE circuits on an unequalized waveform. Notice how the number of abrupt drops in the equalized waveform matches the number of DFE taps used. It is also important to notice that the drops associated with higher tap numbers are smaller. This is due to the fact that the higher order post-cursors are always smaller in magnitude.

III. PROPOSED NEURAL NETWORK ARCHITECTURE

In this work, under the assumption that the DFE effect is K time step dependent, we use an Elman RNN to represent the

978-1-7281-6162-4/20 $31.00 © 2020 IEEE

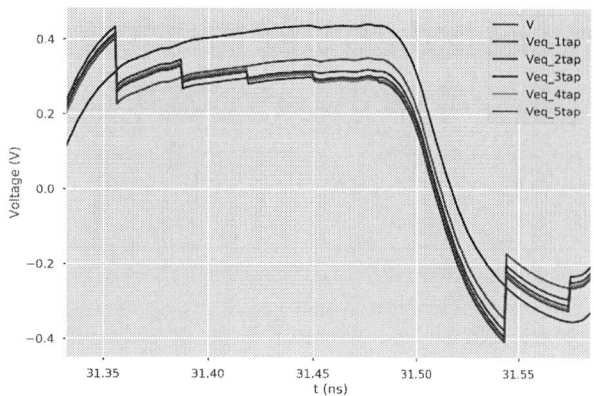

Fig. 2. Example waveform under DFE effect with different number of taps

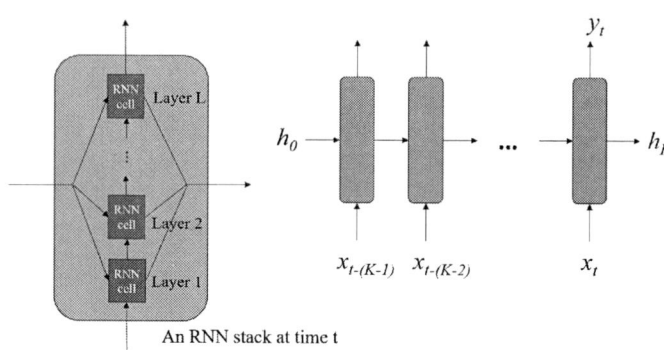

An RNN stack at time t

Fig. 3. K-step RNN unrolled in time

shows a K-step RNN unrolled in time and detailed of a stack L layers of RNN cells.

mapping between a K-step windowed sample of inputs and the next time step output. Mathematically, the K time step truncated full stack RNN is

$$
\begin{cases}
\boldsymbol{h_t} = g_h^{(L)} \circ g_h^{(L-1)} \circ \cdots \circ g_h^{(2)} \circ g_h^{(1)} \left(\boldsymbol{x_t}, \boldsymbol{h_0} \right) \\
y_t = g_o \left(\boldsymbol{h_t} \right),
\end{cases}
\quad (2)
$$

where $A \circ B = A \left(B \left(x \right) \right)$ is the composition operation, $\boldsymbol{x_t} = \begin{bmatrix} x_{t-(K-1)} & x_{t-(K-2)} & \cdots & x_t \end{bmatrix} \in \mathbb{R}^{d_1 \times K}$, and $\boldsymbol{h_t} = \begin{bmatrix} h_{t-(K-1)} & h_{t-(K-1)} & \cdots & h_t \end{bmatrix} \in \mathbb{R}^{l_h \times K}$, $h_t \in \mathbb{R}^{l_h}$ where l_h is the dimension of the hidden state at time step t, $g_h \left(\cdot \right)$ represents the long-short term memory (LSTM) mapping which reads

$$
\begin{cases}
i_t = \sigma \left(W_{ii} x_t + W_{hi} h_{t-1} \right) \\
f_t = \sigma \left(W_{if} x_t + W_{hf} h_{t-1} \right) \\
g_t = \tanh \left(W_{ig} x_t + W_{hg} h_{t-1} \right) \\
o_t = \sigma \left(W_{io} x_t + W_{ho} h_{t-1} \right) \\
c_t = f_t c_{t-1} + i_t g_t \\
h_t = o_t \tanh \left(c_t \right),
\end{cases}
\quad (3)
$$

where h_t is hidden state at time t, c_t is called the cell state, and i_t, f_t, g_t and o_t are the input, forget, cell and output gates respectively. All of the W's are learnable weight matrices., $\boldsymbol{h_0} \in \mathbb{R}^{l_h \times L}$ is the initial state for every K time step dynamic.

The gradient of the loss function of a 1 layer, K time step unrolled RNN at time t w.r.t. any parameter θ can be written as sum of the loss function within the most recent K time steps

$$
\frac{\partial \mathcal{L}}{\partial \theta} = \sum_{\tau = t-(K-1)}^{t} \frac{\partial \mathcal{L}_\tau}{\partial \theta},
\quad (4)
$$

where

$$
\frac{\partial \mathcal{L}_\tau}{\partial \theta} = \sum_{j=t-(K-1)}^{\tau} \frac{\partial \mathcal{L}_\tau}{\partial y_\tau} \frac{\partial y_\tau}{\partial \boldsymbol{h_\tau}} \frac{\partial \boldsymbol{h_\tau}}{\partial \boldsymbol{h_j}} \frac{\partial \boldsymbol{h_j}}{\partial \theta}
\quad (5)
$$

with $\boldsymbol{h_j}$, $\boldsymbol{h_\tau}$ are given by (2) at $t = j$ and $t = \tau$. Each term $\frac{\partial \mathcal{L}_\tau}{\partial \theta}$ represents the partial-time gradient of the error in the past time steps (j's) up to the current time step (τ). Figure 3

In order to represent the effect of the tap values to the RNN dynamic behavior, we use an FNN which takes the tap values as input then take its output to feed into the RNN as $\boldsymbol{h_0}$. An FNN is the composition of multiple weighted nonlinear functions and can be mathematically described as

$$
\hat{y} = f_L \circ W_L f_{L-1} \cdots \circ W_2 f_1 \circ W_1 x
\quad (6)
$$

The input $x \in \mathbb{R}^{d_1}$, at the lth stage, f_l is a non-linear activation function composed with f_{l-1} weighted by the weight matrix $W_l \in \mathbb{R}^{d_{l+1} \times d_l}$, where d_{l+1} and d_l are the dimension of the output of layer $l+1$ and that of layer l, respectively; $l = 1, 2, ..., L$, $\hat{y} \in \mathbb{R}^{d_{L+1}}$ is the prediction of the FNN. Since both input and output are real values, it is most convenient to choose *mean-square error* (MSE) loss function. MSE loss is calculated as the square of 2-norm of the error vector.

$$
\mathcal{L} \left(\hat{y}, y \right) = \text{MSE} \left(\hat{y}, y \right) = \frac{1}{d_{L+1}} \left\| \hat{y} - y \right\|_2^2
\quad (7)
$$

The weights of both networks are learnt at the same time, after backpropagation through time in RNN, the gradient is also backpropagated through the FNN so its weights can also be updated. Figure 4 visualizes how FNN and RNN are combined to make a tunable model in this work.

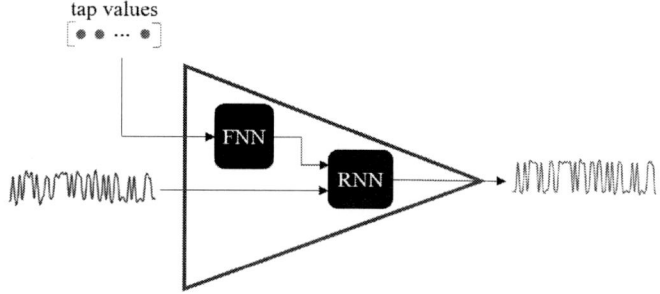

Fig. 4. Proposed architecture

978-1-7281-6162-4/20 $31.00 © 2020 IEEE

IV. Example

In this section, we will demonstrate the proposed method using data from a 2-tap DFE for an about 1.7ns delay differential channel transmitting data at 32 Gbps. The tap values are normalized so that they span from 0 to 1. Negative values of the taps will amplify the post-cursor instead of cancelling it, hence, are excluded. The channel pulse response and the equalized response are shown in Figure 5. It can be seen that there are 1 pre-cursor and 2 significant post-cursors. The effect of DFE is reflected clearly on the equalized waveform which is cancelled out exactly at 2 most significant sampled post-cursor locations.

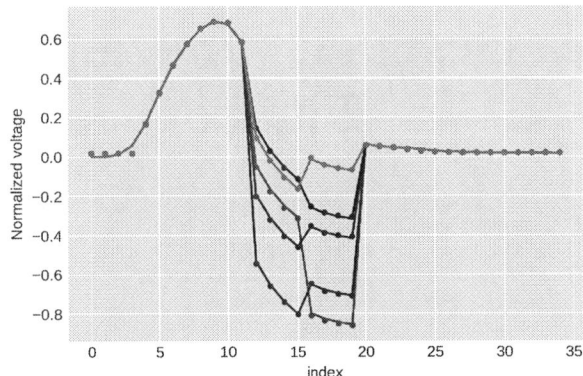

Fig. 6. Prediction (dot) vs. transient waveform (solid) for unseen tap combinations (different colors)

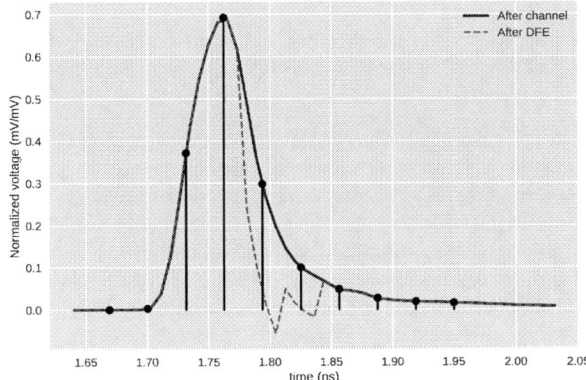

Fig. 5. Single pulse response and DFE effect

To prepare training data, many combinations of tap values are swept, the unequalized and equalized waveform is collected for each combination of tap values. In this example, we chose a 2-layer FNN which has 10 neurons and 20 neurons, respectively and a 6-layer, LSTM-cell RNN whose hidden state is in $\mathbb{R}^{30}$. Adam [4] is used for optimization with initial learning rate being 0.01. Also, RNN when trained was set to start out in *teacher-force* mode but we used a schedule sampling to choose between known data and RNN-generated data such that at the beginning of training, the RNN is fed with more known data (from training data) but as training progresses, the RNN is fed with more of its own generated data. More details about different modes of RNN-based model were presented in [2] and references therein.

After trained, the RNN is set to *read-out* mode and the model is tested with unseen data. Excellent agreement between the proposed model and the transient simulation is observed, the results are shown in Figure 6. Each color in Figure 6 is the output response of DFE model for different tap values.

V. Conclusion and Future work

In this work, we have combined an FNN and an RNN to create a tunable model that can completely replace the transistor model for different values of taps. The validation result on a pulse response matches very well with the transistor simulation result. This model can be exchanged between vendors and designers helping the former to protect their IP

and ensure accuracy for latter at the same time. For EDA vendors, implementation of these kinds of models is straight forward and simple. Exchanging models between vendors and designers would be also simple because all that is needed is the network architectures and its weights. Deep learning community has proposed ONNX [5], an open source framework, built as an effort to offer a unified API to all neural network models so it can be exchanged and used no matter how and by which framework it was developed and trained.

In a future work, we will report how the proposed model would be trained and validated in a channel simulation to generate eye diagrams. A longer term extension of this work is to include the adaptivity of equalization schemes into the model.

Acknowledgement

This material is based upon work supported by the National Science Foundation under Grant No. CNS 16-24810 for Center for Advanced Electronics through Machine Learning (CAEML), the U.S Army Small Business Innovation Research (SBIR) Program office and the U.S. Army Research Office under Contract No.W911NF-16-C-0125.

References

[1] T. Nguyen, T. Lu, J. Sun, Q. Le, K. We, and J. Schut-Aine, "Transient Simulation for High-Speed Channels with Recurrent Neural Network," in *2018 IEEE 27th Conference on Electrical Performance of Electronic Packaging and Systems (EPEPS)*, Oct 2018, pp. 303–305.

[2] T. Nguyen, T. Lu, K. Wu, and J. Schutt-Aine, "Fast Transient simulation of High-Speed Channels using Recurrent Neural Network," 2019. [Online]. Available: https://arxiv.org/abs/1902.02627

[3] J. L. Elman, "Finding structure in time," *COGNITIVE SCIENCE*, vol. 14, no. 2, pp. 179–211, 1990.

[4] D. P. Kingma and J. Ba, "Adam: A method for stochastic optimization," *CoRR*, vol. abs/1412.6980, 2014. [Online]. Available: http://arxiv.org/abs/1412.6980

[5] J. Bai, F. Lu, K. Zhang *et al.*, "ONNX: Open Neural Network Exchange," https://onnx.ai/, 2019.

Via Design Optimization for High Speed Differential Interconnects on Circuit Boards

Armen Vardapetyan
Connectivity Group
Intel Corporation
Santa Clara, U.S.A.
armen.vardapetyan@intel.com

Chong-Jin Ong
Internet-of-Things Group
Intel Corporation
Chandler, U.S.A.
chong-jin.ong@intel.com

Abstract— The unprecedented demand for high bandwidth applications boosts the data rates of major high speed differential interconnect protocols such as PCIe and Thunderbolt/USB. Transmission lines and via transitions form most of the interconnect path between a transmitter and a receiver. To get maximum performance of the system at high signaling rates, the impedance of the interconnect path has to be as uniform as possible to cause minimal signal reflections. While the impedance of transmission lines can be easily controlled, the impedance of vias are much harder to control. In this paper, we use time-domain impedance waveforms in conjunction with channel simulations to optimize the impedance profiles of 3 types of differential vias: through-hole vias, blind vias and buried vias. We do this by varying the via diameter, pad diameter, antipad diameter and via pitch (center-to-center distance). We then show a quick method to optimize the vias for faster turnaround time, depending on whether the via impedance is too capacitive or too inductive. The board designer can use the quick method to try to achieve approximately the impedance profiles we show.

Keywords— *bandwidth, high-speed electronics, impedance matching, printed circuit boards*

I. INTRODUCTION

One of the key challenges of signal integrity engineers is the understanding of how to maximize the electrical performance of high speed differential I/O, which can run at data rates of tens of gigabits per second. It is critical to identify significant factors which can be optimized to stretch the electrical performance of electronic systems in an increasingly competitive business environment.

On circuit boards and packages, vias are impedance discontinuities that are unavoidable as routing needs to change layers. Up to data rates of several gigabits per second, these vias are electrically short, so they do not interfere too much with signal integrity performance. Board designers focus mainly on the maximum length allowable for via stubs to prevent resonance effects. When the data rate goes into tens of gigabits per second as in PCIe Gen. 5, USB 3.2 Gen. 2 and Thunderbolt, the vias could contribute to significant degradations to the signal quality.

The investigation into single ended via modeling and via impedance control has been undertaken by Andreas Hardock et al. [1]. In this paper, the focus will be on differential vias as high speed input-output (HSIO) signaling at high data rates takes place on differential interconnects. In particular, we will focus on PCIe Gen. 5. The methodology is general and applies to high speed differential vias for any signaling protocol.

A lot of researchers have worked on optimizing via electrical performance. Some have tried to optimize ground via placement and patterns [2] [3]. Others have tried to redesign the via structure [4]. In this paper, we will optimize via parameters such as via diameter, pad diameter, anti-pad diameter and via pitch to get optimum signal-to-noise ratio (SNR). (The anti-pads for the differential vias are assumed to be merged.) This involves generating differential via models and performing channel simulations. The optimum impedance profile for 3 types of vias will be shown: the plated through hole (PTH) via, the blind via and the buried via. The blind and buried vias are all assumed to be backdrilled and hence have no stubs. These 3 via types are shown in Fig. 1. We will also show a quick method of optimizing vias, depending on whether the impedance profile is too capacitive or too inductive. The board designer can use the optimized impedance profile as a guide to optimize the via impedance.

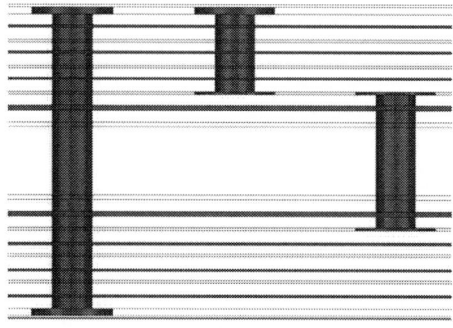

Fig. 1   Three types of vias. From left: through-hole via, blind via and buried via. The stackup is drawn to scale.

II. METHODOLOGY FOR FULL FACTORIAL VIA PARAMETER OPTIMIZATION

The S-parameters and time domain impedance profiles of the PTH vias, blind vias and buried vias were generated. For each via type, the following via parameters were swept: via diameter, pad diameter, antipad diameter and via pitch. Table I shows the via parameter values swept. The minimum values of these parameters were chosen such that the boards can be mass manufactured at reasonable cost.

978-1-7281-6162-4/20 $31.00 © 2020 IEEE

TABLE I. RANGE OF VIA PARAMETERS

Via parameter	Min. Value (mils)	Max. Value (mils)
via diameter	8	12
pad diameter	18	22
antipad diameter	28	32
via pitch	28	36

Each via of the differential pair had a ground via placed 40 mils away (center-to-center). Each ground via had a 10 mil via diameter. The board in total had 18 layers. The through hole via went from the top layer to the bottom layer. The blind via went from the top layer to layer 7. The buried via went from layer 7 to layer 12. See Fig. 1. The ground plane and substrate dimensions for the via model was 150 mils by 150 mils. The total feed length was about 150 mils. Fig. 2 shows the modeled structure for the blind via. The buried and through-hole via were modeled similarly.

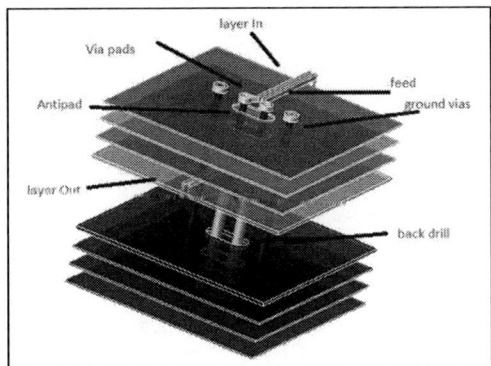

Fig. 2   Via structure for blind via.

Each model for a single set of parameters took about 15 to 20 min. to generate for a frequency range from 1 MHz to 32 GHz. About 240 models were generated for each via type. The machine used to generate the models (and perform the channel simulations) was a 64 bit Windows server with 384 GB of RAM and dual Intel Xeon Gold 6142 CPU's clocked at 2.6 GHz.

The via models were incorporated into channel simulations to generate the eye patterns at the receiver end of a PCIe Gen 5 channel, which had a 32 GT/s data rate. The channel included 2.5 inch long differential transmission lines both before and after the via models. The channel simulation for the blind vias had 2 sets of vias to so that transmitter and receiver were both on the top layer. The channel simulation for the buried vias had 2 sets of buried vias as well as blind vias so that the transmitter and receiver was on the top layer. The channel simulation for the PTH via had only 1 via from top to bottom layer. The transmitter had feed forward equalization (FFE) enabled in the IBIS AMI model. The receiver had decision feedback equalization (DFE) enabled in the IBIS AMI model. No continuous time linear (CTLE) equalization was used. The optimized via parameters were not expected to be heavily influenced by the equalization, as having a better channel should result in better receiver eye parameters irrespective of the equalization scheme.

III. RESULTS

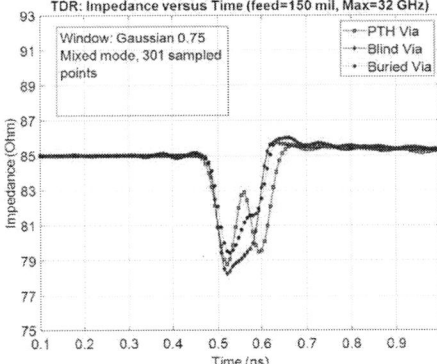

Fig. 3   Impedance profiles of vias with largest SNR for the 3 via types.

Fig. 3 shows the impedance profiles of the vias with the largest SNR for each of the 3 via types. (The differential impedance of the transmission lines was 85 Ω.) These are impedance profiles of the vias with the via parameters optimized such that the channel simulations would give the largest SNR. In general, the via gives optimal SNR when there are minimal impedance changes. This is because reflections will be minimized. Multiple reflections will cause distortions to received waveform. Thus, if the via is capacitive, an inductance 'spike' in the middle of the via is undesirable. We see that for the PTH via, the inductance 'spike' in the center is unavoidable as the via goes through the core where there are no closely spaced ground planes. The buried via goes through the core too but has no inductance spike as the pad capacitance and via pitch can compensate for the inductance due to the via barrel. It also helps that the buried via is less capacitive than the through-hole via. The blind via does not go through the core and is strictly capacitive for maximum SNR.

Table II shows the via parameters corresponding to the impedance profiles shown in Fig. 3.

TABLE II. VIA PARAMETER VALUES FOR VIAS WITH OPTIMIZED SNR

Via Type	Via Diameter (mils)	Pad Diameter (mils)	Antipad Diameter (mils)	Via Pitch (mils)	SNR (dB)
PTH	11	22	28	34	9.05
Blind Via	8	20	32	28	8.50
Buried Via	8	22	30	28	8.98

To illustrate how the optimized via impedance profile differs from a via impedance profile that is not optimized, we show the case of the buried via that was not optimized in Fig. 4, along with the case of the optimized via.

978-1-7281-6162-4/20 $31.00 © 2020 IEEE

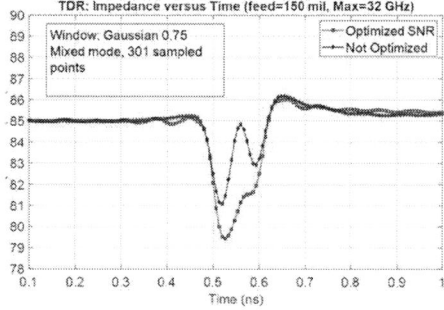

Fig. 4   Impedance profile of buried via that is optimized for SNR and one that has not been optimized. For the latter case, the via diameter is 11 mils, pad diameter is 18 mils, antipad diameter is 28 mils and via pitch is 33 mils. The SNR of the receiver signal was 8.48 dB.

Comparing the buried via impedance profile in Fig. 3 and Fig. 4, it can be seen that maximizing pad diameter and minimizing the via pitch helps to shape the impedance profile to minimize the impedance change for the via optimized for maximum SNR. Fig. 5 shows the receiver eye of the channel with the optimized buried via.

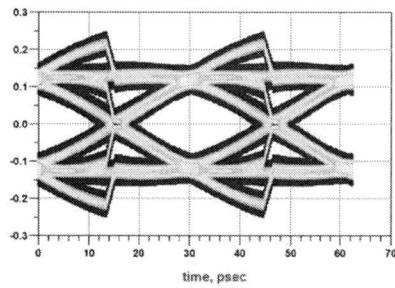

Fig. 5   Receiver eye for channel with optimized buried via. The y-axis has units of volts. The impedance profile of the via is shown in Fig. 3 and the via parameters are shown in table II.

IV. QUICK METHODOLOGY FOR VIA PARAMETER OPTIMIZATION

The results in Fig. 3 were generated using a full factorial parameter sweep and this may take days to accomplish. If such a time period is too prohibitive, an arbitrary via design can be simulated and a sub optimization method can be achieved using tables III and IV.

TABLE III.   QUICK VIA OPTIMIZATION IF VIA IS TOO CAPACITIVE

Action	Rel. Mag. of Impact	Constraints
Decrease via diameter	Highest	Via diameter ≥ 6 mils
Decrease pad diameter	High	Pad diameter – hole diameter ≥ 6 mils
Increase antipad diameter	High	Antipad diameter – via diameter ≥ 16 mils
Increase via pitch	Medium	Via pitch – via diameter ≥ 15 mils

Other methods to lower the capacitance include lowering the dielectric constant of the substrate, increasing dielectric thicknesses and removing non-functional pads.

TABLE IV.   QUICK VIA OPTIMIZATION IF VIA IS TOO INDUCTIVE

Action	Rel. Mag. of Impact	Constraints
Increase via diameter	Highest	Via diameter ≥ 6 mils
Increase pad diameter	High	Pad diameter – hole diameter ≥ 6 mils
Decrease antipad diameter	High	Antipad diameter – via diameter ≥ 16 mils
Decrease via pitch	Medium	Via pitch – via diameter ≥ 15 mils

Other methods for lowering inductance are increasing dielectric constant of the substrate, decreasing dielectric thickness, adding non-functional pads and adding ground vias.

A constraint not in tables III and IV is the maximum aspect ratio. For mechanically drilled vias, it is 10:1 for filled vias and 15:1 for unfilled vias. All the constraints may vary depending on the board manufacturer. The process in tables III and IV is repeated until the via impedance profile approximates the corresponding profile in Fig. 3.

V. CONCLUSION

We have presented a method to optimize differential vias for maximum SNR. We showed a full factorial method with the optimum impedance profiles of 3 types of vias: through hole vias, blind vias and buried vias. We also showed a quick method of optimizing the vias, depending on whether the vias are too capacitive or inductive. The goal for the quick optimization is to achieve an impedance profile that approximates the profiles shown in this paper for the full factorial optimization.

ACKNOWLEDGMENT

The authors would like to thank Benjamin P. Silva of the Intel Corporation for helping to proofread the paper.

REFERENCES

[1] A. Hardock, R. Rimolo-Donadio, S. Muller, Y. H. Kwark and C. Schuster, "Efficient, physics-based via modeling: return path, impedance and stub effect control," *IEEE EMC Magazine*, vol 3., pp. 76-84, 2014.

[2] S. Chen et al., "Via optimization for next generation speeds," *Proc. of IEEE Conf. on EPEPS*, Oct. 2017.

[3] C. Ye, X. Ye and E. L. Miralrio, "Via pattern design and optimization for differential signaling 25Gbs and above," *Proc. of IEEE Int. Symp. on Electromagn. Compat.*, July 2016.

[4] D. Seo, H. Lee, M. Park and W. Nah, "Enhancement of differential signal integrity by employing a novel face via structure," *IEEE Trans. Electromagn. Compat.*, vol. 60, pp. 26-33, Feb. 2018.

978-1-7281-6162-4/20 $31.00 © 2020 IEEE

Causal Transmission Line Geometry Optimization for Impedance Control in PCBs

Zachariah M. Peterson, *Member, IEEE*

Abstract—**PCB transmission lines are designed with broadband impedance matching using differential evolution while including causal dispersion and roughness. Results for striplines show that an optimized geometry can be found with differential evolution.**

Index Terms—**Board-level interconnects, methodologies and algorithms for modeling, simulation and optimization.**

I. INTRODUCTION

TRANSMISSION line behavior is commonplace in printed circuit boards (PCBs) operating with ultrahigh speed digital signals (~10 GHz bandwidth) and analog signals with mmWave frequencies. Dispersion is problematic at these frequencies, and minor impedance mismatches along an interconnect will produce reflections and distortion in various portions of the signal bandwidth. In addition, roughness of copper conductors on PCB laminates creates an additional source of dispersion and losses [1]. These problems of dispersion and roughness lead to signal distortion. A proper description of signal behavior requires enforcing causality in descriptions of dispersion in the substrate's dielectric function $\varepsilon(\omega)$ and copper roughness using Kramers-Kronig relations [2], which is required in equivalent circuit models describing interconnects, e.g., in PCB transmission lines for many high speed standards [2] (e.g., USB 4.0) and 100 Gb/s Ethernet in the IEEE P802.3bj Task Force proposal [3].

A transmission line is designed by adjusting its geometry in CAD software so that its impedance at a specific frequency takes a target value. This method does not consider broadband impedance matching, losses, velocity dispersion, and/or signal distortion. Solving such a design problem is difficult as any change in the line's per-unit-length (p.u.l.) capacitance, inductance, or resistance must be compensated by a change in one of the other quantities, each of which is related to the line's impedance. A simpler method is to use a existing model for the characteristic impedance and incorporate dispersion and roughness throughout the signal bandwidth analytically, which forms a complex nonlinear optimization problem.

As a class of heuristics, evolutionary algorithms can be used to solve nonlinear optimization problems in electromagnetics [4]. This class of algorithms uses iterative solution generation techniques to search a solution space for the global optimal solution. Multi-objective problems can be addressed by solving one objective while holding the others as constraints, and a Pareto surface can be generated through successive iteration.

Previous work has focused on determining circuit model values or dielectric properties using measured S-parameters [2] rather than optimizing the line's geometry. Therefore, a method for designing PCB transmission lines with dispersion and roughness to a target impedance is presented in this paper. Differential evolution was used to determine the geometry that minimizes deviations from a target impedance throughout the signal bandwidth. Analytical expressions for the characteristic impedance of a stripline and a causal model for $\varepsilon(\omega)$ were used to minimize deviations from a target impedance while including dispersion and roughness up to 20 GHz [2, 5]. The method can be applied to other geometries, used with other optimization methods, or reformulated as a multi-objective problem to balance many signal integrity metrics. Some additional objectives include signal distortion metrics, S-parameter values, and crosstalk in coupled transmission lines. The procedure could also be used in modern CAD applications for accurate transmission line design over a broad bandwidth.

II. THEORETICAL MODEL AND METHODS

A. Causality, Dispersion, and Copper Roughness

Building a causal model for a PCB interconnect requires enforcing causal representations for $\varepsilon(\omega)$, conductor roughness, and the line's transfer function [1, 2]. Typical causal dispersion models for PCB substrates are the Lorentzian or wideband Debye models [2]. The RLGC(*f*) model can be used to describe a line's characteristic impedance in terms of lumped circuit elements with dispersion as long as $\varepsilon(\omega) = \varepsilon_R(\omega) + i\varepsilon_I(\omega)$ has a causal representation. The following causal Lorentzian model for striplines on FR4 is valid up to 20 GHz [2, 5]:

$$\varepsilon_R(\omega) = \left(\varepsilon_\infty + \frac{\varepsilon_{s1}-\varepsilon_\infty}{1+(\omega\tau_1)^2} + \frac{\varepsilon_{s2}-\varepsilon_\infty}{1+(\omega\tau_2)^2}\right)\varepsilon_0$$
$$\varepsilon_I(\omega) = \omega\left(\frac{(\varepsilon_{s1}-\varepsilon_\infty)\tau_1}{1+(\omega\tau_1)^2} + \frac{(\varepsilon_{s2}-\varepsilon_\infty)\tau_2}{1+(\omega\tau_2)^2}\right)\varepsilon_0.$$

(1)

Typical values of ε_{s1}, ε_{s2}, ε_∞, τ_1, and τ_2 can be found in [5]. The conductor's RMS roughness H_{RMS} modifies the dielectric constant to $\varepsilon(\omega) \rightarrow \frac{\varepsilon(\omega)T}{T-2H_{RMS}}$ [1]. Dispersion can be placed in

---

Submitted on Jul. 26, 2020, Accepted on Aug. 20, 2020.

Z. M. Peterson is with Northwest Engineering Solutions LLC, Portland, OR 97204 USA (e-mail: zmp@nwengineeringllc.com).

978-1-7281-6162-4/20 $31.00 © 2020 IEEE

standard equations in the causal RLGC(f) model [2] and analytical equations for transmission line impedance determined using conformal mapping [6].

Rather than assume all circuit parameters in the causal RLGC(f) model are known *a priori*, the characteristic impedance $Z_0(\omega)$, DC resistance R_0, and causal $\varepsilon(\omega)$ are considered known in the method shown here. These are used to determine the geometry that most closely matches the line impedance with dispersion and roughness $Z(\omega)$ to a target interconnect impedance Z_T throughout the relevant bandwidth.

In the causal RLGC(f) model, the characteristic impedance $Z_0 = \sqrt{\frac{R+i\omega L}{G+i\omega C}}$ is defined in terms of lumped circuit elements [2]:

$$R(\omega) = R_0 + \sqrt{\omega}R_s \qquad L(\omega) = L_\infty + \omega^{-1/2}R_s$$
$$G(\omega) = \omega C(\omega)\varepsilon_0 \tan\delta(\omega) \qquad C(\omega) = K_g\varepsilon_0\varepsilon_R(\omega) \qquad (2)$$

where R_s is the p.u.l. skin-depth resistance, K_g is a geometry factor, and $\tan\delta(\omega) = \frac{-\varepsilon_I(\omega) - \sigma_{sub}}{\varepsilon_R(\omega)}$ is the frequency-dependent loss tangent [2, 5]. Note that the conductance G_0 of typical PCB substrates is $G_0 \sim 10^{-11}$ S/m, so G_0 and σ_{sub} are ignored in $\tan\delta$. L_∞ is the p.u.l. inductance as $\omega \to \infty$ and is taken as a constant as PCB laminates are non-magnetic [2]. These circuit parameters are related to the characteristic impedance as follows:

$$Z_0(\omega) = \sqrt{\frac{[R(\omega)+i\omega L(\omega)]}{\omega C(\omega)(i+\tan\delta(\omega))}}. \qquad (3)$$

where the propagation constant is written as $\gamma(\omega) = \sqrt{C(\omega)[\omega R(\omega) + i\omega^2 L(\omega)][i + \tan\delta(\omega)]}$.

Accounting for copper roughness on the PCB laminate is accomplished by using a causal roughness correction factor $K(\omega)$. This is incorporated into $R(\omega)$ by applying the linear transformation $R_s \to K(\omega)R_s$. Causal models for $K(\omega)$ (e.g., Hammerstad and Cannonball-Huray models) are found in [1].

B. Geometry Optimization with Differential Evolution

Dispersion and roughness were considered while designing to a target characteristic impedance using the L^2 norm of $Z(\omega) - Z_T$ in ω as the objective function:

$$\min\left[\int_{\omega_1}^{\omega_2}||Z(\omega)-Z_T||^2 d\omega\right]^{0.5},$$
$$\text{subject to: } J < W/H < K, \qquad (4)$$

This formulation is equivalent to minimizing the mean-squared error between $Z(\omega)$ and Z_T. Let W, T, and H be the width, thickness, and distance to the reference plane, respectively; these are the variables used to minimize $L^2(Z(\omega) - Z_T)$, and J and K are constants. The number of variables is reduced from 3 to 2 when W/H and T/H are used as optimization variables. T/H is normally fixed based on the PCB stackup, which reduces the number of variables to 1. This ensures the line is designed

within practical manufacturing constraints on laminate thickness and copper weight.

Algorithm 1 shows pseudocode for the differential evolution algorithm used to solve (4). Differential evolution proceeds by randomly generating a feasible initial solution; trial solutions are generated randomly, and the trial solution that moves the objective function value closer to an optimum in the solution space is accepted as the new solution. Constraint checking for inequalities is implemented using the ConstraintCheck and ConstraintMod functions [7]. GenerateNew mutates the current W/H value using the standard algorithm in [8].

Algorithm 1: Differential evolution pseudocode
Input: K, J, Z_T, N_{max}
00. **While** $N < N_{max}$
01.   **Generate initial solution** $Z(\omega)$ and W/H
02.   **If** $N < N_{max}$
03.   GenerateNew W'/H' and $Z'(\omega)$
04.   **If** $L^2(Z'(\omega) - Z_T) < L^2(Z(\omega) - Z_T)$
and ConstraintCheck = True
05. $Z(\omega) = Z'(\omega), N = 0$
06.   **Go to 02**
07.   **Else**
08.   ConstraintMod
09. $N \to N + 1$
10.   **Go to 03**
11.   **Else**
12.   **End**

Algorithm 1 was executed in Python 3.7 on a PC with a 1.8 GHz quad-core processor and 8 GB RAM. The time required to solve this problem primarily depends on the level of discretization used in the L^2 norm; 400 data points from 0.01 to 20 GHz were used in this method.

III. EXAMPLE FOR A LONG STRIPLINE

In this section, the method outlined in Section II is applied to the stripline shown in Figure 1. Striplines with and without dispersion and roughness will be compared here. The stripline without dispersion and roughness has $\varepsilon = 4.300 + i0.0681$, which is taken at 5 GHz. On the rough line, $H_{RMS} = 0.5$ μm was used with average particle size of 0.2 μm. Wadell's equation [6] for a stripline's characteristic impedance $Z_0(\omega)$ is shown in (5):

$$Z_0(\omega) = \frac{60}{\sqrt{\varepsilon_R(\omega)}}\ln\left(1+\left(\frac{8H}{\pi\widetilde{W}}\right)\left[\left(\frac{16H}{\pi\widetilde{W}}\right)+\sqrt{\left(\frac{16H}{\pi\widetilde{W}}\right)^2+6.27}\right]\right). \qquad (5)$$

In (5), $\widetilde{W} = W + \left(\frac{T}{\pi}\right)\left[1 - \frac{1}{2}\ln\left(\left(\frac{T}{4H+T}\right)^2 + \left(\frac{\pi(T/H)}{4(W/H+1.1T/H)}\right)^m\right)\right]$ and $m = 6H/(3H+T)$ [7].

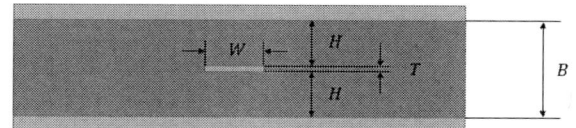

Fig. 1. Stripline transmission line on FR4 showing the definition of the geometric parameters in (5).

Using (3), (5), and the phase velocity $\left(\frac{\omega}{\gamma(\omega)}\right)$ with $R(\omega) = 0$, one can derive equations for $Z(\omega)$ and $\gamma(\omega)$ that include causal dispersion and copper roughness with linear transformations:

$$Z(\omega) = \sqrt{\frac{i\omega Z_0^2 \sqrt{\varepsilon_R(\omega)} + Z_0 c_0 \left(R_0 + (1+i)K(\omega)R_s\sqrt{\omega}\right)}{i\omega\sqrt{\varepsilon_R(\omega)}(1-i\tan\delta)}}, \quad (6a)$$

$$\gamma(\omega) = \sqrt{\frac{-\omega^2 Z_0 \varepsilon(\omega) + i\omega c_0 \sqrt{\varepsilon_R(\omega)}(1-i\tan\delta)\left(R_0 + K(\omega)(1+i)\sqrt{\omega}R_s\right)}{Z_0 c_0^2}}, \quad (6b)$$

where c_0 is the speed of light in vacuum. These equations are applicable to any transmission line if Z_0 and $\varepsilon(\omega)$ are known. In (6), $R_s \approx \sqrt{\frac{\mu_0}{8\sigma(T+W)^2}}$, and $K(\omega)$ obeys the Cannonball-Huray model [1]. Note that $K(\omega)$ could be defined using any other copper roughness model to match the line's morphology.

For the stripline on FR4 shown in Fig. 1, Table 1 shows the values for the parameters in (1) and the conductivity of copper.

TABLE I
CAUSAL TWO-TERM LORENTZIAN MODEL PARAMETERS

ε_{s1}	ε_{s2}	ε_∞	τ_1 (ps)	τ_2 (ps)	σ ($\Omega^{-1}\text{m}^{-1}$)
4.081	4.068	3.95	82.12	5.712	$5.81\cdot10^7$

IV. RESULTS

Fig. 2 shows the impedance spectra of two striplines (25 cm length) designed to a target impedance $Z_T = 50\ \Omega$. The load capacitance is $C_L = 1$ pF with parallel termination at its target impedance; these values collectively determine the impedance seen at the input to the load, which then determines the return loss (S_{11}). The optimized geometry is $W = 0.1774$ mm (6.983 mil) for the rough, dispersive stripline, and $W = 0.1741$ mm (6.856 mil) for the smooth, non-dispersive stripline. The line is placed on an 8-layer PCB stackup with $H = 0.224$ mm and $T = 17.5\ \mu$m (0.5 oz./sq. ft. copper weight).

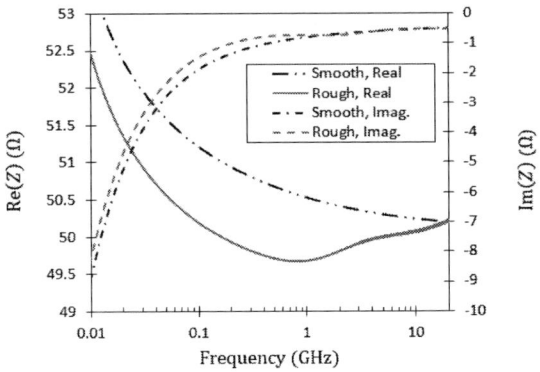

Fig. 2. Optimized impedance spectra for striplines with and without roughness and dispersion.

The proposed procedure gives a design that provides low deviation from the target impedance, despite roughness and dispersion. For electrically short interconnects, return loss is the quantity of concern, while insertion loss is more critical in long interconnects. Fig. 3 compares insertion and return losses on both lines. One can see that making the rough stripline slightly wider provides slightly lower losses than the smooth stripline.

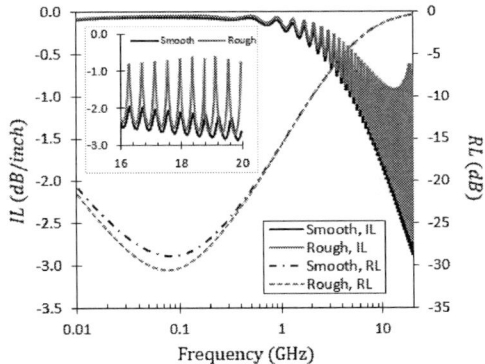

Fig. 3. Insertion loss and return loss in the terminated rough and smooth striplines (1 pF load capacitance). The inset shows a magnified view of insertion loss from 16 to 20 GHz.

Finally, the impulse function can be calculated using the line's causal transfer function for an arbitrary stimulus. Impulse responses for rough and smooth lines are compared in Fig. 4. The input pulse used here matches that used in [2] (see (46)).

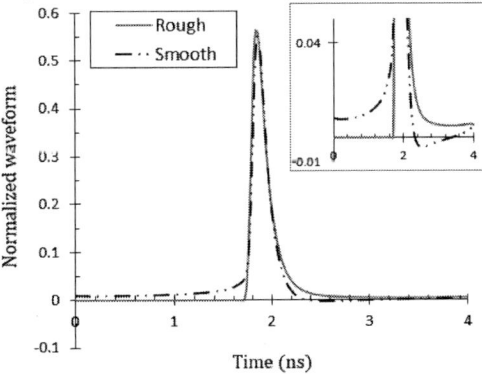

Fig. 4. Impulse responses for rough and smooth transmission lines. The inset shows a magnified view, revealing non-causal artifacts on the smooth transmission line.

By imposing design for manufacturing constraints, the designed stripline meets practical PCB fabrication requirements. Additional objectives could also be used in the L^2 norm in (4), such as wideband S-parameters or insertion loss.

REFERENCES

[1] V. Dmitriev-Zdorov, B. Simonovich, I. Kochikov, "A Causal Conductor Roughness Model and its Effect on Transmission Line Characteristics," *Signal Integrity Journal*, November 2018.

[2] J. Zhang, et al., "Causal *RLGC(f)* Models for Transmission Lines From Measured *S*-Parameters," *IEEE Trans. Electromagn. Compat.*, vol. 52, no. 1, pp. 189-198, Feb. 2010.

[3] A. Healey, C. Moore, R. Mellitz, A. Ran, L. Ben-Artsi, "Proposal for a causal transmission line model," *IEEE P802.3bj Task Force* (Mar 2014).

[4] Y. Rahmat-Samii and E. Michielssen, *Electromagnetic Optimization by Genetic Algorithms*. New York: Wiley, 1999, pp. 1–93.

[5] J. Zhang, et al., "Planar transmission line method for characterization of printed circuit board dielectrics," Progress In Electromagnetics Research, vol. 102, pp. 267-286, 2010.

[6] B. C. Wadell, *Transmission Line Design Handbook*. Norwood, MA: Artech House, 1991.

[7] J. Lampinen, "A constraint handling approach for the differential evolution algorithm," *Proceedings of the 2002 Congress on Evolutionary Computation. CEC'02 (Cat. No.02TH8600)*, Honolulu, HI, USA, 2002, pp. 1468-1473 vol.2, doi: 10.1109/CEC.2002.1004459.

[8] R. Storn, K. Price, "Differential Evolution - A Simple and Efficient Heuristic for Global Optimization over Continuous Spaces," *Journal of Global Optimization*, vol. 11, pp. 341-359, 1997.

Predictor-Corrector Algorithm with Embedded Dimension Reduction for Uncertainty Quantification of MWCNT On-Chip Interconnect Networks

Surila Guglani and Sourajeet Roy
Department of Electronics and Communication Engineering,
Indian Institute of Technology Roorkee, Roorkee, India
Email: sourajeet.roy@ece.iitr.ac.in

Abstract — **This paper presents a novel polynomial chaos (PC) approach for the fast uncertainty quantification of on-chip multi-walled carbon nanotube (MWCNT) interconnect networks. The proposed approach combines the benefits of predictor-corrector algorithms with that of dimension reduction strategies to provide two distinct levels of numerical efficiency when training the PC metamodels. As a result, this approach is even better scalable with respect to the number of problem dimensions than conventional predictor-corrector algorithms and state-of-the-art dimension reduction techniques.**

Keywords — **Interconnect networks, multi-walled carbon nanotubes (MWCNTs), predictor-corrector algorithm, polynomial chaos, uncertainty quantification.**

I. INTRODUCTION

As copper interconnects reach their performance limits at the 22 nm technology node, multi-walled carbon nanotube (MWCNT) interconnects are emerging as their potential replacement for on-chip applications [1], [2]. In particular, MWCNT provide longer mean free path of electrons, consequently smaller scattering resistance, greater current carrying capacity, and greater robustness to thermal breakdown than copper interconnects.

Unfortunately, despite the improved performance of MWCNT interconnect networks, they are highly susceptible to fabrication process variations. Therefore, circuit designers must quantify the impact of fabrication process variations on the performance of MWCNT networks at the earliest design stage. Typically, this is done using the brute-force Monte Carlo (MC) technique. However, due to the slow convergence of MC, its computational cost often becomes prohibitively large. This is especially true for MWCNT interconnect networks where the circuit size can be massively large. In such circumstances, the generalized polynomial chaos (PC) technique has been found to be usually more efficient [3], [4]. The PC technique models the impact of fabrication process variations on the MWCNT network responses as a linear expansion of orthonormal basis functions of random variables. This expansion is referred to as a surrogate model or metamodel. The coefficients of the metamodel are the unknowns of the system which are often evaluated using non-intrusive techniques.

A major drawback of the PC technique is that it suffers from the 'curse of dimensionality'. This means that the number of unknown coefficients in a PC metamodel scales in an exponential manner with respect to the number of random variables (or random dimensions) used to model the fabrication process variations [4], [5]. In turn, the number of deterministic SPICE simulations required to evaluate or 'train' these unknown PC coefficients too scale in a similar exponential manner, thereby making the conventional PC technique computationally intractable for large multidimensional problems.

Recently, multiple approaches have been reported to address the poor scalability of PC specifically for MWCNT interconnect networks [4], [5]. For example, the predictor-corrector algorithm improves the scalability by intelligently combining the numerical efficiency of a low-fidelity compact equivalent single conductor (ESC) model with the accuracy of a high-fidelity rigorous multiconductor circuit (MCC) model of the MWCNT network [5]. Another approach is the reduced dimension technique where the PC metamodel is constructed using only the most important random dimensions while other unimportant dimensions are ignored [4].

This work combines, for the very first time, the benefits of both the predictor-corrector algorithm of [5] and the dimension reduction technique of [4]. The proposed approach begins by constructing a predictor PC metamodel trained using the ESC model simulations of the MWCNT network. Next, in order to correct for the errors in the predictor metamodel arising from the approximations in the ESC model, a new corrector PC metamodel is constructed. Typically, the corrector metamodel is trained using MCC model simulations of the MWCNT network. Crucially, the corrector metamodel exploits the correlation between the results obtained from the predictor metamodel and the true MWCNT network responses in order to minimize the number of MCC training simulations that are required in the first place. In this paper, the required number of MCC training simulations is further decreased by ensuring that the corrector metamodel only includes those dimensions which have a large impact on the network responses as inferred from the already trained predictor metamodel. Thus, there exists two

978-1-7281-6162-4/20 $31.00 © 2020 IEEE

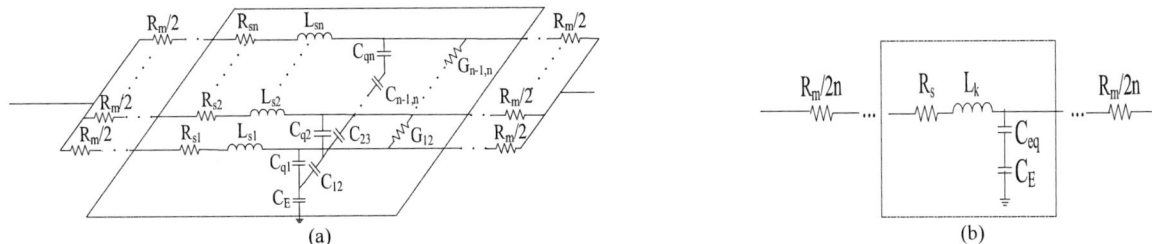

Fig. 1: MCC and ESC model representations of a MWCNT conductor. Note that the circuit elements inside the box represent an elementary part of the interconnect length and have to be repeatedly cascaded to model the entire conductor length. (a) Multiconductor circuit (MCC) model representation of a MWCNT conductor. (b) Equivalent single conductor (ESC) model representation of a MWCNT conductor.

levels of numerical efficiency when training the proposed corrector leading to relatively more time savings compared to both the conventional predictor-corrector algorithm [5] and existing reduced dimension technique [4]. Note that the proposed approach, in general, can be applied to all types of dimension reduction techniques (e.g., those based on ANOVA, active subspaces or sliced inverse regression). For sake of illustration, in this paper the ANOVA technique using Sobol's sensitivity indices is used for dimension reduction.

II. DEVELOPMENT OF PROPOSED APPROACH

Consider a general MWCNT interconnect network consisting of M coupled conductors with N_s number of shells in each conductor. In this case, the electrical behavior of each MWCNT conductor can be accurately represented using the rigorous multiconductor circuit (MCC) model as shown in Fig. 1(a). Let $\lambda = [\lambda_1, \lambda_2, …, \lambda_N]$ be the N-dimensional uncorrelated random variables located within the N-dimensional random space Ω used to represent the fabrication process variations.

A. Developing the Predictor Metamodel

The proposed approach begins by developing a predictor metamodel of the MWCNT network response $x(t,\lambda)$ as

$$x_p(t,\lambda) \approx \sum_{k=0}^{P} x_k^{(p)}(t)\phi_k(\lambda) \qquad (1)$$

where $x_k^{(p)}(t)$ is the k-th predictor coefficient and $\phi_k(\lambda)$ is the kth degree N-dimensional orthonormal polynomial. The number of terms in (1) is truncated to $P+1 = (N+m)!/(N!m!)$, m being the maximum degree of the expansion of (1). The coefficients of (1) are evaluated using SPICE simulations of the ESC model of the MWCNT network. The ESC model is based on the equipotential assumption that has been described in details in [6]. As per the equipotential assumption, all the shells of a conductor will collapse into a single shell as shown in Fig. 1(b). Thus, the ESC model is highly compact when compared to the MCC model and can be solved in SPICE at relatively smaller CPU time costs. In effect, the coefficients of the predictor are trained at very small CPU time costs. However, in reality, the equipotential assumption is an approximation to the MCC model. This means that the basic principle on which the ESC

model is based is not perfectly true. Thus, the ESC model will introduce errors into the coefficients of the predictor.

B. Using Predictor to Identify Important Dimensions

The Sobol's sensitivity index of any general i-th dimension can be evaluated as [4]

$$S_i(t) = \frac{\sigma_i^2}{\sigma^2} = \frac{\sigma_i^2}{\sum_{k=1}^{P}(x_k^{(p)}(t))^2} \qquad (2)$$

where σ_i^2 is the sum of square of those coefficients whose corresponding basis function has the degree of the univariate polynomial of λ_i greater than zero. The Sobol's sensitivity indices of (2) being time varying quantities, they can be integrated over the entire time window of simulation $[0 – T_{max}]$ to reduce them to scalar quantities. These scalar indices reflect the impact of each dimension on the variance of the network response [4]. Thus, those dimensions whose scalar indices is more than a specific threshold ε, they are considered to be the important dimensions while all other dimensions are considered to be unimportant and can be ignored. In this way, λ can be reduced to a vector of N_r important dimensions ξ where $N_r < N$.

C. Developing the Corrector Metamodel

To correct the errors in the predictor metamodel, a corrector PC metamodel is constructed. This corrector metamodel includes only the N_r important dimensions in ξ as

$$f_c(t,\xi) = \sum_{k=0}^{Q} x_k^{(c)}(t)\psi_k(\xi) \approx x(t,\lambda) - x_p(t,\lambda) \qquad (3)$$

where $x_k^{(c)}(t)$ is the k-th corrector coefficient and $\psi_k(\xi)$ is the kth N_r-dimensional orthonormal polynomial. It is known from [5] that the covariance between the predictor output and the true response of the MWCNT network is usually large, thus making the variance of the corrector metamodel small. This means that only a sparse set of $Q+1$ terms in the corrector is sufficient to capture the error term on the right hand side of (3) where $Q+1$ $<< P+1$. This is the main reason why very few MCC model simulations are sufficient to train the corrector coefficients. In this work, however, the value of $Q+1$ is further decreased by including only the most important N_r dimensions in the

978-1-7281-6162-4/20 $31.00 © 2020 IEEE

TABLE I
UNCERTAIN NETWORK PARAMETERS WITH NORMAL DISTRIBUTION

No	Uncertain Network Parameters	Mean	% SD
1	D_{in1} (Inner diameter of conductor 1)	2.28	
2	D_{in2} (Inner diameter of conductor 2)	nm	
3	d_1 (Inter-shell distance of conductor 1)	0.34	
4	d_3 (Inter-shell distance of conductor 3)	nm	
5	Cs_1 (Driver capacitance of conductor 1)	0.14	15 %
6	Cs_2 (Driver capacitance of conductor 2)	fF	
7	C_{L1} (Load capacitance of conductor 1)	0.049	
8	C_{L1} (Load capacitance of conductor 1)	fF	
9	w (Separation between conductors)	22 nm	
10	H (Height of dielectric)	50 nm	

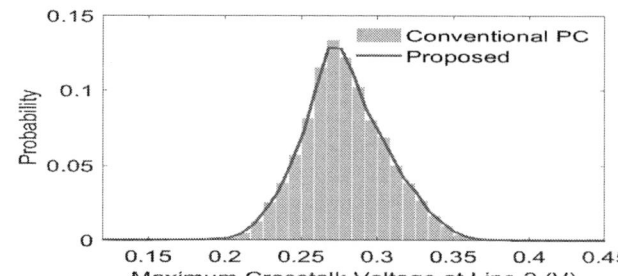

Fig. 2: Histogram depicting the PDF of the peak crosstalk for line 2 obtained using the proposed approach and conventional PC.

TABLE II
CPU TIME FOR NUMERICAL EXAMPLE

	CPU Time (hours)	Speedup w.r.t. Conventional PC
Conventional PC (2002 MCC)	3.03	-
Dimension Reduction PC [4] (654 MCC)	0.99	3.06
Predictor-Corrector Algorithm [5] (2002 ESC + 352 MCC)	0.56	5.41
Proposed (2002 ESC+184 MCC)	0.31	9.83

corrector metamodel instead of the full N dimensions as done before in [5]. Therefore, the dimension reduction technique accelerates the conventional predictor-corrector algorithm of [5]. In fact, the use of the predictor metamodel in (3) ensures that the proposed methodology is also faster than the existing dimension reduction technique of [4].

D. Recovering the PC metamodel

After the predictor and the reduced dimensional corrector are trained, the original PC metamodel of the network response is recovered as

$$
\begin{aligned}
x(t,\boldsymbol{\lambda}) &\approx x_p(t,\boldsymbol{\lambda}) + f_c(t,\boldsymbol{\lambda}) \\
&= \sum_{k=0}^{P} x_k^{(p)}(t)\varphi_k(\boldsymbol{\lambda}) + \sum_{k=0}^{Q} x_k^{(c)}(t)\psi_k(\boldsymbol{\lambda})
\end{aligned}
\tag{4}
$$

The additional terms of the corrector metamodel compensate for the errors in the predictor metamodel arising from the equipotential assumption of the ESC model.

III. NUMERICAL RESULT AND DISCUSSION

In this example, a two conductor ($M = 2$) MWCNT network is considered. The number of shells in each conductor are $N_s = 30$. The uncertainty in the network is represented using the $N = 10$ dimensions described in Table I. Line 1 of the network is excited by a voltage source with a saturated ramp waveform of rise time $T_r = 0.1$ ps and an amplitude of 1 V. Line 2 is victim line. Four approaches are used for the uncertainty quantification of this network – the proposed approach, the reduced dimensional PC scheme of [4], the conventional predictor-corrector algorithm of [5], and the full-blown PC metamodel trained using SPICE MCC model simulations only. All metamodels use a maximum degree of $m = 4$. For this example, all those dimensions whose scalar sensitivity indices is below a threshold of $\varepsilon = 0.05$ is considered to be unimportant and their effects ignored. Thus, for this example $\xi = [Din_1, d_1, d_2, C_{S1}, C_{L1}, w, H]$. The corrector metamodel is constructed in the random space described by these 7 dimensions with the hyperbolic truncation factor $u = 0.7$. For this example, the time cost of an ESC simulation is roughly 53 ms and that of an MCC simulation is 5.45 seconds.

In this example, the accuracy of the proposed approach is validated by comparing the probability density function of the crosstalk response in Fig. 2. Finally, the CPU training cost for all four aforementioned PC metamodels is recorded and listed in Table II. The results of Table II clearly illustrate the efficiency of the proposed approach.

IV. CONCLUSION

In this paper, a novel PC approach for the fast UQ of MWCNT networks is developed. This approach is numerically far more efficient that both conventional predictor-corrector algorithms as well as existing dimension reduction techniques.

REFERENCES

[1] A. Naeemi and J. D. Meindl, "Compact physical models for multiwall carbon-nanotube interconnects," *IEEE Electronic Device Letters*, vol. 27, no. 5, pp. 338-340, May 2006

[2] H. Li, C. Xu, N. Srivastava, and K. Banerjee, "Carbon nanomaterials for next-generation interconnects and passives: Physics, status, and prospects," *IEEE Trans. Electron Devices*, vol. 56, no. 9, pp. 1799–1821, Sep. 2009

[3] I. S. Stievano, P. Manfredi, and F. G. Canavero, "Carbon nanotube interconnects: process variation via polynomial chaos," *IEEE Trans. Electromagn. Compatibility*, vol. 54, no. 1, pp. 140–148, Feb. 2012

[4] A. K. Prasad and S. Roy, "Accurate reduced dimensional polynomial chaos for efficient uncertainty quantification of microwave/RF networks," *IEEE Trans. Microwave Theory and Techn.*, vol. 65, no. 10, pp. 3697-3708, Oct. 2017

[5] Y. Li, S. Bhatnagar, A. Merkley, D. Weber, and S. Roy, "A predictor-corrector algorithm for fast polynomial chaos-based uncertainty quantification of multi-walled carbon nanotube interconnects" *IEEE Trans. Comp., Packag. and Manuf. Technol.*, vol. 9, no. 10, pp. 1963-1975, Oct. 2019

[6] M. S. Sarto and A. Tamburrano, "Single-conductor transmission line model of multiwall carbon nanotubes," *IEEE Trans. Nanotechnol.*, vol. 9, no. 1, pp. 82–92, Jan. 2010

SI Model to Hardware Correlation on a 44 Gb/s HLGA Socket Connector

Pavel Roy Paladhi[1*], Yanyan Zhang[1], Junyan Tang[1], Daniel Rodriguez[1], Jose Hejase[2#], Sungjun Chun[1], Wiren Becker[1], Brian Beaman[1], and Daniel Dreps[1]

1. *IBM Corporation*, Austin, TX 78758
2. *Nvidia Corporation*, Austin, TX 78717
 # Contributions to this work completed prior to April 2020 while with IBM Corporation
* Pavel.Roy.Paladhi@ibm.com

Abstract— **A comprehensive signal integrity model to hardware correlation study on an improved, 44 Gb/s capable, hybrid land grid array (HLGA) socket connector design is presented. The connector only design SI performance is shown through 3D electromagnetic (EM) modelling. Details of the test vehicle designed to carry out the connector hardware evaluation are shown. Simulation modelling and experimental results of the test vehicle inclusive of the connector are presented and compared. The systematic testing ensures that the new component performs up to its required specifications which ensures successful operation at the system level.**

I. INTRODUCTION

Over the past few decades, high speed buses in server computers have evolved drastically with increasing speeds of operation and larger data bandwidths. The trend of achieving higher speeds and bandwidth continues to drive innovation in design and technology in its support. With the emergence of new technologies and component designs, carrying out validation testing of their performance remains an essential part of their realization [1].

To maintain the high speed data signaling integrity, the channel carrying the data between the transmitter and receiver has to achieve acceptable signal integrity (SI) requirements in the form of: insertion loss (IL), return loss (RL), impedance profile and crosstalk isolation properties [2]. The high speed channel topologies can vary largely and depend on the architecture of the high speed bus or the system requirements/implementation. They can have any combination of interconnects such as substrate packages, connectors, cables, single or multiple printed circuit boards (PCB), etc. Different types of connectors are used in channel topologies dependent on the need. Some can connect cables to boards, while others can connect boards to substrate packages etc. Connectors can have deleterious effects on the channel performance if not designed or selected properly. At higher speeds, the challenges increase proportionately.

An important type of connectors is the Hybrid Land Grid Array (HLGA) socket connector which connects a chip substrate package to a PCB, commonly: CPU to a server computer motherboard. HLGA connectors are soldered to the motherboard on the bottom side while the package is connected through a spring-loaded assembly on the top side of the HLGA connector. The advantage of the detachable package is that a defective CPU can be removed without changing the whole motherboard as would be the case in a direct attach BGA soldered package to a board. The industry has seen a considerable drive towards improving existing HLGA connectors to function at higher speeds. As an example, in an earlier work, some of the authors in this work had presented an improved HLGA connector design which can support data rates up to 50 Gb/s [3].

Before system integrating new connectors designed to achieve required SI metrics, the component must be validated through design modelling and lab measurement correlation. To assist in doing this, test vehicles are designed to measure performance of the component and validate the design. In this work: (1) connector only SI metrics are shown for a 44 Gb/s capable new connector from 3D EM modelling, (2) details on the design and development of a test vehicle to validate the HLGA connector design are presented and (3) measurements of the HLGA connector as part of this test vehicle are presented along with their correlation with simulation modelling data and discussion validating the design success.

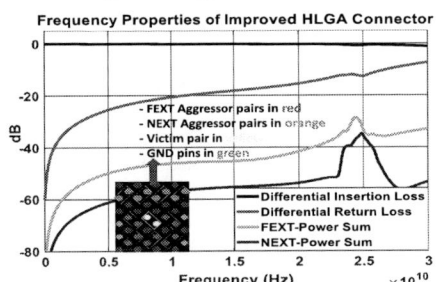

Figure 1. **Improved HLGA Frequency properties from 3D modelling.**

II. IMPROVED HLGA DESIGN

The frequency response of the improved connector design only model is shown in Fig. 1, assuming a 42.5 Ohm port reference impedance. It is worthy to note that this connector was designed for 85 Ohm differential impedance applications but can support up to 100 Ohm impedance applications. Up to 22 GHz, the differential insertion loss (DIL) is less than 0.4 dB and the differential return loss (DRL) is less than 14 dB. The differential near end crosstalk (NEXT) and far end crosstalk (FEXT) power sums, shown for the signal/ground pin distribution in the inset of Fig. 1, are respectively less than -50

978-1-7281-6162-4/20 $31.00 © 2020 IEEE

dB and -38 dB. The frequency responses show that the connector can operate with good signal integrity up to a frequency of 22 GHz, i.e. 44 Gb/s data rate. Based on this connector design, a test vehicle to carry out measurement to hardware correlation validation of the connector was designed.

III. HLGA TEST VEHICLE (TV) PLANNING

An HLGA connector would prevent a direct and isolated measurement of the connector itself on a traditional measurement setup. The TV should be designed in a way so as to be able to measure the properties of the HLGA connector accurately and with minimal interference due to the added traces and connections as part of the TV setup. Furthermore, test points should be efficiently chosen such that all required metrics can be extracted from a minimal set of measurements. With the SI performance evaluation as the primary consideration, the TV is designed to measure IL, RL, impedance and crosstalk metrics of the HLGA connector. A brief description of the TV follows.

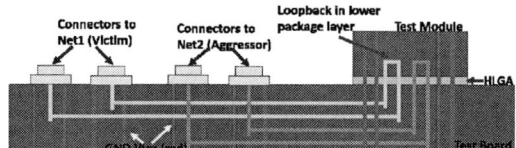

Figure 2. Schematic: HLGA TV showing loopback nets.

The HLGA TV consists of: a PCB board on which the HLGA connector is soldered, the substrate package module which can be mounted on the HLGA connector and the heat sink assembly which provides the mounting capability of the module to the HLGA connector similar to that in a full system.

To measure the frequency domain properties of the HLGA connector, a set of differential-ended loopback nets were designed to be measured with a VNA. An illustrative representation of these test nets in the TV is shown in Fig. 2. The nets originate on the board, outside the HLGA footprint area at SMK connectors where the VNA calibration cable can be directly connected. The general path of the nets is as follows: they breakout from the on-board SMK connectors in the uppermost internal signal layer on the PCB as differential striplines, and then pass through the HLGA connector to the package above. In the package, there is a short trace length in the lower most build up layer which turns back into the HLGA and out into the PCB. The PCB and package wiring layers were chosen to minimize the effect from their respective transition vias from/to the HLGA connector itself. A total of six differential loopback nets were put in the TV. Furthermore, the signal pin locations in the HLGA connector are crucial to get meaningful test data. They were placed in a way to capture representative crosstalk effects which would be observed in an actual system design high speed differential pin distribution. Six differential loopback nets were included in the TV: one victim, two near-end crosstalk aggressors and 3 far-end crosstalk aggressors to evaluate the crosstalk isolation in the HLGA. Fig. 3 shows the top view of the TV. The SMK connectors for the victim net have been circled in the figure. The package (shown in the top right inset) and the heat sink assembly have been removed to show the HLGA connector. The internal wiring for the loopback nets in the pin area of the HLGA connector is shown in the lower inset figure. More details on them are given in the next section. Through the SMK connectors, IL measurement of the six differential pairs and any crosstalk measurement between them can be performed.

Figure 3. Top view of HLGA TV.

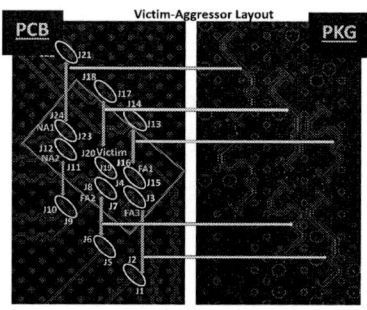

Figure 4: Net layout and grouping for model subdivisions: PCB-HLGA-Package pin transition area subdivided into three sections.

To do a correlation with measurement data, all six loopback nets were needed to be modeled with a 3D EM solver. In order to avoid excessive time and computational complexities, the full structure of the TV was modelled by parts in Ansys HFSS. Then the resulting s-parameters were cascaded to obtain the full channel response. For modelling purposes, the PCB wiring was divided into two zones: open area wiring outside of the HLGA shadow and pin area wiring under the HLGA connector. The other part of the model is the crucial area of the HLGA assembly. In particular, the splitting of the HLGA assembly needed to be done with care to ensure the splitting of the model does not compromise the crosstalk phenomenon in the split models and maintains field continuity. This HLGA assembly contains three main sections: the PCB via, the HLGA connector and the package wiring. It is worthy to note that the HLGA connector model which was used in the modelling of the test nets is the same one for which data was presented for in Section II. This area to be split is shown in Fig. 4 below. The left picture shows the pin area wiring in the PCB section. It shows the transition vias for all the six differential nets. There are six via pairs for going up through the HLGA and six others for coming down (to complete the loopback nets' nature). The upward and downward via pairs for each net have been identified and their connection traces in the package are shown on the right. The rectangular area identified on the left is the key section where the crosstalk phenomena occur. This section is modelled as a single unit to preserve the crosstalk effects. This captures the crosstalk effect from two aggressors separated from the victim with a double-ground isolation (as designed for NEXT in the system) and from three other aggressors with single-ground

978-1-7281-6162-4/20 $31.00 © 2020 IEEE

separation (as designed for FEXT in the system). Within the box, the NEXT aggressors are marked as NA1 and NA2. The FEXT aggressors are marked as FA1, FA2 and FA3. Outside the rectangular box, the 3 via pairs above and 3 pairs below are modelled as two other separate sections. Beside each via in the left figure, the 'J#' number identifies the corresponding connector placed on the PCB, e.g. J17-J18 and J19-J20 are the i/p and o/p connector pairs respectively for the victim net. Measurements were taken in the TV to measure the cross talk between the victim and all the aggressors.

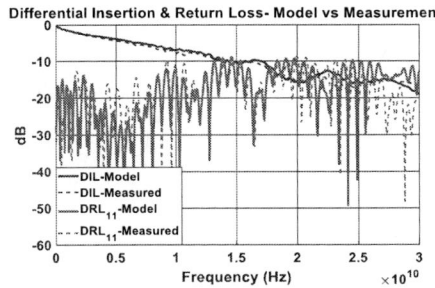

Figure 5. Victim DIL and DRL comparison between measured and modelled data.

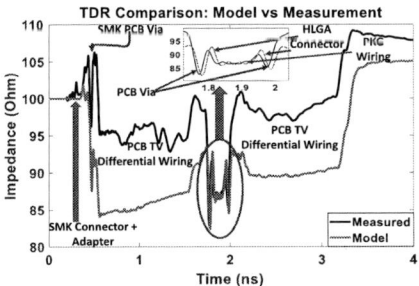

Figure 6. Impedance profile comparison between measured and modeled data of the victim net. TDR is computed from the s-parameters.

IV. MODEL TO MEASUREMENT CORRELATION

S-parameter measurements on the TV were made using Anritsu's VectorStar VNA. DIL, DRL, impedance profile (calculated from s-parameter), FEXT and NEXT measurements were carried and compared to modelling data. The VNA calibration plane is at the top of the 50 Ohm SMK connectors. The models include the full path for each of the nets starting and ending at the SMK connector tips. Hence, de-embedding of the traces and connectors in the PCB was not needed, in part helping minimize de-embedding inaccuracies, though de-embedding traces existed on the TV PCB. Since the SMK connectors are 50 Ohm and the purpose of this section is solely for comparison, the data is presented for 50 Ohm port reference impedance as opposed to converting to 42.5 Ohm port reference impedance. Fig. 5 shows a comparison of the DIL and DRL in the TV with modelled data for the victim net. A very good correlation is observed for DIL. For DRL, the correlation is good with the peaks and dips matching up well. In Fig. 6, an impedance comparison of the victim channel is shown through TDR plots calculated from the measured and modelled s-parameters using a 25 ps rise time. The different regions of the TV are identified in the figure. The middle section with the two HLGA connector sections connected by package wiring shows very good

correlation. The deviation observed in the PCB wiring region can be alluded to fabrication tolerances resulting in deviations in trace dimensions from design values intended to produce 85 ohms differential impedance. Figs. 7 and 8 respectively show FEXT and NEXT comparisons. For FEXT, the nearest and farthest aggressors' responses are compared and clearly can be distinguished by the difference in their crosstalk energy levels. In both FEXT and NEXT plots, the correlation is good. The deviations in the dB though observable are considerably low taking into consideration the absolute dB levels.

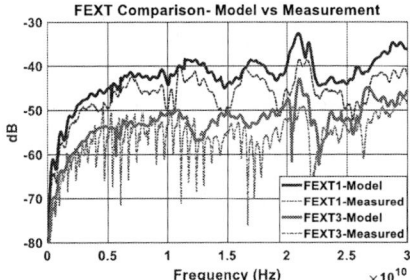

Figure 7. FEXT comparison between measured and modelled data.

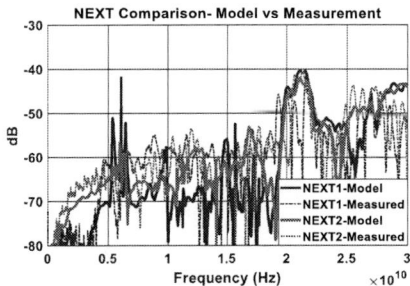

Figure 8. NEXT comparison between measured and modelled data.

V. CONCLUSION

In this paper, a model to hardware correlation study, including test vehicle design and implementation, to validate the performance of a new and improved 44Gb/s capable HLGA connector design. Modelling studies and measurements showed good correlation for the tested SI parameters, building confidence on the validity of the connector design, for which its good performance in isolation from simulation was shown in the Section II. The comparisons clearly show that the new connector design manufactured are consistent with the good performance in simulation. With the need to perform at higher speeds from one generation to another, studies focusing on validating new and/or improved designs/technologies prior to system implementation are becoming more important.

REFERENCES

[1] K.J. Han et al., "Modeling On-Board Via Stubs and Traces in High-Speed Channels for Achieving Higher Data Bandwidth", *IEEE Transactions on Components, Packaging and Manufacturing Technology*, vol. 4, no. 2, pp. 268-278, Feb. 2014.

[2] E. Bogatin, Signal and Power Integrity Simplified, 2nd ed.. New York, USA: Prentice Hall, 2004.

[3] J.A. Hejase et al., "A Hybrid Land Grid Array Socket Connector Design for Achieving Higher Signalling Data Rates", 2017 IEEE 26th Conference on Electrical Performance of Electronic Packaging and Systems (EPEPS), San Jose, CA, 2017, pp. 1-3.

ANN Performance for the Prediction of High-Speed Digital Interconnects over Multiple PCBs

Katharina Scharff, Christian Morten Schierholz, Cheng Yang, Christian Schuster

Institute of Electromagnetic Theory, Hamburg University of Technology, Hamburg, Germany

Email: katharina.scharff@tuhh.de

Abstract—In this paper the performance and the accuracy of artificial neural networks for the prediction of high-speed digital interconnects up to 100 GHz on printed circuit boards are analyzed and evaluated. The prediction accuracy is evaluated both for scattering parameters in frequency domain as well as weighted power sums thereof. The interconnects considered all contain a backplane connected to a daughtercard, showing two via arrays each. Several parameter variations of the basic setup lead to a wide range of possible transmission and crosstalk parameters. Training data sets are obtained using physics-based via modeling up to 100 GHz. Approximately 7000 data sets were made available in total for this study. Neural networks are able to predict the overall link behavior.

Index Terms—Signal Integrity, Machine Learning, High-Speed Links, Physics-Based Modeling

I. INTRODUCTION

With increasing data rates of modern high-speed links come new requirements for the simulation environments that are used for their design. Even though hardware capabilities have increased, it is often not feasible to use time-consuming full-wave simulation methods. In this context, Machine learning (ML) can become an important element in the link design toolbox.

Recently, several works have studied the applicability of ML. It has been shown how to predict properties of the passive link such as via impedances ([1]) or stripline parameters ([2], [3]). In [4] the eye opening is predicted. In [5] eye opening predictions for a SATA 3.0 example up to 9 GHz are shown. Another example of eye opening prediction is given in [6].

From these previous works it is clear that artificial neural networks (ANNs) can indeed be used for tasks in signal integrity engineering. Still, ANNs are not yet able to replace physics-based simulation tools, such as full-wave solvers. The aim of this contribution is to explore the applicability of ANNs for a more complicated high-speed interconnect that shows full-wave effects and therefore high parameter sensitivities. An example of such a link is shown in Fig. 1. It consists of a backplane that is connected to a daughtercard via a connector. In this work different variations of this link are simulated up to 100 GHz with an efficient modeling technique called the physics-based via modeling (PBV) [7]. Based on the PBV simulations, an ANN is trained to predict the S-parameters. The frequency-dependent S-parameter analysis of the SI behavior is extended by an analysis of frequency domain figures of merit (FOMs), the weighted power sums [8].

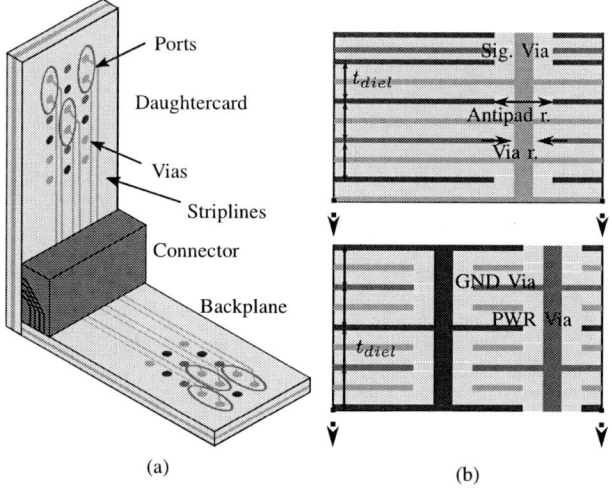

Fig. 1. Complete model consisting of backplane and daughtercard. Only a section of the via arrays is shown. The connector is modeled as a thru, connecting backplane and daughtercard. (a) Model. (b) Upper part of the stackup of backplane and daughtercard. The lower part is symmetrical to the upper half.

II. LINK MODEL AND PHYSICS-BASED SIMULATION

The complete link model consists of a backplane and an attached daughtercard (see Fig. 1). The connector is not included in the modeling process, instead both parts are simulated separately and concatenated directly based on network parameters. The vertical stack-ups are shown in Figs. 1 (b) and (c), respectively. The individual board layouts are shown in Figs. 2 (a) and 2 (b) respectively. The differential stripline pairs are routed to avoid skew that could impair the differential transmission. On both boards perfectly matched layer boundaries are assumed.

The boards are simulated with the physics-based via modeling [7], and a 2D trace model. For similar links a good correlation with full-wave simulations up to 100 GHz has been observed [9].

Both boards have variable parameters, whose ranges are given in Table I. To match their footprints, both boards have the same pitch inside the via arrays. Via and antipad radius are the same for backplane and daughtercard. This yields a total of 13 variable parameters. The transmission line width and the separation inside a differential pair are chosen to achieve a differential impedance of 100Ω. Parameter combinations that are impossible, such as a via radius larger than the antipad

978-1-7281-6162-4/20 $31.00 © 2020 IEEE

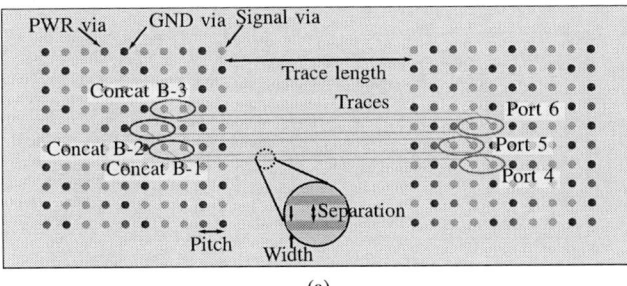

(a)

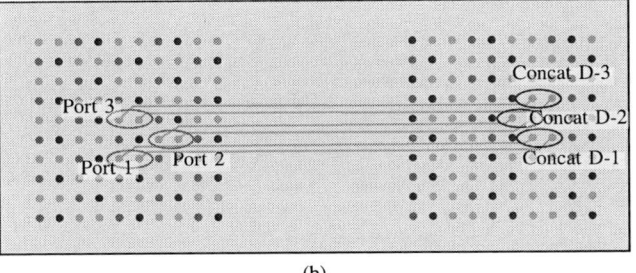

(b)

Fig. 2. Top view of the backplane and daughtercard boards. Differential ports are indicated with a circle. The blue ports "Concat B-1" to "B-3" are concatenated with the blue ports "Concat D-1" to "D-3" of the daughtercard. The port numbers of the red ports refer to the port labels of the concatenated link. The dimensions are not drawn to scale. (a) Backplane. (b) Daughtercard.

Table I
RANGE OF GEOMETRICAL PARAMETERS OF THE INTERCONNECT MODEL.

Parameter	Range Backpl.	Range Daughterc.
pitch	[40,80] mil	*same as backplane*
via r.	[3,9] mil	[3,9] mil
antipad r.	[6,18] mil	[6,18] mil
ε_r	[3.6,4.4]	[3.6,4.4]
tan δ	[0.0,0.02]	[0.0,0.02]
diel. h.	[6,14] mil	[6,14] mil
trace len.	[500,5000] mil	[500,2000] mil

radius, are excluded. The exception is the stripline separation. If the calculated separation is not achievable due to the via pitch, the separation is set to the maximum possible value.

The S-parameters under consideration are the transmission and the far-end crosstalk (FEXT). The transmission is taken between ports 5 and 2 (S_{d5d2}) and the FEXT is taken between ports 5 and 1 (S_{d5d1}) and ports 5 and 3 (S_{d5d3}).

Besides S-parameters, the links are also evaluated based on the weighted power sums, which are FOMs in frequency domain. They can represent transmission, crosstalk, and also a weighted signal to crosstalk ratio. Details can be found in [8].

III. DESIGN OF ANN MODELS

The design of the ANN can be found manually with a grid search or with an optimization algorithm. The ANNs found with the Bayesian Optimization (BO) do not perform significantly better than the ANNs from the manual search.

Both the S-parameters as well as the weighted power sums could be predicted with an ANN model. In case of the S-parameters, the target is the magnitude vector which

Table II
ANN CONFIGURATIONS FOR DIFFERENT PREDICTION TARGETS.

Parameter	Trans.	FEXT	WPSXT	WPT	WSXTR
N. of hidden layers	4	4	3	3	4
N. of neurons per layer	2000	2000	2000	2000	2000
Activation function	relu	relu	relu	relu	relu
Learning rate	0.001	0.0005	0.001	0.001	0.005
Train RSME	0.0211	0.0040	2.5962	141.6727	6.06405
Test RSME	0.0258	0.0043	10.3185	194.2862	17.4680
Training time (s)	202.75	165.16	263.30	511.37	1094.70

is sampled at 200 frequency steps. Only the magnitude is predicted because this part is required for the calculation of the weighted power sums. To predict the phase, an extended ANN would be required. In case of the FOM, the target is a 6 element vector, where the sampling points correspond to 6 different bitrates.

The samples are simulated with the PBV. Samples that are not physical realizable are excluded. In total 7030 samples are generated. Each sample is a concatenation of backplane and daughtercard model. The simulation of one sample requires approximately $8\,m\,56\,s$. The total simulation time for all samples (without parallelization) would be approximately 44 days. The parameters are chosen within the ranges specified in Table I. The samples are chosen based on latin hypercube sampling (LHS) to achieve an even distribution of parameters within the parameter space. The data split into an 80% training batch and a 20% test batch. Only the training samples are seen by the ANN model during the training. The final model quality is determined by the test batch.

The ANNs are implemented with the open source python package *scikit-learn* which uses the Adam solver [10]. The hyperparameters are tuned by manually testing different parameter combinations (grid search). This requires a lot of manual tuning but if similar problems have been studied before, the knowledge of the solutions can be applied. The final ANN design for all five predictions can be found in Table II.

IV. EVALUATION OF ANN PERFORMANCE

This section analyzes the errors of the predictions, both for S-parameter and FOM prediction. The ANN topologies are given in Table II.

Fig. 3 shows the S-parameters that are obtained from the PBV, and the predicted S-parameters from the ANN model. Figs. 3 (a) and (b) show the predictions with the lowest mean squared error (MSE) for transmission and FEXT. The predicted results show large variations and are less smooth than the original data. A few datasets show an unexpected S-parameter behavior that also lead to a very poor prediction performance. This could be due to unfavorable link parameter combinations and is still under investigation. Furthermore, the predicted S-parameters are not guaranteed to be passive.

ANNs can also predict the FOMs. Their configurations for the prediction of weighted power sum of crosstalk (WPSXT),

978-1-7281-6162-4/20 $31.00 © 2020 IEEE

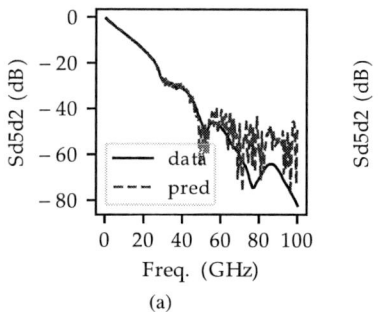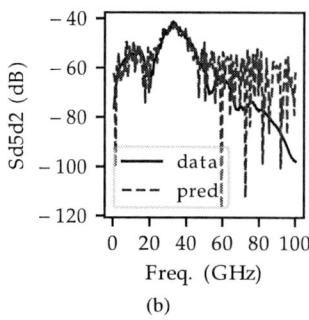

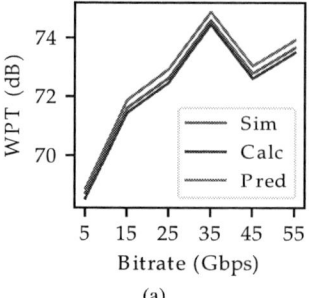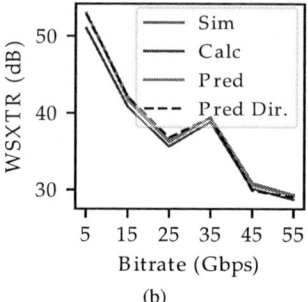

Fig. 3. Comparison of Transmission and FEXT with best prediction results. (a) Transmission. (b) FEXT.

Fig. 5. Comparison of FOMs generated by different means. The simulated parameters are calculated from S-parameters simulated with the PBV ("Sim"). "Calc." indicates that the FOM is calculated from predicted S-parameters. "Pred." indicates that FOMs are predicted from the link geometry with the ANN. In case of the WSXTR the FOM can be predicted directly ("Dir. Pred.") or with the WPT and WPSXT ("Pred."). (a) WPT. (c) WSXTR.

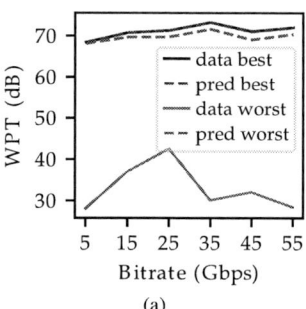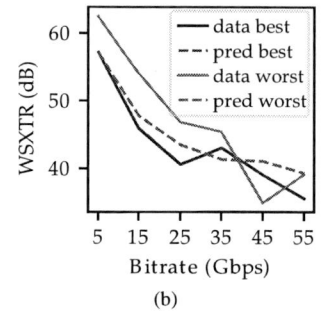

Fig. 4. Prediction results for FOMs. (a) WPT best and worst prediction. (b) WSXTR best and worst prediction.

weighted power sum of transmission (WPT), and weighted signal to crosstalk ratio (WSXTR) are given in Table II. Fig. 4 shows the samples with the best prediction results for WPT and WSXTR. The best predictions match the simulated data almost perfectly. The worst predictions (not shown here) show very large deviations. For the prediction of the WSXTR the worst prediction still represents the main behavior of the data. As the WSXTR is the division of WPT and WPSXT, it is possible that this division cancels out some differences.

Fig. 5 shows the FOMs generated by different means for one link variation. The references are the FOMs calculated from simulated S-parameters ("Sim"). The second set is calculated from S-parameters that are predicted with an ANN from the link parameters ("Calc"). The third set is directly predicted from the link geometry ("Pred"). The WSXTR can be calculated by dividing the predicted WPT and WPSXT ("Pred.") or it can be predicted directly ("Dir. Pred."). The presented results show little differences between the different methods. Greater variations can be observed for other link variations (not shown here). It appears that directly predicting the WSXTR leads to worse results than predicting WPT and WPSXT separately. Even though the calculation of the power sums cancel out a part of the noise of S-parameters, there appears to be no clear advantage in predicting the S-parameters instead of directly predicting the FOMs.

V. CONCLUSION

It was found that ANNs of modest complexity can predict interconnect performance for connected PCBs up to 100 GHz.

ANN designs found with an optimization lead to a similar performances than manual designs. Both S-parameters and figures of merit can be predicted. A further investigation should focus on extending the range of link models.

REFERENCES

[1] K.-T. Hsu, W.-D. Guo, G.-H. Shiue, C.-M. Lin, T.-W. Huang, and R.-B. Wu, "Design of reflectionless vias using neural network-based approach," *IEEE Transactions on Advanced Packaging*, vol. 31, no. 1, p. 211–218, Feb. 2008.

[2] H. Kim, C. Sui, K. Cai, B. Sen, and J. Fan, "Fast and precise high-speed channel modeling and optimization technique based on machine learning," *IEEE Transactions on Electromagnetic Compatibility*, vol. 60, no. 6, pp. 2049–2052, Dec. 2018.

[3] K. Roy, M. A. Dolatsara, H. M. Torun, R. Trinchero, and M. Swaminathan, "Inverse design of transmission lines with deep learning," in *IEEE Conference on Electrical Performance of Electronic Packaging and Systems (EPEPS)*, Montreal, Canada, Oct. 2019.

[4] T. Lu, J. Sun, K. Wu, and Z. Yang, "High-speed channel modeling with machine learning methods for signal integrity analysis," *IEEE Transactions on Electromagnetic Compatibility*, vol. 60, no. 6, pp. 1957–1964, Dec. 2018.

[5] N. Ambasana, G. Anand, D. Gope, and B. Mutnury, "S-parameter and frequency identification method for ANN-based eye-height/width prediction," *IEEE Transactions on Components, Packaging and Manufacturing Technology*, vol. 7, no. 5, pp. 698–709, May 2017.

[6] R. Trinchero and F. G. Canavero, "Modeling of eye diagram height in high-speed links via support vector machine," in *IEEE Workshop on Signal and Power Integrity (SPI)*, Brest, France, May 2018.

[7] R. Rimolo-Donadio, X. Gu, Y. Kwark, M. Ritter, B. Archambeault, F. de Paulis, Y. Zhang, J. Fan, H.-D. Brüns, and C. Schuster, "Physics-Based Via and Trace Models for Efficient Link Simulation on Multilayer Structures Up to 40 GHz," *IEEE Transactions on Microwave Theory and Techniques*, vol. 57, no. 8, pp. 2072–2083, Aug. 2009.

[8] K. Scharff, H.-D. Brüns, and C. Schuster, "Performance metrics for crosstalk on printed circuit boards in frequency domain," in *IEEE Workshop on Signal and Power Integrity (SPI)*, Chambéry, France, June 2019.

[9] ——, "Efficient crosstalk analysis of differential links on printed circuit boards up to 100 GHz," *IEEE Transactions on Electromagnetic Compatibility*, vol. 61, no. 6, pp. 1849–1859, Dec. 2019.

[10] F. Pedregosa, G. Varoquaux, A. Gramfort, V. Michel, B. Thirion, O. Grisel, M. Blondel, P. Prettenhofer, R. Weiss, V. Dubourg, J. Vanderplas, A. Passos, D. Cournapeau, M. Brucher, M. Perrot, and E. Duchesnay, "Scikit-learn: Machine learning in Python," *Journal of Machine Learning Research*, vol. 12, pp. 2825–2830, 2011.

Cost-Effective Implementation of Air-Filled Waveguides on Printed Circuit Boards

Felix Sepaintner
THD Technische Hochschule
Deggendorf
Deggendorf, Germany
felix.sepaintner@th-deg.de

Andreas Scharl
THD Technische Hochschule
Deggendorf
Deggendorf, Germany
andreas.scharl@th-deg.de

Johannes Jakob
THD Technische Hochschule
Deggendorf
Deggendorf, Germany
johannes.jakob@th-deg.de

Florian Keck
THD Technische Hochschule
Deggendorf
Deggendorf, Germany
florian.keck@stud.th-deg.de

Kevin Kunze
THD Technische Hochschule
Deggendorf
Deggendorf, Germany
kevin.kunze@stud.th-deg.de

Franz Xaver Röhrl
Rohde & Schwarz GmbH & Co. KG
Teisnach, Germany
franz.roehrl@rohde-schwarz.com

Werner Bogner
THD Technische Hochschule
Deggendorf
Deggendorf, Germany
werner.bogner@th-deg.de

Stefan Zorn
Rohde & Schwarz GmbH & Co. KG
Teisnach, Germany
stefan.zorn@rohde-schwarz.com

Abstract— This paper presents a new cost-effective method to produce air-filled waveguides out of a printed circuit board (PCB) with a milling machine and how to implement them with standard PCB technology. For this purpose, low cost substrates like FR-4 are used. For the baseboard to waveguide transition in the E-Band (60 - 90 GHz) a WR12 waveguide connector [1] was used. The WR12 waveguide was manufactured and analyzed on mechanical deviations. The RF performance in the E-Band was measured and compared to common PCB waveguides like microstrip lines (MS) and grounded coplanar waveguides (GCPW).

Keywords—low loss transmission lines, air-filled waveguide, submount, WR12, PCB

I. INTRODUCTION

Millimeter wave systems nowadays have to deal with high data rates and continuously increasing integration density. Besides the steady progressing miniaturization in electronics industry, low fabrication costs and high quality are required simultaneously. Moreover, for millimeter wave applications low loss transmission lines are a key technology. In [2], a GCPW with reduced attenuation by partially removed substrate was analyzed. The measurement showed an improved attenuation of 0.25 dB/cm at 90 GHz. Another technology of great interest for many practical applications in standard low cost PCB manufacturing processes is the substrate integrated waveguide (SIW), because of its robustness against production tolerances [3]. With regard to quality factor and losses, these devices are a good compromise between the performance of planar circuits and classical waveguides. However, conventional rectangular waveguides are often preferred, due to the low insertion loss as they are completely filled with air (dielectric constant $DK_{air} \approx 1$ and dissipation factor $DF_{air} \approx 0$). Therefore, high research effort is taken to implement waveguides on PCBs to combine the benefits of the SIW and the classical rectangular waveguide. Devices like the air-filled substrate integrated waveguide (AFSIW) already showed, that passive structures (e.g. waveguides or filters) produced in AFSIW technology have a better RF performance than the conventional SIW ([4] - [5]). In [4] and [5] the AFSIW was realized as planar structure, analyzed and compared to conventional SIWs up to 60 GHz. Whereas in this paper, the air-filled waveguides are realized in submount technology and their RF performance were analyzed in the full E-Band. A great advantage of using waveguides as submounts is the fact that the necessary manufacturing techniques are available in conventional circuit board technology. In addition, the submounts can be fully assembled in SMD production lines due to their smooth surface.

II. AIR-FILLED WAVEGUIDES

A. Manufacturing of Submount Waveguides

Producing a waveguide as a submount allows the use of cost effective substrates (e.g. FR-4) as they also can be used for the production of the baseboard. For this purpose, a cavity depth milling technology was used to create the waveguide structures. Subsequently, the structures were coated with copper (25 µm) and silver (250 nm), to create a conductive surface. The produced WR12 waveguide submounts are shown in Fig. 1. The milling machine has to deal with composed materials in a multilayer PCB (consisting of different substrates and prepregs), which have various physical, thermal and mechanical properties. Furthermore, the end of a milling cutter normally has different sharp edges. Those sharp edges and high process dynamics usually lead to work piece vibrations. Therefore, locally different cutting forces affect the processed surface, leading to variable surface qualities (see Chapter II B). Fig. 2 shows a close-up view of the produced waveguide. The notches and scratches on the surface are caused by an inappropriate milling strategy and by worn milling cutters. This causes mechanical deviations which will also influence the RF properties. Milling processes have to be optimized individually, depending on which PCB materials and cutters are involved.

978-1-7281-6162-4/20 $31.00 © 2020 IEEE

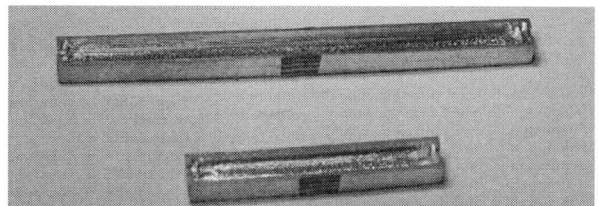

Fig. 1 Manufactured 22 mm and 44 mm long WR12 waveguide submounts

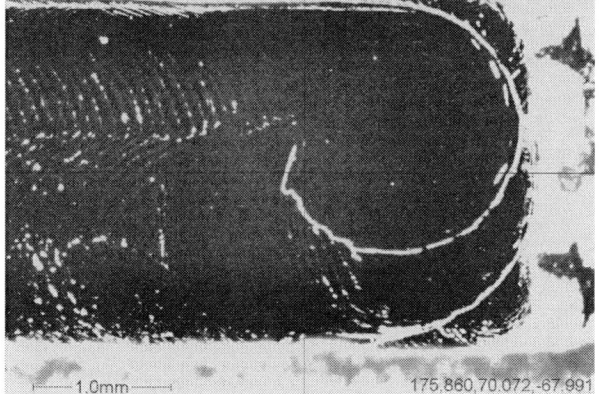

Fig. 2 Irregularities on the surface of the WR12 waveguide submount

B. Influence of Mechanical Deviations

As the quality of the inner surfaces is affected by nearly every parameter of the milling process and the used substrate materials, roughness may vary at every surface. Especially the copper roughness of the different waveguide walls has big influence on the RF properties in the upper two-digit GHz range. Both signal phase delay and attenuation are significantly increased, when dealing with rough surfaces in the millimeter wave range [6]. Therefore, it is important to fabricate the inner walls of the waveguide as smooth as possible. The area based RMS roughness S_q of the waveguide was deduced with a laser scanning microscope. Each surface was measured five times and then averaged. Fig. 3 shows a cutout of three different waveguide surfaces.

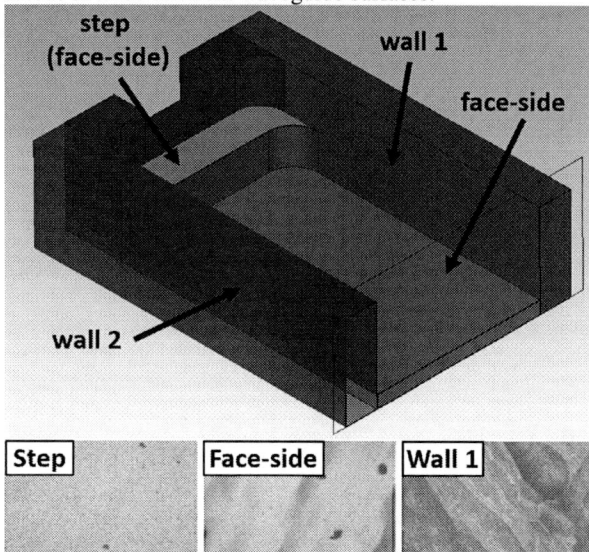

Fig. 3 Model of the waveguide and laser scanning microscope images of the different surfaces

It has been found that $S_{q,wall\ 1}$=2.77 µm, $S_{q,wall\ 2}$=1.67 µm, $S_{q,step}$=0.19 µm and $S_{q,face-side}$=1.15 µm. The roughness of the baseboard copper was considered with the values denoted in [2]. Two different milling cutters were used for the step and the face-side, leading to different $S_{q,step}$ and $S_{q,face-side}$ values. For the face-side a worn milling cutter was used, and a new one for the step. Therefore, the use of worn milling cutters should be avoided. Furthermore, the roughness of the two walls varies by about 1.1 µm. This is due to the different cavity milling processes (parallel and reverse milling). It has been found, that reverse milling (used for Wall 1) causes rougher surfaces than parallel milling (used for Wall 2). Besides the copper roughness, the manufacturing deviation (±50 µm) of the used cavity milling machine also influences the quality of the transition from the WR12 connector to the WR12 waveguide submount. Table 1 shows the mechanical dimensions of the produced WR12 waveguide.

Table 1 Mechanical dimensions of the WR12 waveguide

Mechanical deviation	WR12 waveguide 22 mm	
	Target value (mm)	Measured (mm)
a (waveguide width)	3.10	3.12
b (waveguide height)	1.52	1.53
Total length	22	22
Step length	0.9	0.87
Step depth	1	1.01

Deviations in the dimensions of the step (length/depth) lead to worse impedance matching and decreased quality of the transition. Also the dimensions (height (b) and width (a)) of the waveguide submount can be affected. The conductor losses of a waveguide can be calculated as [7]:

$$\alpha_C = \frac{R_S}{b \cdot \frac{\mu_0}{\varepsilon_0}} \cdot \frac{(1 + \frac{2b}{a} \cdot \frac{\omega_C^2}{\omega^2})}{\sqrt{1 - \frac{\omega_C^2}{\omega^2}}} \qquad (1)$$

With R_S denoting the surface resistance, ω_c the angular cutoff frequency, μ_0 the permeability constant and ε_0 the permittivity constant. At maximal deviation of a and b (-50 µm), the insertion loss is increased by about 0.18 dB/m at 75 GHz and, therefore, has less impact on the attenuation. Additionally, very precise milling is necessary for the step.

III. DESIGN AND MEASUREMENT

For the measurement of the air-filled waveguide submounts a WR12 waveguide connector [1] was used. The PCB waveguide transitions was designed and analyzed with the 3D EM field simulator CST Microwave Studio®. This has been done with a WR12 rectangular waveguide connector as described in [1].

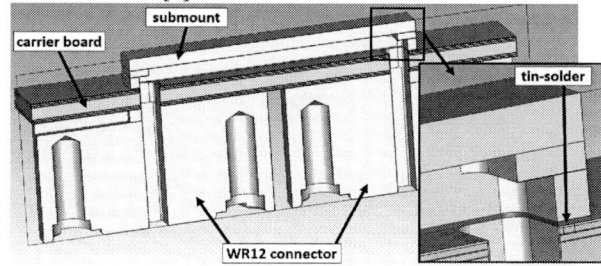

Fig. 4 WR12 waveguide connector to WR12 waveguide submount transition

978-1-7281-6162-4/20 $31.00 © 2020 IEEE

The WR12 waveguide connector can easily be attached on the backside of the baseboard. For this purpose, an additional milling process is necessary to connect the WR12 connector with the waveguide submount. To eliminate parasitic propagation of the electric field into the substrate material of the baseboard, the milled hole was coated with copper through electroplating (Fig. 4). The corresponding simulation and measurement results are illustrated in Fig. 5. The measurement of the 22 mm long waveguide submount has an insertion loss of $0.45 - 1.36$ dB in the full E-Band ($60 - 90$ GHz).

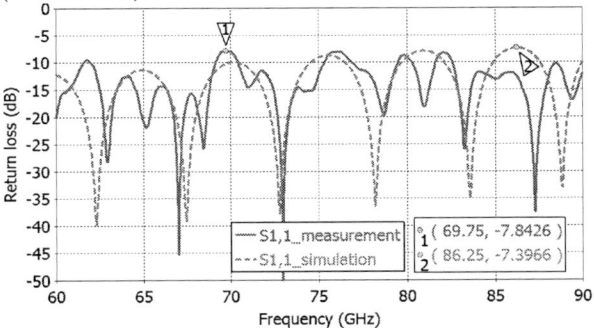

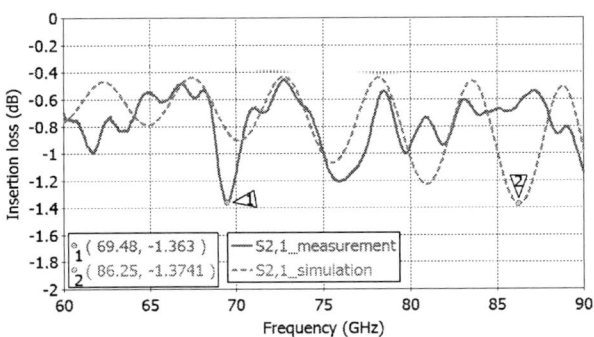

Fig. 5 Simulation and measurement of the WR12 connector to WR12 waveguide submount transition

Due to mechanical deviations, the waveguide has slightly larger dimensions than the WR12 standard. This decreases the attenuation of the measurement compared to the simulation.

IV. Potential of Air-Filled Waveguides on PCBs

In order to show the great potential of an air-filled waveguide in terms of insertion loss, a WR12 waveguide with 1 m length was measured directly with WR12 connectors. To get a compact design, the delay line was realized spiral-shaped which can easily be implemented on PCBs. Fig. 6 shows the delay line soldered onto the baseboard and the corresponding X-ray picture for evaluating the mechanical quality of the waveguide. Fig. 7 compares the insertion loss of the air-filled waveguide to common waveguides on PCBs. In contrast to MS and GCPW the attenuation of the WR12 waveguide is improved by 73 dB and 53 dB at 75 GHz, respectively (Fig. 7).

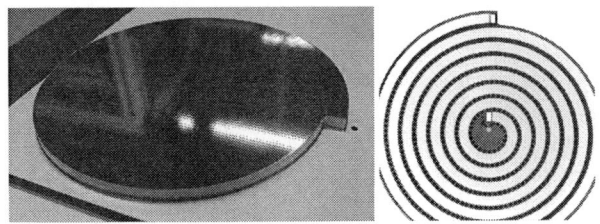

Fig. 6 Mounted WR12 delay line (left) and corresponding X-ray view (right)

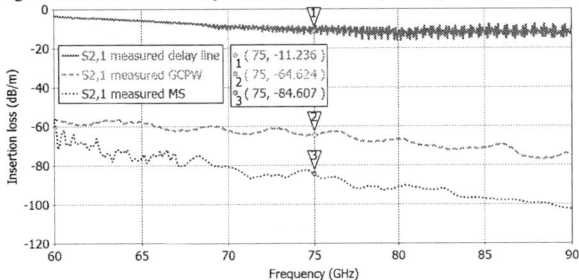

Fig. 7 Comparison between common waveguides (MS and GCPW with 5 mil substrate: DF=0.0015) and an air-filled WR12 waveguide

Acknowledgment

The Project PaGAnIni is financially supported by the Bavarian Ministry of Economic Affairs, Regional Development and Energy. All kinds of support are gratefully acknowledged. The authors also like to thank Rohde & Schwarz Teisnach for all the fabrication and measurement support.

References

[1] J. Jakob, R. Sammer, F. X. Röhrl, W. Bogner and S. Zorn, "WR12 to planar transmission line transition on organic substrate," in *49th European Microwave Conference (EuMC)*, Paris, 2019.

[2] F. Sepaintner, A. Scharl, F. X. Röhrl, W. Bogner and S. Zorn, "Simulation and Manufacturing of Low Loss PCB Structures with Additional Electromagnetic Field in Air," in *IEEE MTT-S International Microwave Workshop Series on Advanced Materials and Processes for RF and THz Applications (IMWS-AMP)*, Bochum, 2019.

[3] F. X. Röhrl, R. Sammer, J. Jakob, W. Bogner, R. Weigel, U. Hassel and S. Zorn, "Cost-Effective SIW Band-Pass Filters for Millimeter Wave Applications," in *47th European Microwave Conference (EuMC)*, Nuremberg, 2017.

[4] A. Ghiotto, F. Parment, F. Martin, T. P. Vuong and K. Wu, "Air-filled substrate integrated waveguide — A flexible and low loss technological platform," in *13th International Conference on Advanced Technologies, Systems and Services in Telecommunications (TELSIKS)*, Nis, 2017.

[5] F. Parment, A. Ghiotto, T.-P. Vuong, J.-M. Duchamp and K. Wu, "Air-Filled SIW Transmission Line and Phase Shifter For High-Performance and Low-Cost U-Band Integrated Circuits and Systems," in *Global Symposium on Millimeter-Waves (GSMM)*, Montreal, 2015.

[6] G. Gold and K. Helmreich, "A Physical Surface Roughness Model and Its Applications," *IEEE Transactions on Microwave Theory and Techniques*, pp. 3720-3732, May 2017.

[7] S. J. Orfanidis, "www.ece.rutgers.edu," 1999. [Online]. Available: https://www.ece.rutgers.edu/~orfanidi/ewa/ewa-1up.pdf. [Accessed 30 June 2020].

Thermal Sensitivity of Dielectric Materials in High-Speed Designs

Sunil Pathania[1], Bhyrav Mutnury[2], Mallikarjun Vasa[3], Vijender Kumar[3], Sukumar Muthusamy[3], Seema P K[3] and Rohit Sharma[1]

[1]Department of Electrical Engineering, Indian Institute of Technology Ropar, INDIA
[2]Dell Enterprise Server Group, Round Rock, TX, USA
[3]DellEMC Enterprise Server Group, Bangalore, INDIA
Email: rohit@iitrpr.ac.in

Abstract—**At high-speeds, careful analysis is required at the design stage to ensure robust signal integrity (SI) in high-speed printed circuit boards (PCBs). Signal loss in PCBs is predominantly due to conductor loss, dielectric loss and impedance mismatch.**

In this paper, thermal impact on loss and impedance is studied. In that, thermal sensitivity for standard loss, mid-loss, low-loss and ultra-low loss dielectric materials is studied. It is observed that ultra-low loss materials are less sensitive as compared to standard loss materials. Also, thermal impact on impedance and loss in transmission lines, vias and SMT pads is analyzed.

Keywords—Impedance, Thermal sensitivity, PCB

I. INTRODUCTION

Thermal impact on signal integrity (SI) is becoming dominant in high-speed printed circuit boards (PCBs). Heat generated from components can degrade the electrical performance of traces and lead to SI failure [1]. Electrical characteristics of copper traces and dielectric materials vary with temperature, which degrades the performance of signals. In the case of high-speed signals, where the signal-to-noise margin is less, the probability of signal failure is high. Therefore, care should be taken in the modeling phase to account for thermal impact on high-speed signals [2].

Some work is done in the literature on electro-thermal co-simulation of 3D integration system for power delivery networks [3]. This work addressed the importance of electro-thermal co-simulation for on-chip scenario only. Authors in [4] presented the characterization of insertion loss in striplines on various substrate materials as function of temperature. Laboratory measurements for standard, mid-, low- and ultra-low loss materials were performed and it was concluded that temperature sensitivity reduces as the material loss decreases. Authors in [5] discussed experimental thermal sensitivity of dielectric constant and dissipation factor. In [6], thermal sensitivity on PCB cross-section was discussed.

In this paper, the work done in [4]-[6] is extended by taking measured dielectric constant and loss tangent values across various materials and extending them to model insertion loss and impedance for traces, vias and surface mount (SMT) pads. The impact of temperature on impedance and loss is studied across various material classifications through simulations.

This paper is organized as below. Section II explains temperature impact on dielectric constant and loss tangent using vendor measurement data and theoretical postulations. Section III presents thermal sensitivity of impedance and loss for transmission lines, vias and SMT pads. Section IV concludes the paper.

II. TEMPERATURE IMPACT ON MATERIALS

In this section, thermal sensitivity of different PCB materials is discussed. In addition, the theory for thermal sensitivity is also presented.

A. Thermal sensitivity of PCB materials

Dielectric materials are classified based on losses. Major classifications typically include ultra-low loss, low-loss, mid-loss and standard loss. These materials are sensitive to temperature. Table 1 shows that variation of Dk and Df with respect to temperature [7]. It can be seen that high Dk materials exhibit larger temperature variation compared to low Dk materials.

TABLE I: THERMAL SENSITIVITY DATA OF DIELECTRIC MATERIALS [7]

Materials		0°C	20°C	60°C	100°C	Delta (20-100)°C
Ultra-Low Loss	Dk	3.88	3.88	3.89	3.9	0.52 %
	Df	0.0059	0.0064	0.0068	0.0074	15.63%
Low Loss	Dk	3.93	3.94	3.96	3.97	0.76 %
	Df	0.0083	0.0091	0.0102	0.0113	24.18 %
Mid Loss	Dk	4.21	4.23	4.26	4.3	1.65 %
	Df	0.0113	0.0124	0.0146	0.0173	39.52 %
Standard Loss	Dk	4.23	4.26	4.32	4.4	3.29 %
	Df	0.0164	0.0184	0.0234	0.0279	51.63 %

B. Theory for thermal sensitivity of materials

Dielectric constant of materials is defined by equation (1). Here, E represents the applied electric filed to material and E_{in} is induced electric field [8].

$$Dk = \frac{E}{E-E_{in}} \qquad (1)$$

When the molecules are polarized by applying the electric field, E, it generates a dipole moment. Due to dipole moment of molecules, induced electric filed, E_{in}, is formed inside the dielectric material. The polarization vector per unit volume is given by equation (2), where N represents the number of molecules per unit volume, which contributes to dipole moment. Note that, α_e is polarizability and E is applied electric field [8]. Corresponding to the dipole moment, P, the charge density and induced electric field are given by equations (3) and (4) [9], where, ρ represents the charge density and $\hat{n}$ is unit vector in equation (3). In equation (4), ∇ (del) operator represents the divergence of induced electric field.

978-1-7281-6162-4/20 $31.00 © 2020 IEEE

$$P = N\alpha_e E \qquad (2)$$

$$\rho = \mathbf{P}.\hat{n} \qquad (3)$$

$$\nabla.\mathbf{E}_{in} = \rho \qquad (4)$$

Usually, the dipole moment does not reach the saturation level on applying the electric filed. Few molecules have an orientation such that they could cancel the dipole movement of others. With increase in temperature these molecules get the energy and begin to orient in the direction of electric field. Hence, number of molecules that contribute to the dipole moment are increased. Based on the above explanation, equations (2), (3) and (4) can be modified as equations (5), (6) and (7) as the temperature increases. Here, N' is the number of molecules per unit volume after increase in temperature and N' is greater than N.

$$P' = N'\alpha_e E \qquad (5)$$

$$\rho' = P'.\hat{n} \qquad (6)$$

$$\nabla.E'_{in} = \rho' \qquad (7)$$

Thus, the polarization vector magnitude increases, which results in increased charge density and increase in magnitude of induced electric field as per equations (6) and (7). Induced electric field (E_{in}) increases as per equation (1) that results in higher dielectric. Thus, we can conclude that the temperature sensitivity is more dominant on materials with higher dielectric constant.

III. TEMPERATURE IMPACT ON SIGNAL IMPEDANCE AND LOSS

In this section transmission lines, vias, SMT pads are considered to quantify the impact due to thermal variations in a PCB.

a) Transmission Line

To study the thermal impact on transmission lines, various material classes from standard loss to ultra-low loss are considered. Transmission lines are modeled as per Dk and Df numbers from Table 1. Geometry for microstrip line includes trace-width of 4.5 mils, spacing of 6 mils, trace thickness of 1.9 mils, prepreg thickness of 3 mils and solder mask of 0.5 mil. The geometry considered for striplines includes trace-width of 4.5 mils, spacing of 6 mils, trace thickness of 1.3 mils, core thickness of 3 mils and prepreg of 6 mils. It is observed that ultra-low loss material is less sensitive to thermal impact as compared to standard loss material. As shown in Fig. 1, ultra-low loss material shows 4.7% and 3.7% change in loss at 16GHz for a temperature change from 20°C to 100°C for stripline and microstrip line, respectively. A standard loss material shows 30.3% and 25.2% change in loss at 16GHz for a temperature change from 20°C to 100°C for stripline and microstrip line, respectively. Thermal sensitivity of impedance is not significant as shown in Fig. 2. As standard loss material is more sensitive to temperature, via and SMT pad are analyzed with standard loss material only for insertion loss and impedance.

b) Via

Vias produce impedance discontinuity when the signal transits from one layer to another. To maintain required impedance different aspects of vias should be considered such as material of the stackup, anti-pad, spacing between the signal vias and signal-to-ground as shown in the Fig. 3. To minimize the impedance discontinuity of vias, designers should clearly understand their thermal sensitivity.

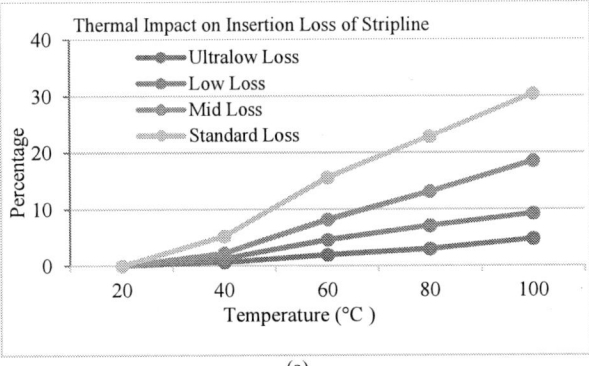

(a)

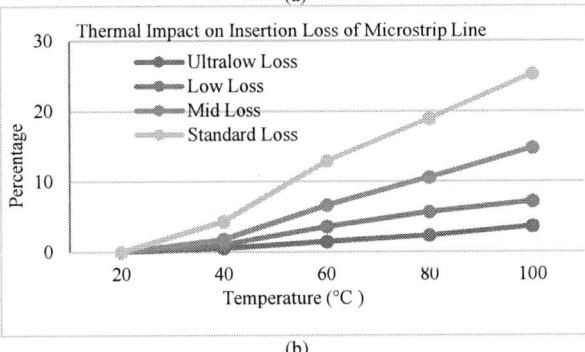

(b)

Fig. 1. Percentage change of insertion loss with temperature (a) Stripline (b) Microstrip line

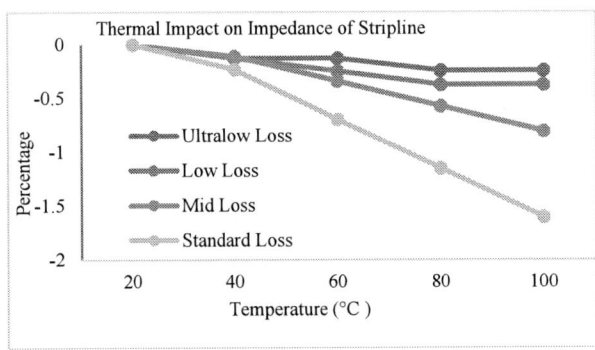

(a)

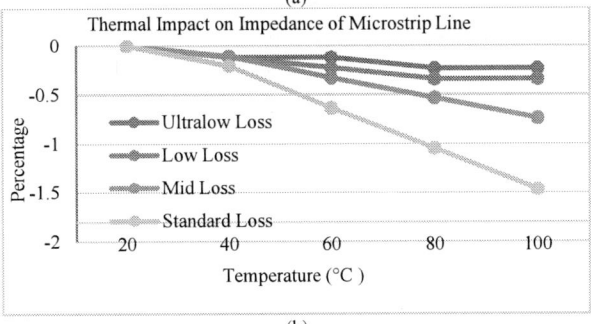

(b)

Fig. 2. Percentage change of impedance with temperature (a) Stripline (b) Microstrip line

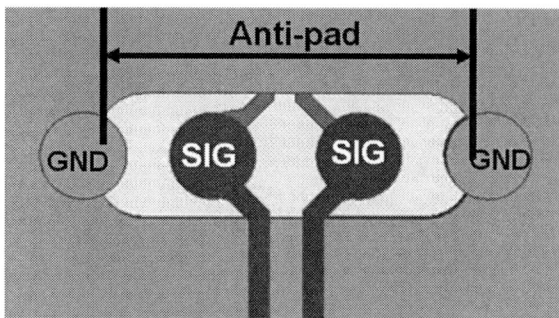

Fig. 3. Top view of a differential via

TABLE III: SMT IMPEDANCE AND LOSS DIFFERENT Dk AND Df

Dk	Df	Impedance@ 5ps	% change in impedance	Insertion loss(dB) @16GHz	% change in insertion loss
4.1	0.0075	65.6-91.2	--	0.41	--
4.2	0.0142	65.3-90.7	0.55	0.45	9.75
4.3	0.0217	64.3-90.1	1.2	0.48	17.03
4.4	0.0279	64.1- 89.8	1.53	0.59	31.7

From Section II, one can observe that the *Dk* of material increases with increasing temperature. Table 2 shows the variation in impedance and insertion loss for a range of *Dk* and *Df* values for a stack-up of 100 mils thickness with signal transiting from top to bottom layer. It is observed that there is 3% variation in impedance and 31% variation in loss at 16 GHz. Thus, thermal sensitivity of vias is similar to that of transmission lines.

TABLE II: VIA IMPEDANCE AND LOSS FOR DIFFERENT Dk AND Df

Dk	Df	Impedance@ 5ps	% change in impedance	Insertion loss (dB) @16GHz	% change in insertion loss
4.1	0.0075	74-88	--	0.48	--
4.2	0.0142	73.1-87	1.14	0.52	8.34
4.3	0.0217	72.2-85.6	2.72	0.57	18.75
4.4	0.0279	72.1-85.3	3.06	0.63	31.25

c) Surface mount (SMT) pads:

By using SMT pads, electrical components are mounted directly onto the surface of PCB. SMT pads offer higher component density and provide more routing space for inner layers. However, the landing pads in SMT connectors offer large capacitance that can significantly affect their impedance. Table 3 shows the impedance and loss variations for various *Dk* and *Df* for standard loss material. From our analysis, it can be seen that impedance of SMT connector pads varies by 1.5% and insertion loss by ~32%.

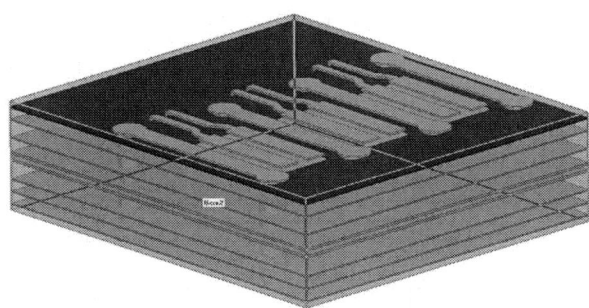

Fig. 4. Surface mount connector pads

IV. CONCLUSION

In this paper, thermal sensitivity of dielectric materials and its effect on loss and impedance for high-speed PCBs are discussed. The theory behind thermal sensitivity of materials is also discussed, which helps in providing physical insights into this phenomenon. Thermal sensitivity study is done for PCB transmission lines, vias and SMT pads. It is observed that standard loss material is more sensitive to thermal variations as compared with mid-loss, low-loss and ultra-low loss materials. Insertion loss variation of ~30% is observed in transmission lines, vias and SMT pads at 16GHz with standard loss material. Standard loss materials are thermally insensitive for impedance with a ~2-3% variation across transmission lines, vias and SMT pads. Based on our analysis, we conclude that high-speed PCBs need to take thermal sensitivity into account in the early design stages.

V. ACKNOWLEDGEMENTS

Authors would like to thank ITEQ Corporation for providing temperature dependent properties of dielectric materials. This work was supported by the Ministry of Electronics and IT under the SMDP program (*Grant no: 9(1/2014-MMD)*).

REFERENCES

[1] K. Banerjee, A. Mehrotra, A. Sangiovanni-Vincentelli, and H. Chenming, "On thermal effects in deep sub-micron VLSI interconnects," in *Proceedings 1999 Design Automation Conference*, 1999, pp. 885-891.

[2] S. Pathania *et al.*, "Multiphysics Approach Using Computational Fluid Dynamics for Signal Integrity Analysis in High Speed Serial Links," in *2019 Electrical Design of Advanced Packaging and Systems*, 2019, pp. 1-3.

[3] J. Xie and M. Swaminathan, "Electrical-Thermal Co-Simulation of 3D Integrated Systems With Micro-Fluidic Cooling and Joule Heating Effects," *IEEE Transactions on Components, Packaging and Manufacturing Technology,* vol. 1, no. 2, pp. 234-246, 2011.

[4] J. Hsu, C.-J. Ong, and X. Ye, "*Characterization of Insertion Loss of Striplines on Various Substrate Materials as a Function of Temperature,*" in *2019 Joint International Symposium on Electromagnetic Compatibility, Sapporo and Asia-Pacific International Symposium on Electromagnetic Compatibility,* Sapporo, Japan, 2019, pp. 96-99.

[5] S. Pathania, M. Vasa, B. Mutnury, and R. Sharma, "Thermal Impact on High Speed PCB Interconnects," in *2019 IEEE 28th Conference on Electrical Performance of Electronic Packaging and Systems*, 2019, pp. 1-3.

[6] S. Hinaga, M. Koledintseva, J. Drewniak, A. Koul, and F. Zhou, "Thermal Effects on PCB Laminate Material Dielectric Constant and Dissipation Factor," in *IPC APEX EXPO Technical Conference 2010*, vol. 2, 2010.

[7] http://www.iteq.com.tw/?lang=en

[8] J. Nakamura, W. Sadakazu, and N. Akiko, "First-principles evaluations of dielectric constants," in *2006 International Workshop on Nano CMOS*, 2006, pp. 236-249.

[9] S. Kim, J. Ha, and J.-B. Kim, "Calculation of polarization and bound charge density inside a dielectric material in triboelectric nanogenerators: Analytical and numerical study," *The European Physical Journal Plus*, vol. 131, no. 11, p. 382, 2016.

978-1-7281-6162-4/20 $31.00 © 2020 IEEE

An Inspection Based Method to Analyse Deterministic Noise in N-port Circuits

Vijender Kumar Sharma[1], Jai Narayan Tripathi[2] and Hitesh Shrimali[1]

[1]Indian Institute of Technology Mandi, India, [2]Indian Institute of Technology Jodhpur, India.

Email: [1]vijender_s@students.iitmandi.ac.in, [2]jai@iitj.ac.in, [1]hitesh@iitmandi.ac.in

Abstract—**This paper proposes the estimation by inspection method to analyse the impact of deterministic supply noise on the design specifications of the analog and mixed signal (AMS) systems. The method is based on the indefinite admittance matrix (IAM) method. The voltage gain, phase and input-output impedance have been considered as the design specifications. To validate the method, two examples of output stages for analog and digital blocks have been simulated in standard 0.18 μm technology with 1.8 V of supply and same geometric area. The proposed models using the inspection method and the SPICE based simulations shows maximum mean percentage error (MPE) of 3% for all the examples.**

Index Terms—**Two-port network, indefinite admittance matrix (IAM), deterministic supply noise, signal integrity.**

I. INTRODUCTION

Power supply noise is one of the dominant factors which degrades the performance of any high-speed AMS system [1]. The common causes of power supply noise are simultaneous switching noise, insertion-loss, reflection, electromagnetic interference, etc. [2]. The variations in the supply voltage, known as power supply noise (PSN), may affect the signal integrity (SI) of the system. The SI of analog and digital blocks of the AMS system depends on the primary design parameters such as gain, phase and input-output (I/O) impedance. In case of noisy power supply, deviations in these parameters may occur and hence the performance of the system will be degraded. Therefore, the causes and impacts of these quantities under supply/bulk fluctuations need to be investigated for deep sub-micron technologies.

For the estimation of the aforementioned-design parameters, a two port network analysis is widely used to describe the I/O behavior of a system [3]. In the recent years, the two-port network theory was used in many complex systems to analyse the input-output behavior of the systems [4], [5]. In the literature, the modeling of two-port network parameters including the effects of PSN, ground supply noise (GSN) and bulk-supply noise (BSN) has not been reported yet.

In this paper, the standard two-port network analysis using the IAM is extended and a new method, estimation-by-inspection for multi-input port is proposed to understand the effects in the presence of supply noise. The initial results of this work are presented in [6]. The inspection method derives crucial SI metrics such as gain, phase, I/O impedance and output transient response in the closed-form expressions. These expressions are derived to form a generalised two-port network for an active circuit having N-terminals.

II. PROPOSED ANALYSIS FOR AN N-PORT NETWORK

A. Estimation by Inspection Method

The IAM, is a zero sum matrix ($n \times n$ singular matrix), the reference terminal of which is at an arbitrary potential which lies outside the network. The procedure to write IAM and design parameters for two-port networks are discussed in [7]. However, the standard IAM method is not valid for the estimation of design parameters in case of multiple terminals.

Algorithm 1 Proposed algorithm to calculate two-port parameters for the generalised multi-port system.

Input: v_{na}, v_{in}, v_b, v_{bn}, v_{bp} and angular frequency (ω)

Output: A_{v_i}, $\tilde{A}_v$, Z_i, $\tilde{\phi}$

1: Initialise : p = Number of input terminals, o = output terminal (v_{out}), i_m = main input terminal of a circuit.

2: Draw the small-signal model and assign the node numbers.

3: Compute the required IAM elements for a circuit with all the deterministic noise sources.

4: **for** $i = 1$ to p **do**

4: $\quad A_{v_i} = \frac{-|Y_{o,i}|}{|Y_{o,o}|}$

4: $\quad Z_i = \dfrac{Y_{o,o}}{\begin{vmatrix} Y_{i,i} & Y_{i,o} \\ Y_{o,i} & Y_{o,o} \end{vmatrix}}; \quad Z_{out} = \dfrac{|Y_{i_m,i_m}|}{\begin{vmatrix} Y_{i_m,i_m} & Y_{i_m,o} \\ Y_{o,i_m} & Y_{o,o} \end{vmatrix}}$

4: $\quad i = i + 1$

5: **end for**

6: $\tilde{A}_v = \sum\limits_{i=1}^{p} A_{v_i}$

7: $\tilde{\phi} = \pi - \angle \tilde{A}_v(j\omega)$

8: **return** Two port parameters

For multi-input terminals, the estimation-by-inspection method, based on the IAM approach is used to compute the two-port parameters. Here, multiple inputs refer to the different deterministic supply noise sources. The abbreviations used for the PSN, p-bulk, n-bulk, ground, common-mode noise and input voltage sources are v_{na}, v_{bp}, v_{bn}, GND, v_b and v_{in}, respectively. The first step of the method is to draw the small-signal model of the circuit with all the deterministic supply noise sources. In the next step, the node numbers need to be assigned in the small-signal model. After that, the required IAM ($Y_{i,i}$, $Y_{o,o}$, $Y_{o,i}$, $Y_{i,o}$) elements are extracted for the circuit including all the nodes. The $Y_{i,i}$ and $Y_{o,o}$ refer to the self admittance parameters of input and output nodes, respectively. $Y_{o,i}$ and $Y_{i,o}$ are the negative signed mutual admittance parameters between output-to-input and input-to-

---

978-1-7281-6162-4/20 $31.00 © 2020 IEEE

output nodes, respectively. Next, to determine the closed-form equations of the required parameters, the superposition theorem has been applied on the circuit by selecting one excited input node at a time. Finally, the transfer function (TF), impedance, overall gain and the overall phase expressions can be calculated using the Algorithm 1. The derivations of these formulae are skipped for the interest of readers.

B. Validation of the Proposed Method

The proposed method of multi-port analysis by inspection is validated using two examples viz. a CMOS inverter and a chain of inverters, in the presence of PSN and BSN. For the sake of simplicity and a proof of concept, the GSN has not been considered. In this method, the circuits are analysed with respect to the lowest potential i.e. GND.

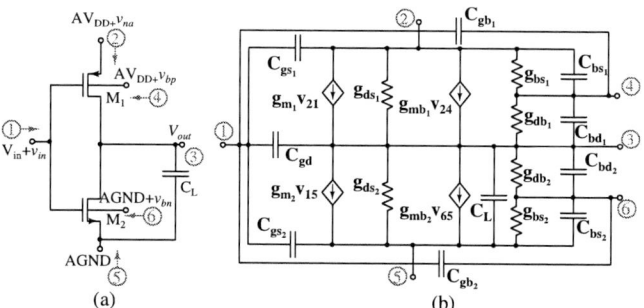

Fig. 1: The CMOS inverter (a) schematic diagram and (b) equivalent small-signal model.

Example-I: CMOS inverter

The CMOS inverters have diverse applications in AMS system as a class-AB push-pull amplifier, high-speed buffer, output driver and a transmission gate. The schematic and the small-signal diagram of an inverter is depicted in Fig. 1. The possible paths of supply noise in the circuit are denoted by red circled arrows in the Fig. 1(a). The required IAM parameters to estimate the TFs and I/O impedance are mentioned in Table I.

TABLE I: The required IAM parameters for the inverter.

Y_{ij}	Expressions	Y_{ij}	Expressions
Y_{11}	$s(C_{gd}+C_{gs}+C_{gb})$	Y_{13}	$-sC_{gd}$
Y_{22}	$s(C_{gs_1}+C_{bs_1})+\gamma_3$	Y_{23}	$-g_{ds_1}$
Y_{31}	$g_{m_1}+g_{m_2}-sC_{gd}$	Y_{32}	$-g_{m_1}-g_{mb_1}-g_{ds_1}$
Y_{33}	$g_{ds}+g_{bd}+sC_1$	Y_{34}	$g_{mb_1}-g_{bd_1}-sC_{bd_1}$
Y_{35}	$-g_{ds_2}-g_{mb_2}-sC_L$	Y_{36}	$g_{mb_2}-g_{bd_2}-sC_{bd_2}$
Y_{43}	$-g_{bd_1}-sC_{bd_1}$	Y_{44}	$g_{bs_1}+g_{bd_1}+s(C_{bd_1}+C_{bs_1}+C_{gb_1})$
Y_{63}	$-g_{bd_2}-sC_{bd_2}$	Y_{66}	$g_{bs_2}+g_{bd_2}+s(C_{bd_2}+C_{bs_2}+C_{gb_2})$

The voltage gain, phase and the impedance expressions for the inverter can be formulated as:

$$A_{v1_{inv}} = \frac{-g_{m_1}-g_{m_2}+sC_{gd}}{\gamma_1+sC_1}, \quad A_{v2_{inv}} = \frac{g_{m_1}+g_{mb_1}+g_{ds_1}}{\gamma_1+sC_1}, \quad (1)$$

$$A_{v4_{inv}} = \frac{-g_{mb_1}+g_{bd_1}+sC_{bd_1}}{\gamma_1+sC_1}, \quad A_{v6_{inv}} = \frac{-g_{mb_2}+g_{bd_2}+sC_{bd_2}}{\gamma_1+sC_1} \quad (2)$$

$$Z_{1_{inv}} = \frac{\gamma_1+sC_1}{s^2(C_2C_1-C_{gd}^2)+s(C_2\gamma_1+g_mC_{gd})}, \quad (3)$$

$$Z_{2_{inv}} = \frac{\gamma_1+sC_1}{\alpha_1 s^2+\alpha_2 s+\alpha_3}, \quad (4)$$

$$Z_{3_{inv}} = Z_{out_{inv}} = \frac{sC_2}{s^2(C_1C_2-C_{gd}^2)+s(g_mC_{gd}+C_2\gamma_1)}, \quad (5)$$

$$Z_{4_{inv}} = \frac{\gamma_1+sC_1}{\alpha_4 s^2+\alpha_5 s+\alpha_6}, \quad Z_{6_{inv}} = \frac{\gamma_1+sC_1}{\alpha_7 s^2+\alpha_8 s+\alpha_9}, \quad (6)$$

where, $C_{gs} = C_{gs_1}+C_{gs_2}$, $C_{gd} = C_{gd_1}+C_{gd_2}$, $C_{db} = C_{db1}+C_{db2}$, $C_1 = C_{gd}+C_{db}+C_L$, $C_2 = C_{gd}+C_{gs}+C_{gb}$, $\gamma_1 = g_{ds}+g_{db}$, $g_{ds} = g_{ds_1}+g_{ds_2}$, $g_m = g_{m_1}+g_{m_2}$, $\gamma_2 = g_{ds_1}+g_{bs_1}+g_{m_1}+g_{mb_1}$, $\alpha_1 = C_1(C_{gs_1}+C_{bs_1})$, $\alpha_2 = C_1\gamma_2+\gamma_1(C_{gs_1}+C_{bs_1})$, $\alpha_3 = \gamma_1\gamma_2-g_{ds_1}(g_{m_1}+g_{mb_1}+g_{ds_1})$, $\alpha_4 = C_1(C_{bd_1}+C_{bs_1})-C_{bd_1}^2$, $\alpha_5 = C_1(g_{bs_1}+g_{bd_1})+\gamma_1(C_{bd_1}+C_{bs_1})-C_{bd_1}(2g_{bd_1}-g_{mb_1})$, $\alpha_6 = \gamma_1(g_{bs_1}+g_{bd_1})-(g_{bd_1}-g_{mb_1})g_{bd_1}$, $\alpha_7 = C_1(C_{bd_2}+C_{bs_2})-C_{bd_2}^2$, $\alpha_8 = C_1(g_{bs_2}+g_{bd_2})+\gamma_1(C_{bd_2}+C_{bs_2})-C_{bd_2}(2g_{bd_2}-g_{mb_2})$, $\alpha_9 = \gamma_1(g_{bs_2}+g_{bd_2})-(g_{bd_2}-g_{mb_2})g_{bd_2}$.

Here, the subscript in the gain (A) and the impedance (Z) expressions refer to the input terminal. For example, $A_{v1_{inv}}$ is the input-output voltage gain and $Z_{1_{inv}}$ is impedance at terminal ① of an inverter circuit. The overall voltage gain ($\tilde{A}_{v_{inv}}$) of the inverter is:

$$\tilde{A}_{v_{inv}} = G_{m_{inv}} Z_{out_{inv}}, \quad (7)$$

$$G_{m_{inv}} = \frac{A_{v1_{inv}}(1+\zeta)[s^2(C_1C_2-C_{gd}^2)+s(g_mC_{gd}+C_2\gamma_1)]}{s(C_{gd}+C_{gs}+C_{gb})}, \quad (8)$$

$$\zeta = \frac{g_{mb_2}-g_{m_1}-g_{ds_1}-g_{bd}-sC_{bd}}{g_m-sC_{gd}}. \quad (9)$$

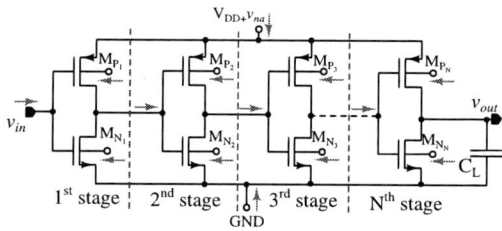

Fig. 2: Schematic diagram of a chain of CMOS inverters.

The overall output response ($\tilde{V}_{o_{inv}}$) of CMOS inverter due to deterministic noise fluctuations is as follows:

$$\tilde{V}_{o_{inv}} = A_{v1_{inv}}\left(v_{in}+\frac{Y_{32}}{Y_{31}}v_{na}+\frac{Y_{34}}{Y_{31}}v_{bp}+\frac{Y_{36}}{Y_{31}}v_{bn}\right). \quad (10)$$

Example-II: N-stages Inverter Chain

Fig. 2 shows such N-stages of chain considering the supply/bulk/common-mode noise (denoted by red colored arrows). The required two-port model expressions for the chain can be written in a similar way as discussed in the previous example. The impedance and phase expressions for the respective terminals are the same as derived for the inverter. The overall gain ($A'_{v_{ch}}$) with all the supply noise sources for an N-stages of inverter chain is:

$$A'_{v_{ch}} = \sum_{n=2}^{N}\left(\prod_{k=n}^{N}\left(\frac{Y_{o1}}{Y_{oo}}\right)_k \left(\frac{\sum_{i=1}^{p-1}(Y_{oi})}{(Y_{oo})}\right)_{n-1}\right)$$
$$+ \prod_{n=1}^{N}\left(\frac{Y_{o1}}{Y_{oo}}\right)_n + \left(\frac{\sum_{i=1}^{p-1}(Y_{oi})}{(Y_{oo})}\right)_N \quad (11)$$

where, N is total number of stages in an inverter chain and p refers to the input noise sources in a single stage. The derivation of the $A'_{v_{ch}}$ and the other details of two-parameters for the inverter chain are omitted due to space constraints.

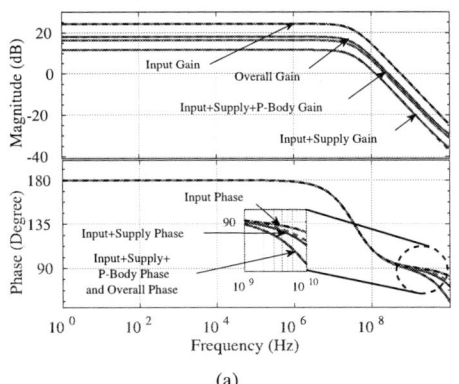
(a)

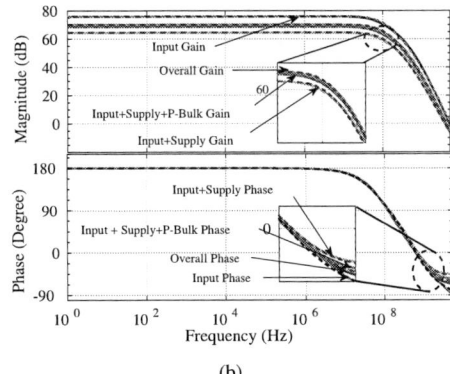
(b)

Fig. 3: Frequency and phase response due to the combination of noise sources for: (a) the CMOS inverter and; (b) the cascaded three stage CMOS inverter.

III. RESULTS AND DISCUSSIONS

In order to verify the proposed approach, two example circuits are designed in a standard 180 nm CMOS technology with supply voltage of 1.8 V and a load capacitor of 100 fF. The DC model parameters have been calculated using the SPICE based simulations and BSIM model.

Fig. 3(a) and (b) compare the analytical and simulation results of the single inverter and three-stage inverter, respectively. The analytical results are shown by dashed line whereas, the solid lines are for the simulation results. The impact of combination of different noise sources on the I/O gain and the phase behavior for the CMOS inverter and the three stage inverter are shown in Fig. 3(a) and (b), respectively. These different combinations of noise sources include the effect of supply, common-mode and bulk noises. The combined response of pull-up transistor due to supply and input node is the difference of the individual gain of these two node. This composite response in the presence of the PSN and input node noise is lowest among the other combinations as they are out-of-phase from each other. The analytical plot matches accurately with the simulation results with MPE of 3%.

IV. INSIGHTS AND USAGE OF THE PROPOSED ANALYSIS

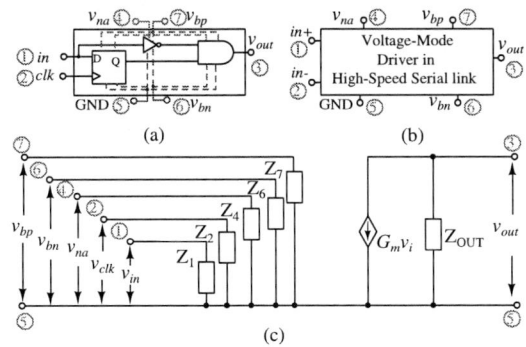

(c)

Fig. 4: (a) Edge detector circuit and, (b) voltage-mode driver circuit [8] with its (c) equivalent N-port network.

Fig. 4 shows the generalised two-port network for high-speed AMS block when the excited signals are applied at gate, supply and bulk nodes. Here, the word *generalised* means that Fig. 4(a) and (b) or any other system can be realised in the form of a two-port network as shown in Fig. 4(c). Consider

an example of an edge detector circuit, depicted in Fig. 4(a) which has multiple-terminals viz. in, clk, v_{na}, GND, v_{bn}, v_{bp} and v_{out}. The deterministic noise may disturb the DC voltage at these terminals and may degrade the SI of the system. Therefore, the SI effects due to supply noise can be analysed using the modified two-port network, shown in Fig. 4(c). Note that, the outputs of an AMS circuits are generally taken from drain to drain abutted output of nMOS and pMOS. This makes the circuit with high output impedance and hence, the circuits are modeled as transconductance amplifier with the PSN, GSN and the BSN. Here, Z_i is the impedance at the node i. The G_m depends on number of input ports and it can be changed by varying the input sources (v_i). These metrics are estimated using the proposed method. Moreover, the other circuits can be modelled using the proposed method in the similar fashion.

V. CONCLUSIONS

A method, estimation-by-inspection, is proposed to model the design parameters for a multi-port AMS system in the presence of noisy supplies. The mathematical model of the performance metrics viz. A_v, Z_i and ϕ, under the influence of PSN, GSN, BSN and common-mode noises are derived. On the basis of derived TFs, the gain cross over frequency, gain margin, phase margin and, subsequently, stability can be predicted. The method is not only limited to simple two transistors based circuits, but, can be extended for the system level blocks such as op-amp, flip-flops, data converters, etc.

REFERENCES

[1] K. Arabi *et al.*, "Power supply noise in SoCs: Metrics, management, and measurement," *IEEE Des. Test*, vol. 24, no. 3, pp. 236–244, 2007.
[2] J. N. Tripathi *et al.*, "A review on power supply induced jitter," *IEEE Trans. CPMT*, vol. 9, no. 3, pp. 511–524, 2018.
[3] P. R. Gray *et al.*, *Analysis and design of analog integrated circuits*. Wiley, 2001.
[4] A. S. Elwakil, "On the two-port network classification of Colpitts oscillators," *IET circ, devices & syst*, vol. 3, no. 5, pp. 223–232, 2009.
[5] P. Mlynek *et al.*, "Two-port network transfer function for power line topology modeling," *Radioengineering*, vol. 21, no. 1, 2012.
[6] V. K. Sharma *et al.*, "Deterministic noise analysis for single-stage amplifiers by extension of indefinite admittance matrix," *IEEE Open Journal of Circuits and Systems*, vol. 1, no. 1, pp. 1–16, 2020.
[7] W.-K. Chen, *Active Network Analysis*. River Edge, NJ, USA: World Scientific Publishing Co., Inc., 1991.
[8] J. N. Tripathi *et al.*, "Efficient modeling of power supply induced jitter in voltage-mode drivers (EMPSIJ)," in *IEEE Trans. CPMT*, vol. 7, no. 10, pp. 1691–1701, 2017.

978-1-7281-6162-4/20 $31.00 © 2020 IEEE

Accelerated Boundary Element Modeling of Lossy Conductors in Layered Media with a Single-Source Surface Impedance Operator

Shashwat Sharma, Piero Triverio

Department of Electrical & Computer Engineering, University of Toronto, Canada

shash.sharma@mail.utoronto.ca, piero.triverio@utoronto.ca

Abstract—**A new boundary element formulation is presented for the electromagnetic analysis of interconnects in layered media, which accurately captures skin effect in conductors of arbitrary shape and size. Results show superior CPU-time and memory performance compared to several state-of-the-art techniques.**

Index Terms—**Surface integral equations, layered media, lossy conductors**

I. Introduction

The accurate electromagnetic modeling of skin effect over a broad frequency range is essential in the design of integrated circuits, for the quantification of signal integrity phenomena in high-speed on-chip interconnects. Volume-based numerical methods, such as the finite element method (FEM) [1] or volume integral equations [2], require an extremely fine volumetric mesh for the structure to resolve the small skin depth at high frequencies, which is expensive. The boundary element method (BEM) is an appealing alternative, since it requires only a surface-based discretization of the objects in the structure [3].

Conductor modeling with the BEM requires formulating an interior problem to capture the skin effect inside objects, and an exterior problem to model coupling between them. The generalized impedance boundary condition (GIBC) [4] is an accurate and well-conditioned formulation for skin effect modeling, but requires both single- and double-layer potential operators [3] for the exterior problem. The computation of both operators is expensive for objects embedded in layered media, since the expensive multilayer Green's function and its curl are both required [5]. Multi-region formulations like [6] suffer from the same drawback, and also require expensive dual basis functions to achieve good conditioning [7].

A single-layer impedance matrix (SLIM) formulation was recently proposed [8]. The SLIM method is formulated in terms of a single source [9] and does not require the double-layer potential operator for the exterior problem, and therefore avoids the computational and implementation costs associated with the curl of the MGF. Unlike preceding single-source formulations [10], the SLIM method is well conditioned even without dual basis functions [8]. However, the SLIM approach

This work was supported by the Natural Sciences and Engineering Research Council of Canada, by Advanced Micro Devices, and by CMC Microsystems.

requires factorizing two dense matrices per object, which is only feasible for small objects.

In this work, an accelerated SLIM formulation is proposed, which avoids the assembly and factorization of dense matrices in the interior problem and allows handling large objects efficiently. This is achieved with the use of an object-wise adaptive integral method (AIM) [11] to accelerate matrix-vector products associated to the interior problem [12].

II. Proposed Formulation

We consider a homogeneous object bounded by a surface $\mathcal{S}$ with outward unit normal vector $\hat{n}$. The object has permittivity ε', conductivity σ, and $\varepsilon \triangleq \varepsilon' - j\sigma/\omega$, where ω is the cyclical frequency. A triangular mesh is generated for $\mathcal{S}$. The object is embedded in a layered medium, where the lth layer has permittivity ε_l and permeability μ_l.

A. Interior Problem

1) Original Configuration: The tangential electric and magnetic fields on $\mathcal{S}$ are related via the magnetic field integral equation (MFIE) [3], which is discretized and tested using RWG and $\hat{n} \times$ RWG basis functions [3], respectively, to get

$$-j\omega\varepsilon\, \mathbf{L}\mathbf{E} - \left(\mathbf{K} - \frac{1}{2}\mathbf{I}_\times\right)\mathbf{H} = \mathbf{0}, \qquad (1)$$

where the entries of matrices $\mathbf{L}$ and $\mathbf{K}$ may be found in literature [3], and involve the homogeneous Green's function of the object's material. Matrix $\mathbf{I}_\times$ is the identity operator obtained when RWG functions are tested with $\hat{n} \times$ RWG functions. Column vectors $\mathbf{E}$ and $\mathbf{H}$ contain the coefficients of the basis functions associated with the tangential electric and magnetic fields, respectively. Vector $\mathbf{E}$ can be expressed in terms of $\mathbf{H}$ with a rearrangement of (1),

$$\mathbf{E} = \frac{-1}{j\omega\varepsilon}\left(\mathbf{L}\right)^{-1}\left(\mathbf{K} - \frac{1}{2}\mathbf{I}_\times\right)\mathbf{H} = \mathbf{Z}\mathbf{H}, \qquad (2)$$

where $\mathbf{Z}$ may be interpreted as the surface impedance operator associated to the object [4].

2) Equivalent Configuration: Next, we use the surface equivalence principle to replace the conductive object with the background material in which it resides, while requiring that the tangential electric field remains unchanged for $\vec{r} \in \mathcal{S}$ [9]. An equivalent electric current density, $\mathbf{J}_\Delta = \mathbf{H} - \mathbf{H}_{\text{eq}}$, must be introduced on S to keep fields exterior to S unchanged, where $\mathbf{H}_{eq}$ is the discretized tangential magnetic field on S in the equivalent configuration.

In the equivalent configuration, the fields tangential to S can also be related via the discretized electric field integral equation (EFIE) [3]

$$j\omega\mu_l \, \mathbf{L}_l \mathbf{H}_{eq} - \left(\mathbf{K}_l - \frac{1}{2}\mathbf{I}_\times\right)\mathbf{E} = \mathbf{0}, \qquad (3)$$

where $\mathbf{L}_l$ and $\mathbf{K}_l$ involve the homogeneous Green's function of the medium just outside S. The equivalent tangential magnetic field $\mathbf{H}_{eq}$ may be expressed in terms of $\mathbf{E}$ with a rearrangement of (3),

$$\mathbf{H}_{eq} = \frac{1}{j\omega\mu_{eq}}\left(\mathbf{L}_{eq}\right)^{-1}\left(\mathbf{K}_{eq} - \frac{1}{2}\mathbf{I}_x\right)\mathbf{E} = \mathbf{Y}_{eq}\mathbf{E}. \qquad (4)$$

where $\mathbf{Y}_{eq}$ may be interpreted as the surface admittance operator of the object filled with the surrounding medium [10].

B. Exterior Problem

The augmented EFIE (AEFIE) [13] is employed to model the exterior problem over a broad frequency range. The discretized charge density $\rho_\triangle$ associated to $\mathbf{J}_\triangle$ is introduced as an additional unknown [13] to yield

$$jk_0\mathbf{L}_m^{(A)}\mathbf{J}_\triangle + c_0\mathbf{D}^T\mathbf{L}_m^{(\phi)}\mathbf{B}\rho_\triangle + \eta_0^{-1}\mathbf{I}_\times\mathbf{E} = \eta_0^{-1}\mathbf{E}_{inc}, \quad (5)$$

where $\mathbf{L}_m^{(A)}$ and $\mathbf{L}_m^{(\phi)}$ are, respectively, the discretized vector and scalar potential parts of the single-layer operator for the exterior problem, and involve the MGF. Column vector $\mathbf{E}_{inc}$ is related to the incident electric field. Quantities k_0, η_0, and c_0 are, respectively, the wave number, wave impedance, and speed of light in free space. Definitions of the sparse matrices $\mathbf{D}$ and $\mathbf{B}$ may be found in [13].

Equations (2), (4) and the definition of $\mathbf{J}_\triangle$ are then used in (5) to obtain the final system of equations

$$\begin{bmatrix} jk_0\mathbf{L}_m^{(A)} + \mathbf{C} & \mathbf{D}^T\mathbf{L}_m^{(\phi)}\mathbf{B} \\ \mathbf{FD}\left(\mathbf{I} - \mathbf{Y}_{eq}\mathbf{Z}\right) & jk_0\mathbf{I} \end{bmatrix}\begin{bmatrix} \mathbf{H} \\ c_0\rho_\triangle \end{bmatrix} = \begin{bmatrix} \mathbf{E}_{inc}/\eta_0 \\ \mathbf{0} \end{bmatrix}, \quad (6)$$

where $\mathbf{C} = \left(-jk_0\mathbf{L}_m^{(A)}\mathbf{Y}_{eq} + \frac{1}{\eta_0}\mathbf{I}_\times\right)\mathbf{Z}$. The second row of (6) is the discretized continuity equation relating $\mathbf{J}_\triangle$ and $\rho_\triangle$ [13]. Matrix $\mathbf{F}$ is defined in [13], and $\mathbf{I}$ is the identity matrix. Equation (6) is the SLIM formulation [8], which is well-conditioned and avoids the double-layer potential operator for the exterior problem, unlike the GIBC [4]. To solve (6) iteratively, the matrix-vector products involving $\mathbf{L}_m^{(A)}$ and $\mathbf{L}_m^{(\phi)}$ are accelerated with a multilayer AIM [14].

C. Accelerated Modeling of the Interior Problem

Solving (6) iteratively requires computing the two following matrix-vector products at each iteration k:

$$\mathbf{a}^{(k)} = \mathbf{Z}\mathbf{H}^{(k)} = \left(\frac{-\left(\mathbf{L}\right)^{-1}}{j\omega\varepsilon}\left(\mathbf{K} - \frac{1}{2}\mathbf{I}_\times\right)\right)\mathbf{H}^{(k)}, \qquad (7)$$

$$\mathbf{b}^{(k)} = \mathbf{Y}_{eq}\mathbf{a}^{(k)} = \left(\frac{\left(\mathbf{L}_{eq}\right)^{-1}}{j\omega\mu_{eq}}\left(\mathbf{K}_{eq} - \frac{1}{2}\mathbf{I}_\times\right)\right)\mathbf{a}^{(k)}. \qquad (8)$$

For large objects, the matrix inversions in (7) and (8) are not feasible. Instead, with a rearrangement of the matrices to be inverted, (7) and (8) are rewritten as two systems of equations,

$$\mathbf{L}\mathbf{a}^{(k)} = \frac{-1}{j\omega\varepsilon}\left(\mathbf{K} - \frac{1}{2}\mathbf{I}_\times\right)\mathbf{H}^{(k)}, \qquad (9)$$

$$\mathbf{L}_{eq}\mathbf{b}^{(k)} = \frac{1}{j\omega\mu_{eq}}\left(\mathbf{K}_{eq} - \frac{1}{2}\mathbf{I}_\times\right)\mathbf{a}^{(k)}. \qquad (10)$$

Since systems (9) and (10) are "nested" into (6), their solution must be computed at every iteration k, and can seriously burden CPU time. To avoid this bottleneck, we propose an efficient preconditioning and solution strategy for (9) and (10). The AIM is used to accelerate all matrix-vector products involving $\mathbf{L}$, $\mathbf{K}$, $\mathbf{L}_{eq}$ and $\mathbf{K}_{eq}$. An independent AIM grid is generated for each object [12]. We can write $\mathbf{A} \approx \mathbf{A}_{NR} + \mathbf{A}_{FR}$, where $\mathbf{A} \in \{\mathbf{L}, \mathbf{K}, \mathbf{L}_{eq}, \mathbf{K}_{eq}\}$, $\mathbf{A}_{NR}$ is sparse and contains the near-region entries of $\mathbf{A}$ which have been pre-corrected to account for the AIM grid contributions [15], and $\mathbf{A}_{FR}$ contains far-region interactions. Matrix-vector products involving $\mathbf{A}_{FR}$ are computed with the fast Fourier transform (FFT) [15]. The near-region entries of $\mathbf{L}$ and $\mathbf{L}_{eq}$ are used as preconditioners for the efficient solution of (9) and (10), respectively. The proposed acceleration scheme allows the SLIM method to capture all electromagnetic phenomena inside conductors, both small and large, in an efficient, accurate and broadband way.

III. RESULTS

We consider two examples: an on-chip inductor coil [15], and a portion of an interposer with 80 copper signal lines and a finite ground plane (courtesy of Dr. Rubaiyat Islam, Advanced Micro Devices). Both structures are embedded in a dielectric substrate ($\varepsilon_r = 2.1$ and 4, respectively) backed by a silicon layer over an infinite ground plane. The geometry and electric surface current densities are shown for the inductor and the interposer in the top panels of Fig. 1 and Fig. 2, respectively. For the inductor, the proposed method is compared against the DSA [10], GIBC [4] and SLIM [8] formulations, which all require the factorization of dense matrices for the interior problem. Results are also compared against a commercial finite element solver, Ansys HFSS. The scattering (S) parameters in Fig. 1 confirm the accuracy of the proposed method. Table I shows that the proposed method is between 1.5 and $16\times$ faster than all other methods, and among the most memory efficient. The SLIM formulation also requires, on average, $31\times$ fewer iterations than the DSA method. For the interposer, the proposed method is compared to HFSS, and to a recently-proposed accelerated GIBC formulation [12], where matrix factorizations in the interior problem are avoided. A comparison against DSA [10] and GIBC [4] was not feasible within the available 256 GB of memory. Similarly, HFSS was not able to accurately resolve skin effect beyond about 30 GHz without running out of memory, so a coarser mesh had to be used. The coarse HFSS mesh has 2–4 volume elements spanning the conductor cross section, while the skin depth becomes over $10\times$ smaller than the cross sectional dimensions. The S parameters reported in the bottom panel of Fig. 2

TABLE I: Performance comparison for the examples in Section III, on a 3 GHz Intel Xeon CPU, single-threaded.

	Inductor (4,470 triangles, 22 frequency points)					Interposer (156,820 triangles, 31 frequency points)			
	HFSS	DSA [10]	GIBC [4]	SLIM [8]	Proposed	HFSS	HFSS (coarse)	Accel. GIBC [12]	Proposed
Total CPU time (hours)	2.5	10.2	1.0	4.1	0.6	N/A	41.0	212.1	104.4
Peak memory (GB)	16.7	7.2	1.0	6.8	1.2	> 256	189.2	30.5	28.3

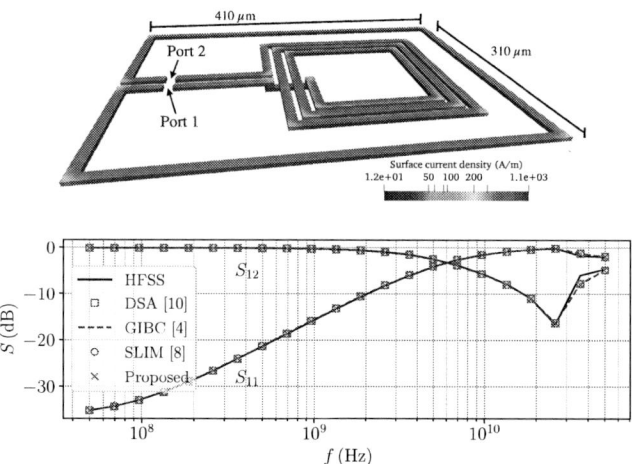

Fig. 1: Top panel: geometry of the inductor in Section III, with the electric surface current density at 100 MHz. Bottom panel: comparison of scattering parameters.

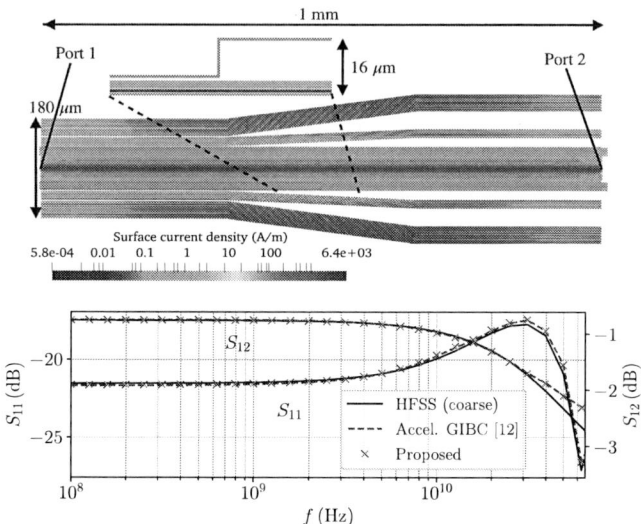

Fig. 2: Top panel: geometry of the interconnect network in Section III, with the electric surface current density at 1 GHz. Bottom panel: comparison of scattering parameters.

show an excellent match between the two BEM approaches. Results are also in agreement with HFSS up to about 30 GHz. Beyond this frequency, HFSS results start deviating due to the shrinking skin depth. Table I confirms that the proposed method is faster than the accelerated GIBC by a factor of over $2\times$, since the double-layer potential is avoided in the exterior problem. The proposed method is also $6.7\times$ more memory efficient than HFSS with the coarse mesh, while providing better accuracy beyond 30 GHz. In conclusion, the numerical examples considered demonstrate that the proposed method can efficiently model large realistic lossy conductors in layered media, thanks to an object-wise application of the AIM, and is significantly more efficient than representative BEM and FEM techniques from the state of the art.

REFERENCES

[1] J.-M. Jin, *The Finite Element Method in Electromagnetics*, 3rd ed. Hoboken, NJ, USA: Wiley, 2014.

[2] A. E. Ruehli, G. Antonini, and L. Jiang, *Circuit Oriented Electromagnetic Modeling Using the PEEC Techniques*. IEEE Press, 2017.

[3] W. C. Gibson, *The Method of Moments in Electromagnetics*. Boca Raton, FL, USA: CRC press, 2014.

[4] Z. G. Qian, W. C. Chew, and R. Suaya, "Generalized impedance boundary condition for conductor modeling in surface integral equation," *IEEE Trans. Microw. Theory Tech.*, vol. 55, no. 11, pp. 2354–2364, Nov. 2007.

[5] K. A. Michalski and J. R. Mosig, "Multilayered media Green's functions in integral equation formulations," *IEEE Trans. Antennas Propag.*, vol. 45, no. 3, pp. 508–519, Mar. 1997.

[6] T. Xia, H. Gan, M. Wei, W. C. Chew, H. Braunisch, Z. Qian, K. Aygün, and A. Aydiner, "An integral equation modeling of lossy conductors with

the enhanced augmented electric field integral equation," *IEEE Trans. Antennas Propag.*, vol. 65, no. 8, pp. 4181–4190, Aug. 2017.

[7] A. Buffa and S. H. Christiansen, "A dual finite element complex on the barycentric refinement," *Math. Computation*, vol. 76, pp. 1743–1769, 2007.

[8] S. Sharma and P. Triverio, "SLIM: A well-conditioned single-source boundary element method for modeling lossy conductors in layered media," *IEEE Antennas Wireless Propag. Lett.*, 2020 (accepted, axXiv: 2007.07378).

[9] M. Huynen, K. Y. Kapusuz, X. Sun, G. Van der Plas, E. Beyne, D. De Zutter, and D. Vande Ginste, "Entire domain basis function expansion of the differential surface admittance for efficient broadband characterization of lossy interconnects," *IEEE Trans. Microw. Theory Tech.*, vol. 68, no. 4, pp. 1217–1233, Jan. 2020.

[10] U. R. Patel, S. Sharma, S. Yang, S. V. Hum, and P. Triverio, "Full-wave electromagnetic characterization of 3D interconnects using a surface integral formulation," in *IEEE EPEPS*, San Jose, CA, Oct. 2017.

[11] E. Bleszynski, M. Bleszynski, and T. Jaroszewicz, "AIM: Adaptive integral method for solving large-scale electromagnetic scattering and radiation problems," *Radio Sci.*, vol. 31, no. 5, pp. 1225–1251, Sep. 1996.

[12] S. Sharma and P. Triverio, "A fully-accelerated surface integral equation method for the electromagnetic modeling of arbitrary objects," *IEEE Trans. Antennas Propag.*, 2020 (submitted, arXiv: 2003.11679).

[13] Z.-G. Qian and W. C. Chew, "Fast full-wave surface integral equation solver for multiscale structure modeling," *IEEE Trans. Antennas Propag.*, vol. 57, no. 11, pp. 3594–3601, Nov. 2009.

[14] S. Sharma, U. R. Patel, S. V. Hum, and P. Triverio, "A complete surface integral method for broadband modeling of 3D interconnects in stratified media," *arXiv e-prints*, p. arXiv: 1810.04030, Oct. 2018.

[15] T. Moselhy, X. Hu, and L. Daniel, "pFFT in FastMaxwell: A fast impedance extraction solver for 3D conductor structures over substrate," in *Proc. Conf. Des., Automat. Test*, 2007.

978-1-7281-6162-4/20 $31.00 © 2020 IEEE

Estimating Per-Unit-Length Resistance Parameter in Emerging Copper-Graphene Hybrid Interconnects via Prior Knowledge based Accelerated Neural Networks

Rahul Kumar[*], S. S. Likith Narayan[#], Somesh Kumar[†], Sourajeet Roy[#], Brajesh K. Kaushik[#], Ramachandra Achar[§] and Rohit Sharma[*]

[*] Department of Electrical Engineering, Indian Institute of Technology Ropar, Ropar, India
[#] Department of Electronics and Communication Engineering, Indian Institute of Technology Roorkee, Roorkee, India
[†] Department of Electronics and Communication Engineering, ABV-Indian Institute of Information Technology & Management, Gwalior, India
[§] Department of Electronics, Carleton Univeristy, Ottawa, Canada
Email: rohit@iitrpr.ac.in

Abstract — **In this paper, an artificial neural network (ANN) is developed to model how the geometrical parameters of hybrid copper-graphene interconnects affect the per-unit-length resistance values. The proposed ANN is intelligently trained using large amounts of data representing the prior knowledge about the interconnects, extracted from an analytical model and sparse amount of data extracted from a rigorous full-wave electromagnetic solver. In this way, the training of the ANN model is accelerated without significant loss in accuracy.**

Keywords— **Artificial neural networks (ANN), copper-graphene hybrid interconnects, per-unit-length resistance, variability analysis, training data.**

I. INTRODUCTION

In order to meet the high data rates and low energy-per-bit standards of sub-22nm technology nodes, various hybrid interconnect networks combining the electrical properties of multiple materials are currently being investigated. Among them, hybrid copper-graphene interconnects have shown some promising results for on-chip applications [1], [2]. In such interconnects, multiple layers of graphene act as barrier layers encapsulating the copper conductor. Due to the relatively large mean free paths of electrons in graphene and its high current carrying capacity, the resistance of the barrier layers come in parallel to that of the copper conductor, thereby effectively lowering the overall per-unit-length (p. u. l.) resistance of the interconnect. In order to quantify the performance of such novel hybrid interconnects, electronic design automation (EDA) tools need to be capable of quickly but accurately determining their p. u. l. parameters from their geometrical values.

Typically, full-wave electromagnetic (EM) solvers are used to extract the p. u. l. parameters from the geometrical and physical description of interconnect structures [3]. However, as the geometry of the interconnect structures change during design iterations, the EM solver has to be rerun. Given the exorbitant CPU run time for each EM solver run, this task quickly becomes computationally intractable. To mitigate this problem, surrogate models such as artificial neural networks (ANNs) have been employed [4]. ANNs employ flexible and nonlinear activation functions to characterize the functional mapping between the values of the geometrical parameters and the p. u. l. parameters of the interconnects [5]. Once trained, the ANN can be stored as a closed-form surrogate to the EM solver. Subsequently, ANN coefficients can then be repeatedly used to efficiently estimate the p. u. l. parameter values for different design scenarios. Unfortunately, ANNs require large sets of data extracted from full-wave EM solvers at excessive computational time cost for training.

In this paper, a methodology to accelerate the training of ANNs in order to quickly quantify the p. u. l. resistance of hybrid copper-graphene interconnects from the knowledge of the geometrical interconnect structures is developed. The proposed methodology begins by constructing an initial ANN that is trained using the data from an analytic model. The analytic model used is based on the Fuchs-Sondheimer and Mayadas-Shatzkes bulk, surface, and grain boundary scattering resistance models of copper combined with the scattering resistance model of graphene nanoribbons [6]. This analytic model represents the circuit designer's prior knowledge regarding the interconnect p. u. l. resistance and so the ANN is referred to as a prior knowledge ANN (PK-ANN). Importantly, this model being analytic, it can be quickly solved as compared to a full-wave EM solver to extract the requisite training data. Furthermore, in the work of [6], it has been shown that this analytic model is able to quantify the p. u. l. resistance of hybrid copper-graphene interconnects with a reasonable degree of accuracy. The ability to do so at extremely small time costs is the key attribute of the proposed methodology.

Next, in order to further enhance the accuracy of the PK-ANN, a second ANN is constructed. This ANN, referred to as a corrector ANN (C-ANN), models the error between the results of the PK-ANN and the EM solver. This error quantity, typically being small, ensures that the C-ANN can be trained using a small set of EM solver runs compared to the large number of runs required to train a conventional ANN. Finally, the sum of the outputs from the PK-ANN and C-ANN together provide an

978-1-7281-6162-4/20 $31.00 © 2020 IEEE

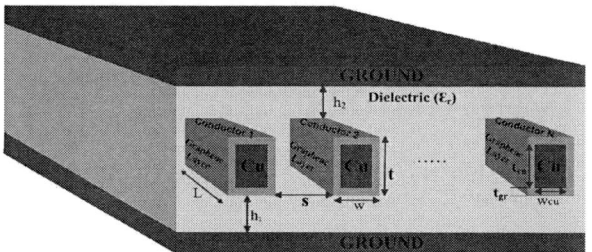

Fig. 1. Cross-section of copper-graphene hybrid interconnect network.

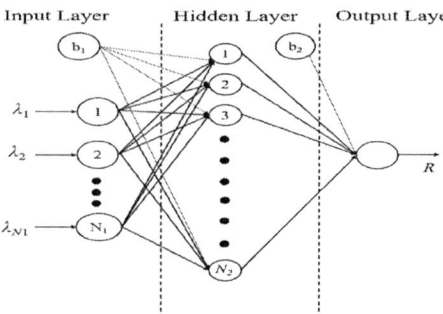

Fig. 2. Neural network architecture for PK-ANN showing input, hidden, and output layers.

accurate estimate of the p. u. l. resistance of hybrid copper-graphene interconnects. Thus, in this paper, prior knowledge of the interconnects is leveraged to expedite the training of ANNs without much loss in accuracy.

II. DEVELOPMENT OF THE PROPOSED PRIOR KNOWLEDGE BASED ACCELERATED ANN

Consider a hybrid copper-graphene interconnect network as shown in Fig. 1. Each conductor consists of a copper core and multiple graphene barrier layers encapsulating this core. Now, in order to model the impact of the geometrical parameters of the structure of Fig, 1 on the p. u. l. resistance, the proposed prior knowledge acclereated ANN model is adopted.

A. Constructing Prior Knowledge ANN (PK-ANN)

The proposed approach begins by constructing the prior knowledge ANN (PK-ANN). The PK-ANN has a three-layer structure as depicted in Fig. 2. The input variables $\lambda = [\lambda_1, \cdots, \lambda_{N_1}]$ represent the N_1 different geometric parameters of interest. In the second (hidden) layer, the weighted values of the inputs are gathered and processed using N_2 neurons. Finally, in the output layer, the weighted output of the neurons in the hidden layers are gathered and again processed using a single neuron. The output R represents the estimated p. u. l. resistance of the conductors of Fig. 1. Overall, the output R is mathematically expressed as

$$R = \sigma_{out}\left(\sum_{i=1}^{N_2}\omega_{i1}^{(2)}\sigma_{hi}\left(\sum_{j=1}^{N_1}\omega_{ji}^{(1)}\lambda_j + \omega_{0i}^{(1)}b_1\right) + \omega_{01}^{(2)}b_2\right) \quad (1)$$

where σ_{out} and σ_{hi} are the activation functions in the output layer and the i-th hidden layer neurons respectively, $\omega = [\omega_{11}^{(1)}, \cdots, \omega_{N_21}^{(2)}]$ represents all the synaptic weights, and the general synaptic weight $\omega_{ij}^{(k)}$ is the weight between the i-th neuron in the k-th layer and j-th neuron in the $k+1$-th layer. Moreover, the variables b_1 and b_2 refer to the bias values. The weights of (1) are optimized so that the result of the ANN closely matches the training data within a set error limit [5].

For the PK-ANN, the initial training data is extracted from an analytic model based on the Fuchs-Sondheimer and Mayadas-Shatzkes bulk, surface, and grain boundary scattering resistance models of copper and the scattering resistance of graphene nanoribbons [6]. As per this model, the equivalent p. u. l. resistance of the interconnects is given as [6]

$$\frac{1}{R} = \frac{1}{R_{Cu}} + \sum_{k=1}^{N_b}\frac{1}{R_{Gk}};$$

$$R_{Cu} = \frac{\rho_{bulk}(F_{FS}+F_{MS})}{wt}; \quad R_{Gk} = \frac{\square}{2e^2 N_b N_{c\square}}\left(1 + \frac{L}{\xi_{eff}}\right) \quad (2)$$

In (2), R_{Cu} and R_{Gk} are the p. u. l. resistance due to the copper core and the k-th graphene layer, ρ_{bulk} is the resistivity of bulk copper, w and t are the interconnect width and thickness respectively, N_b is the number of graphene barrier layers, N_{ch} is the number of conduction channels per layer and ξ_{eff} is the effective mean free path of graphene barrier layer. The functions F_{FS} and F_{MS} are the Fuchs-Sondheimer and Mayadas-Shatzkes models for the Surface and grain boundary scattering mechanisms in copper expressed as

$$F_{FS} = 0.45\xi_{Cu}\left(1-\rho_{Cu}\right)\left(\frac{w+t}{wt}\right);$$

$$F_{MS} = \left(1 - 1.5\phi + 3\phi^2 - 3\phi^3\ln\left(1+\frac{1}{\phi}\right)\right)^{-1}; \phi = \frac{\xi_{Cu}R_f}{D_g(1-R_f)} \quad (3)$$

where ξ_{Cu} is the mean free path of electrons in copper, R_f is the reflection coefficient, and D_g is the grain size of copper. The main advantage of the analytic model of (2) is that it offers reasonably accurate results but requires virtually negligible time costs to solve compared to a full-wave EM solver. Thus, it is possible to extract large amounts of data from (2) in order to train a reasonably accurate PK-ANN at very low time costs.

B. Constructing Corrector ANN (C-ANN)

In this work, to enhance the accuracy of the PK-ANN, a corrector ANN (C-ANN) is constructed. The C-ANN too has the same basic structure as shown in Fig. 2 with the exception that the output is in the form

$$\hat{R} = R_{EM} - R \quad (4)$$

where R_{EM} is the p. u. l. resistance value obtained from the EM solver and R is the estimate of the same obtained from the PK-ANN. It is understood from (4) that the variance of the C-ANN output can be expressed as [7]

978-1-7281-6162-4/20 $31.00 © 2020 IEEE

TABLE I: VALUE OF CORRELATION COEFFICIENT OF OUTPUT OF PK-ANN

	Value of ANN training + testing samples (N_s)					
	100	200	300	500	600	800
Correlation coefficient	0.782	0.971	0.967	0.956	0.961	0.958

TABLE 2: CPU COST COMPARISON

	Proposed PKA Estimator	Conventional ANN	EM Solver
CPU Time	~35 hours	50 hours	50 hours

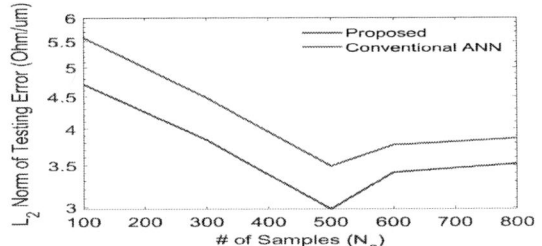

Fig. 3. L_2 norm of the testing error for the proposed PKA estimator and the conventional ANN with respect to the number of samples.

$$Var\left(\hat{R}\right) = Var\left(R_{EM}\right) + Var\left(R\right) - 2Cov\left(R_{EM}, R\right) \qquad (5)$$

Now, because the error of (4) is typically small, the correlation coefficient of R_{EM} and R tends to be closer to 1. Under this condition, based on (5) the variance of the C-ANN output will approach zero. Because of this reason, the C-ANN can be trained using very few runs of the full-wave EM solver.

C. Combining PK-ANN and C-ANN

Based on (4), an estimator of the p. u. l. resistance of hybrid copper-graphene interconnects is designed to be the sum of the outputs from the PK-ANN and C-ANN. This estimator is called the prior knowledge based accelerated (PKA) estimator. Notably, the total time cost required to train the PKA estimator is the sum of the time cost required to train the PK-ANN and C-ANN. Given that both these individual time costs are very small due to the analytic nature of the model of (2) and the small variance of the output of C-ANN in (5), their sum is expected to be small as well. Thus, the overall time cost to train the PKA estimator is usually smaller than that required to train a conventional ANN.

III. NUMERICAL RESULTS AND DISCUSSIONS

In this section, a 3-line hybrid copper-graphene interconnect network is considered. The geometrical and physical parameters of interest are [w, t, s, h_1, t_{gr}] (see Fig. 1). The nominal values considered for these parameters are [13, 27.3, 12, 27, 0.8] (all in nm), with an assumed variation of +/- 10%.

In this example, the proposed PKA estimator and a conventional ANN model are constructed. When constructing the PKA estimator, the PK-ANN is trained using 4000 data samples extracted from the analytic model of (2) in order to achieve L_2 training and testing error norms of below 0.15 $\Omega/\mu m$. Only a single hidden layer with 10 neurons is used with the activation functions as the hyperbolic tangent function. The time cost of extracting each data sample from the analytic model of (2) is only 50 μs (which is more than four orders of magnitude faster compared to the EM solver (Ansys Q3D Extractor)). For training the C-ANN and the conventional

ANN, data samples are extracted from the EM solver. The number of data samples used is varied as N_s = [100, 200, 300, 500, 600, 800] where half of the samples served as training samples while the other half as testing samples. As before, only a single hidden layer is required for the C-ANN and conventional ANN. For the different values of N_s, the correlation coefficient between the output from PK-ANN and the output from the conventional ANN is calculated using 5000 Monte Carlo samples and the results are listed in Table I. It is seen from Table I that when sufficient samples are taken (i.e., N_s > 100), the correlation coefficient is very close to 1, thereby indicating the reasonable accuracy provided by the PK-ANN. Notably, the PK-ANN achieves accurate results at a combined training and testing CPU cost of only 0.2 seconds whereas the conventional ANN requires roughly 10 hours to extract and fit data for N_s = 100 samples. Next, the scaling of the L_2 testing error norms w. r. t. N_s for the proposed PKA estimator and the conventional ANN is shown in Fig. 3. From Fig. 3, it is noted that the proposed PKA estimator is consistently able to achieve a lower error for same N_s than the conventional ANN. In fact, to reach the same error norm of 3.5 $\Omega/\mu m$, the proposed PKA estimator needs roughly 350 full-wave EM solver runs compared to the 500 needed by the conventional ANN. Beyond 500 samples, both the PKA and the conventional ANN overfit the data as seen by the increasing error.

IV. CONCLUSION

In this work, a prior knowledge driven approach to accelerate the training of ANNs is proposed. This approach efficiently estimates the p. u. l. resistance of hybrid copper-graphene interconnects using less expected EM solver runs as compared to traditional ANN based approaches.

REFERENCES

[1] W.-S. Zhao, D.-W. Wang, G. Wang, and W.-Y. Yin, "Electrical modeling of on-chip Cu-graphene heterogeneous interconencts," *IEEE Electronic Devices Letters*, vol. 36, no. 1, pp. 74-76, Jan. 2015.

[2] Z.-H. Cheng *et. al.*, "Analysis of Cu-graphene interconnects," *IEEE Access*, vol. 6, pp. 53499-53508, Oct. 2018.

[3] R. C. Paul, *Analysis of Transmission Lines*, New York: Wiley 1994.

[4] A. Veluswami, M. S. Nakhla, and Q.-J. Zhang, "The application of neural networks to EM based imualtion and optimization of interconencts in high-speed VLSI circuits," *IEEE Trans. Microwave Theory and Techn.*, vol. 45, no. 5, pp. 712-723, May 1997.

[5] Q.-J. Zhang and K. C. Gupta, *Neural Networks for RF and Microwave Design*, Norwood, Massachusetts: Artech House 2000.

[6] R . Kumar *et al.*, "Role of grain size on the effective resistivity of cu-graphene hybrid interconnects," in *Proc. IEEE Electronic Components and Tech. Conf.*, May 2020.

[7] Y. Li *et al.*, "A predictor–corrector algorithm for fast polynomial chaos-based uncertainty quantification of multi-walled carbon nanotube interconnects," *IEEE Trans. Comp., Packag. and Manuf. Tech.*, vol. 9, no. 10, pp. 1963-1975, Oct. 2019.

Variational Inference approach to Jitter decomposition in High-speed Link

Bobi Shi, Thong Nguyen and Jose Schutt-Aine
Department of Electrical and Computer Engineering.
University of Illinois Urbana - Champaign
{bobishi2, tnnguye3, jesa} @illinois.edu

Abstract—**Jitter, a timing deviation from the ideal edge position, is an unwanted phenomenon in high-speed link systems. Decomposing jitter into its components and identifying each type of jitter are beneficial to diagnose the root causes of jitter, thereof improving the system design. In recent years, variational bayes inference (VBI) has made substantial progress towards improving the efficiency of statistical modeling. This paper proposes a jitter approximation method using VBI. Applications approximating different mixtures of well-known components of jitter show good results. The approximated distribution is much closer to the true jitter distributions as compared to traditional methods.**

Index Terms—**jitter decomposition, high-speed link system, Gaussian mixture model, stochastic variontional inference**

I. INTRODUCTION

Over the past decades, signal analysis has been aggressively pursued in order to meet urgent demands for multi-gigabit data rate transfer. As data rates keep increasing, jitter, defined as the deviation of timing edges from their ideal positions, becomes a crucial element affecting the performance of overall high-speed link systems. In the high-speed link system, a small timing deviation might be a significant portion of the signal interval because of the fast transition and short unit interval. Then the jitter-induced errors corrupt the signals and even the clock, causing failure of the proper signal transmission. As a consequence, tighter control over jitter for high-speed link systems is needed to prevent signal failure or non-ideality, and understanding amount of jitter generated by different jitter sources is vital for high-speed link system designer to meet the requirements.

Generally, an observed total jitter can be classified into two categories: deterministic jitter (DJ) and random jitter (RJ). Deterministic jitter follows the bounded distribution and is normally separated into various components to investigate different root causes. It is further divided into periodic jitter (PJ), data-dependent jitter (DDJ) and bounded uncorrelated jitter (BUJ). PJ repeats in a sinusoidal fashion and external deterministic noise sources such as switching power supply noise can lead to PJ. DDJ is the data pattern related to jitter. The rest of the bounded jitter goes into the BUJ category. Random jitter is unpredictable jitter and is caused by unbounded jitter sources. Random jitter normally has a Gaussian distribution model because the sources of random jitter are thermal noise, 1/f flicker noise or shot noise, which fit to a bell curve.

In order to separate jitter into its components, various methods have been developed. The tail-fitting algorithm is one of the prominent jitter separation method [1]. Two Gaussian distribution curves are found to fit the tails of total jitter probability distribution. Then, the quantity of RJ and DJ can be determined through tail-fitting. However, the drawback is that a large amount of jitter samples is required to identify the fit tail part of the distribution. In [2] and [3], a better jitter decomposition method based on Gaussian mixture model (GMM) was promoted. For comparison, in this preliminary work on jitter extraction using stochastic variation inference (SVI) framework, we aslo implemented a GMM model but reformulated the maxmimum likelihood problem into a SVI problem. Convergence rate is about the same between the two methods but accuracy is improved by using SVI. In addition, rather than restricting to a limited class of models that allows an analytical solution, SVI provides a general tool to define any statistical model,regardless of the complexity.

In the next section, the Dual-Dirac model of jitter is reviewed at first. Section III presents the GMM and variational GMM algorithm to segregate the total jitter. Examples of decomposing a mixture of RJ and DJ jitters with different type of DJ are presented in Section IV. Section V concludes the paper.

II. DUAL-DIRAC MODEL OF JITTER

As mentioned in Section I, jitter can be mainly classified into RJ and DJ, Dual-Dirac model [4] assumes that the total jitter can be explained by 2 components whose probability density functions (pdf) are

$$
\begin{aligned}
p\left(DJ\right) &= 0.5\delta\left(t-\mu_L\right)+0.5\delta\left(t-\mu_R\right) \\
p\left(RJ\right) &= \mathcal{N}\left(0,\sigma^2\right)
\end{aligned}
\tag{1}
$$

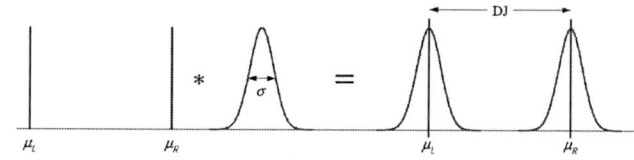

Fig. 1. Dual-Dirac model.

Illustration of total jitter distribution is shown in Figure 1. Fitting jitter data to dual-dirac model means finding σ, μ_L and μ_R to maximize the likelihood observing the data. In most cases, PJ, an example of whose distribution is shown in Figure 2, can also be approximated by 2 Dirac functions as well.

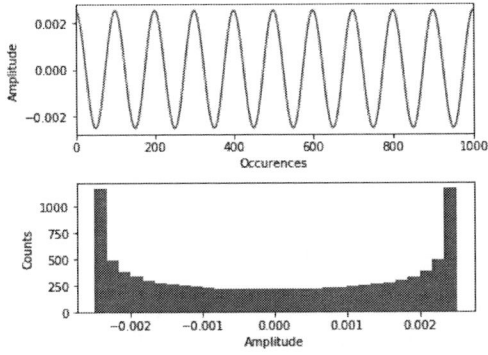

Fig. 2. Periodic jitter.

An advantage of the dual-dirac model is that it is simple and analytical. However, the trade-off is that the accuracy is low and jitter decomposed by the dual-dirac model can only be separated into either DJ or RJ, which is not very useful for debugging the circuit of interest.

III. STATISTICAL MODELING OF JITTER WITH GMM

Due to the oversimplicity of dual-dirac model, many attempts to improve jitter approximation have been proposed in the literature. Among those, standing out is the work in [2], a GMM is used to approximate the total jitter distribution. It is useful because GMM is a powerful and flexible model that can accurately approximate a smooth pdf.

Consider a statistical model with the latent variable z, observed variable x, let $\boldsymbol{\theta}$ denote the set of parameters. The log-likelihood of observing a set of data X is

$$\log p\left(X \mid \boldsymbol{\theta}\right) = \int \log p\left(X \mid \boldsymbol{z}, \boldsymbol{\theta}\right) p\left(\boldsymbol{z} \mid \boldsymbol{\theta}\right) \mathrm{d}\boldsymbol{z} \quad (2)$$

Statistical inference involves finding the maximum likelihood (ML) solution, $\boldsymbol{\theta}_{ML}$, that maximizes the (log-)likelihood in (2). The expectation maximization (EM) algorithm is an iterative algorithm converging the $\boldsymbol{\theta}$ to the ML solution and is especially useful when no closed-form solution is available.

A GMM of n clusters is formulated using a latent variable z representing the Gaussian cluster from which a sample x comes from. Thus, z is sampled from a categorial, or multinomial, distribution with the mass probability $\boldsymbol{\pi} = [\pi_1, \pi_2, \ldots, \pi_n]^T$ such that $\pi_i \geq 0$, $\sum_{i=1}^{n} \pi_i = 1$, i.e. $p\left(z = k\right) = \pi_k$.

$$\begin{aligned} \boldsymbol{z}, \boldsymbol{\theta} &\sim \text{Categorical}\left(\boldsymbol{\pi}\right) \\ \boldsymbol{x} \mid \boldsymbol{z}, \boldsymbol{\theta} &\sim \textstyle\sum_{k=1}^{K} \pi_k \mathcal{N}\left(\boldsymbol{\mu}_k, \boldsymbol{\Sigma}_k\right) \end{aligned} \quad (3)$$

In this case, $\boldsymbol{\theta}$ are the set of unknown parameters such as $\pi_1, \boldsymbol{\mu}_1, \boldsymbol{\Sigma}_1$, etc. Given a set of data $\{\boldsymbol{x}_i\}$, $i = 1, 2, ..., N$, the model log-likelihood is

$$\log p\left(X \mid \boldsymbol{\theta}\right) = \sum_{i=1}^{N} \log \left[\sum_{k=1}^{K} \pi_i p\left(\boldsymbol{x}_i \mid \boldsymbol{z}_k\right)\right] \quad (4)$$

The update formulas for k^{th} Gaussian component at iteration j are given by [5]:

$$\left\langle z_{k,i}^{(j)}\right\rangle = \frac{\pi_k \mathcal{N}\left(\boldsymbol{x}_i \mid \boldsymbol{\mu}_k^{(j)}, \boldsymbol{\Sigma}_k^{(j)}\right)}{\sum_{m=1}^{M} \pi_m \mathcal{N}\left(\boldsymbol{x}_i \mid \boldsymbol{\mu}_m^{(j)}, \boldsymbol{\Sigma}_m^{(j)}\right)} \quad (5a)$$

$$\pi_k^{(j+1)} = \frac{1}{N}\sum_{i=1}^{N} \left\langle z_{k,i}^{(j)}\right\rangle \quad (5b)$$

$$\boldsymbol{\mu}_k^{(j+1)} = \frac{\sum_{i=1}^{N} \left\langle z_{k,i}^{(j)}\right\rangle \boldsymbol{x}_i}{\sum_{i=1}^{N} \left\langle z_{k,i}^{(j)}\right\rangle} \quad (5c)$$

$$\boldsymbol{\Sigma}_k^{(j+1)} = \frac{1}{\sum_{i=1}^{N} \left\langle z_{k,i}^{(j)}\right\rangle}\sum_{i=1}^{N} \left\langle z_{k,i}^{(j)}\right\rangle \left(\boldsymbol{x}_i - \boldsymbol{\mu}_k^{(j)}\right) \left(\boldsymbol{x}_i - \boldsymbol{\mu}_k^{(j)}\right)^T$$
$$(5d)$$

In variational inference view, let $q\left(\boldsymbol{h}, \boldsymbol{\phi}\right)$ be the probability distribution on some hidden variables $\boldsymbol{h} \in \mathbb{R}^d$ with some parameters $\boldsymbol{\phi}$, for simplicity, the $\boldsymbol{\phi}$-dependency is implicitly acknowledged, $\boldsymbol{\phi}$ will be dropped from the notation. Since $\int q\left(\boldsymbol{z}\right) \mathrm{d}\boldsymbol{z} = 1$, the log-likelihood can be re-written as [6]

$$\log p\left(\boldsymbol{x} \mid \boldsymbol{\theta}\right) = \mathcal{L}\left(q, \boldsymbol{\theta}\right) + KL\left(q \parallel p_{\boldsymbol{h}\mid\boldsymbol{x}}\right) \quad (6)$$

with

$$\mathcal{L}\left(q, \theta\right) = \int q\left(\boldsymbol{h}\right) \log \left(\frac{p\left(\boldsymbol{x}, \boldsymbol{h} \mid \theta\right)}{q\left(\boldsymbol{h}\right)}\right) \mathrm{d}\boldsymbol{h} \quad (6a)$$

and

$$KL\left(q \parallel p_{\boldsymbol{h}\mid\boldsymbol{x}}\right) = -\int q\left(\boldsymbol{h}\right) \log \left(\frac{p\left(\boldsymbol{h} \mid \boldsymbol{x}, \theta\right)}{q\left(\boldsymbol{h}\right)}\right) \mathrm{d}\boldsymbol{h} \quad (6b)$$

The term $KL\left(q \parallel p\right) \geq 0$, $\forall p, q$ is known as the Kullback-Leibler divergence which measures how much q differs from p. Hence, (6) becomes

$$\log p\left(\boldsymbol{x} \mid \boldsymbol{\theta}\right) \geq \mathcal{L}\left(q, \boldsymbol{\theta}\right) \quad (7)$$

$\mathcal{L}\left(q, \boldsymbol{\theta}\right)$ is called the evidence lower bound (ELBO). As can be seen from (7), the ELBO provides a lower bound to the likelihood, which makes maximizing it is as good as maximizing the likelihood. The gap (difference) between the log-likelihood and the ELBO is exactly the KL divergence, if $KL\left(q \parallel p\right) = 0$, the ELBO hits the log-likelihood, q is identical to $p_{\boldsymbol{h}\mid\boldsymbol{x}}$. Variational inference focuses on choosing a suitable variational distribution, i.e. the form of q, that best approximates the true distribution $p\left(\boldsymbol{h} \mid \boldsymbol{x}, \theta\right)$ by optimizing the ELBO through coordinate ascent [7].

One main focus of variational inference approach is to choose the form of q. In this work, we use a mean-field approximation [5] to $q\left(\boldsymbol{h}\right)$, i.e. the variational variables are factorizable, each variational distribution q_i is choosen from the exponential family

$$q\left(\boldsymbol{h}\right) = \prod_{t=1}^{d} q_i\left(h_i\right) \qquad (8)$$

The ELBO can be written as [7]

$$\mathcal{L}\left(q, \boldsymbol{\theta}\right) \sim \sum_{t=1}^{d} \mathbb{E}_q\left[\log p\left(h_t \mid h_{\bar{t}}, X\right)\right] - \mathbb{E}_{q_t}\left[\log q_t\left(h_t\right)\right] \qquad (9)$$

where the subscript $\bar{t}$ means all indices other than t. Setting the derivative of $\mathcal{L}\left(q, \theta\right)$ to 0, we arrive at coordinate ascent update

$$q^*\left(h_t\right) \sim \mathbb{E}_{q_{\bar{t}}}\left[\log p\left(h_t \mid h_{\bar{t}}, X\right)\right] \qquad (10)$$

Once a conjugate prior is put over the parameters, the update rule in (10) is tractable and analytical.

IV. EXAMPLES

In order to test the validity of our method, two types of total jitter construction are performed: one consists of PJ and RJ, and another has DDJ and RJ. Since the pdf of a typical, single frequency PJ has highest density at its bounds, it is reasonable to approximate the pdf of PJ as dual-dirac. As shown in Table I, there is a total of 7 jitter combinations based on the selection of RJ and PJ/DDJ. The term PJ/DDJ in Table 1 is used to describe $|\mu_L - \mu_R|$ while the term RJ tells the size of σ. The amount of 10,000 total jitter samples are collected for extraction. The jitter estimated value and jitter error are recorded.

TABLE I
JITTER STUDY CASES

Cases	Jitter Added (ps)
Case 1	RJ = 10, PJ = 5
Case 2	RJ = 15, PJ = 5
Case 3	RJ = 5, PJ/DDJ = 5
Case 4	RJ = 10, PJ/DDJ = 10
Case 5	RJ = 15, PJ/DDJ = 15
Case 6	RJ = 5, DDJ = 10
Case 7	RJ = 5, DDJ = 15

Comparing the DJ error in Figure 3, SVI shows less error in both PJ and DDJ cases than EM. As expected, in cases that have PJ, when approximating the PJ as dual-delta, DJ estimation error is larger than that in cases that have DDJ. It is worth noting that this approximation gets worse (over 100% error) when RJ is the dominant contributor in the underlying jitter. The reason could be the distinction between the dual-RJ in total jitter (TJ) is ambiguous but SVI is still advanced by the half error rate of EM method.

In Figure 4, SVI method performs significantly better than the EM method when estimating RJ.

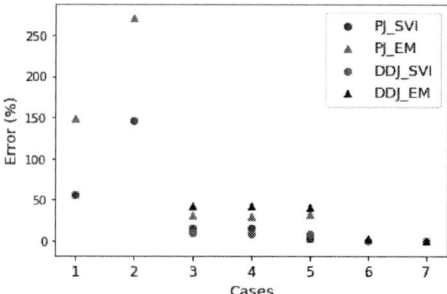

Fig. 3. DJ error between true and extracted values

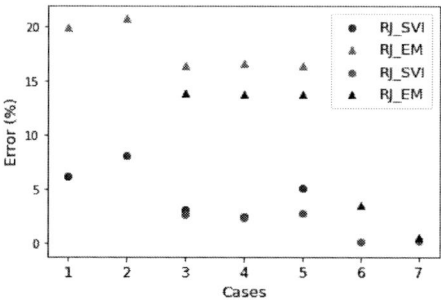

Fig. 4. RJ error between true and extracted values

V. CONCLUSION

An efficient and accurate jitter decomposition method that separates periodic jitter or data-dependent jitter and random jitter is presented. This method is based on stochastic variational inference of Gaussian mixture model to update and estimate jitter parameters. The comparison between this novel method and traditional GMM demonstrates the accuracy of the proposed jitter separation method. In future work, we will demonstrate the robustness of SVI in compiling more complicated graphical models to extract more complicated jitter mixtures.

REFERENCES

[1] M. P. Li, J. Wilstrup, R. Jessen, and D. Petrich, "A new method for jitter decomposition through its distribution tail fitting," in *International Test Conference 1999. Proceedings (IEEE Cat. No.99CH37034)*, pp. 788–794, 1999.

[2] F. Nan, Y. Wang, F. Li, W. Yang, and X. Ma, "A better method than tail-fitting algorithm for jitter separation based on gaussian mixture model," *J. Electronic Testing*, vol. 25, pp. 337–342, Dec 2009.

[3] D. Mistry, S. Joshi, and N. Agrawal, "A novel jitter separation method based on gaussian mixture model," in *2015 International Conference on Pervasive Computing (ICPC)*, pp. 1–4, 2015.

[4] R. Stephens, "Jitter analysis: The dual-dirac model, RJ/DJ, and Qscale." White Paper, Agilent Technologies, Dec 2004.

[5] C. M. Bishop, *Pattern Recognition and Machine Learning (Information Science and Statistics)*. Berlin, Heidelberg: Springer-Verlag, 2006.

[6] R. Neal and G. Hinton, "A view of the EM algorithm that justifies incremental, sparse, and other variants," in *Learning in Graphical Models* (M. I. Jordan, ed.), pp. 355–368, MIT Press, 1999.

[7] D. M. Blei, A. Kucukelbir, and J. D. McAuliffe, "Variational inference: A review for statisticians," *Journal of the American Statistical Association*, vol. 112, pp. 859–877, Feb 2017.

Assessment of 2x Thru De-embedding Accuracy for Package Transmission Line DUTs

Stephen A. Smith, Zhichao Zhang, Kemal Aygün

Assembly Test and Technology Development, Intel Corporation

Chandler, Arizona, US

stephen1.a.smith@intel.com, zhichao.zhang@intel.com, kemal.aygun@intel.com

Abstract—This paper examines 2x Thru de-embedding accuracy for uniform package transmission line devices under test (DUTs). Accuracy is first assessed using 3D modeled data, with and without 2x Thru impedance mismatch. The learnings from this analysis are then applied to improve the match between de-embedded measurement data and modeled data.

Keywords—De-embedding; 2x Thru; TDR

I. INTRODUCTION

2x Thru de-embedding is a broadband, time-domain-based de-embedding method that is gaining popularity in the industry as a convenient alternative to Thru-Reflect-Line (TRL) de-embedding, offering a similar level of accuracy with reduced design complexity [1]. In the case of symmetric fixtures, 2x Thru de-embedding requires a single characterization fixture (called the 2x Thru) to determine the test fixture response. This approach utilizes both the frequency-domain and time-domain responses of the characterization fixture; the calculation is described in detail in [2]. Several commercially available tools implement 2x Thru de-embedding, as well as open-source code developed as part of the IEEE P370 standard [3].

The success of a de-embedding method depends upon several factors: complexity of design, complexity of the measurement process, accuracy (including resilience to mismatch between the calibration structure and test fixture), effective bandwidth, and applicability to a wide range of DUTs. The simplicity of the 2x Thru design and calibration process, as well as its broad applicability, are widely recognized benefits. Further, various means of evaluating the accuracy have been proposed in the literature: analytical [2], [4], comparison to TRL [1], use of synthesized S-parameters [5], and use of NIST-traceable test coupons [6].

However, some gaps remain. Most of the papers addressing accuracy use complex or poorly matched DUTs, thereby masking more subtle de-embedding effects [1]; and few address the impact of impedance mismatch between the 2x Thru and test fixture [5]. This work is intended to address these gaps by examining the accuracy of 2x Thru de-embedding for a uniform package transmission line DUT across a wider range of metrics and in the presence of 2x Thru impedance mismatch. The accuracy is first evaluated using simulated data as in [5]; the learnings from this analysis are then applied to a fabricated package transmission line. All de-embedded results are obtained using a commercially available 2x Thru algorithm.

II. DE-EMBEDDING ACCURACY FOR IDEAL SIMULATED DATA

To establish the baseline accuracy of the de-embedding algorithm for package transmission lines, data for the 2x Thru, fixtured DUT, and true DUT were generated through 3D full-wave electromagnetic (EM) simulation; the fixtured DUT was then de-embedded and compared with the true DUT. The DUT comprises a 5 mm, 50 Ω single-ended stripline structure. The fixture comprises a transmission line segment terminating in surface-layer probe pads at one end. The fixture length is varied between 2.5 mm and 7.5 mm. The 2x Thru structures mirror the test fixture geometry and vary between 5 mm and 15 mm in length. Fig. 1 illustrates the general model geometry.

Fig. 2 shows the insertion loss of the true DUT overlaid by the de-embedded result. The maximum $\Delta S_{21} < 0.003$ from DC to 110 GHz for all fixture lengths. Since this is significantly below the convergence criterion of the field solver ($\Delta S < 0.01$ at 110 GHz), we conclude that de-embedding of insertion loss magnitude is highly accurate, conditioned upon the equivalence of 1x fixture and DUT test fixture loss. Similarly, the de-embedded insertion phase (not shown) is within 2° (0.4%) of the true value for all cases; we conclude that insertion phase de-embedding is also highly accurate, conditioned upon the equivalence of 1x fixture and DUT test fixture insertion phase.

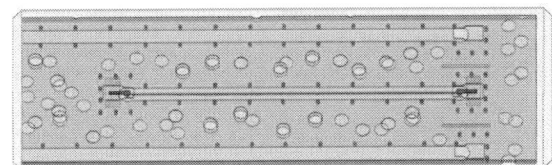

Figure 1. 3D model of the 5 mm 2x Thru structure.

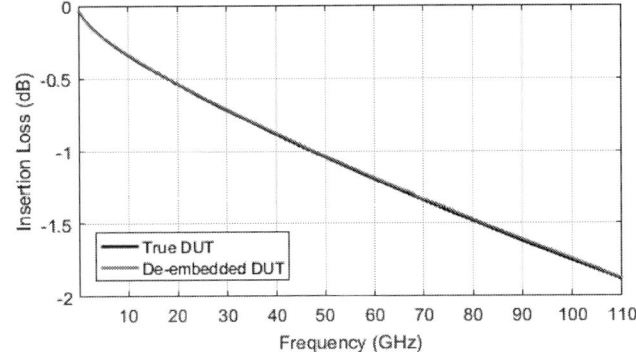

Figure 2. Insertion loss comparison between de-embedded and true DUT.

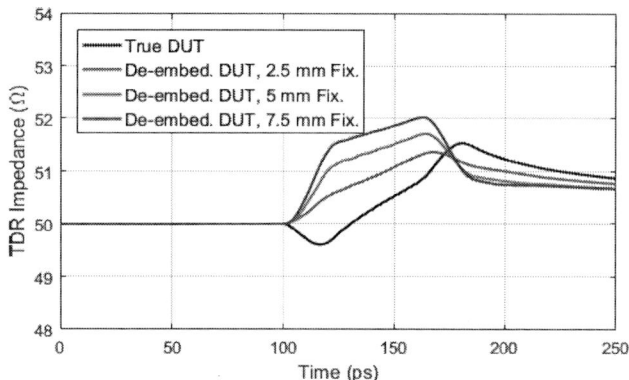

Figure 3. TDR profiles of the 5 mm true DUT and de-embedded DUTs with varying lengths of fixture removed: 2.5 mm, 5 mm, and 7.5 mm.

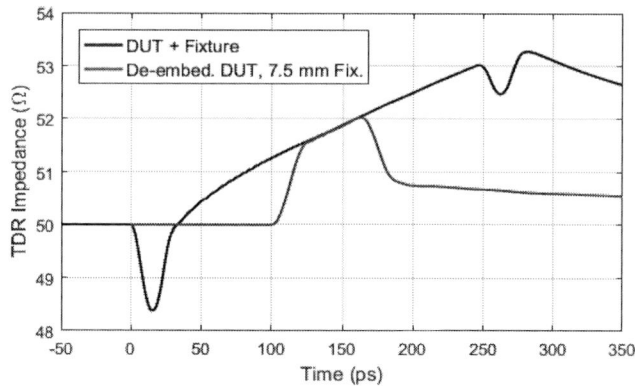

Figure 4. TDR profile of fixtured DUT (20 mm) and de-embedded DUT (5 mm). The de-embedded DUT is shifted to align in time with the fixtured DUT.

Examining time-domain reflectometry (TDR) impedance, however, reveals an unintuitive result: there is a systematic offset between the de-embedded TDR impedance and true DUT impedance, proportional to the fixture length. Fig. 3 shows the impedance gap growing from 0.6 Ω for the 2.5 mm fixture case, to 1.5 Ω for the 7.5 mm fixture case. The source of this discrepancy is evident in Fig. 4. Despite a uniform cross-section, there is an apparent increase in impedance along the line, owing to attenuation of the TDR pulse. When the de-embedded DUT response is shifted to align in time with the fixtured DUT, it is evident that the two impedance profiles agree very well. It can be concluded that 2x Thru de-embedding successfully removes the fixture portions of the embedded DUT but does not remove the artificial impedance shift induced by the fixture loss.

The gap in TDR impedance leads to a gap in return loss as well. The discrepancy is roughly proportional to the fixture loss, and can cause the de-embedded return loss to be either better or worse than the true DUT result, depending on whether the true impedance is lower or higher than the reference impedance, respectively. The magnitude of the discrepancy is probably small enough to be ignored for most applications, but must be taken into account for optimal model-to-measurement correlation of transmission lines or other very well-matched DUTs.

III. DE-EMBEDDING ACCURACY FOR NON-IDEAL DATA

De-embedding theory rests on the assumption that the test fixture and characterization fixture are identical. However, this assumption is never perfectly true in the real world due to dimensional variation in any fabricated structure. Therefore, to build confidence in the de-embedding algorithm for real-world applications, one must assess the sensitivity of the de-embedded result to 2x Thru–test fixture mismatch.

The conventional 2x Thru algorithm calculates the fixture model S_{11} term by time-gating the first half of the time-domain 2x Thru S_{11} term, and converting back to the frequency domain. The impedance-corrected algorithm, however, calculates the fixture model S_{11} by time-gating the DUT test fixture response itself. This method significantly reduces the sensitivity of the return loss result to 2x Thru impedance mismatch. The extent of improvement is illustrated in Fig. 5.

For the results in Fig. 5, the fixtured DUT is a 10 mm single-ended stripline with 50 Ω impedance, while the 2x Thru is a 5 mm stripline with trace width adjusted to achieve 53 Ω impedance (6% mismatch). Without impedance correction, the de-embedded return loss is much worse than would be expected from a simple impedance discontinuity. De-embedding with a 53 Ω 2x Thru yields return loss peaks around −18 dB, whereas a 53 Ω − 50 Ω discontinuity gives a return loss of −30.7 dB.

Impedance-corrected return loss is significantly improved, but does not perfectly agree with the true DUT behavior. While the fixture model S_{11} term is now independent of 2x Thru mismatch, the S_{21} and S_{11} terms of the mismatched 2x Thru are still used to calculate the other fixture model parameters, and thus contribute to inaccuracy in the final result. 2x Thru impedance mismatch imposes a kind of noise floor on the de-embedded return loss, that can be calculated similarly to a simple reflection coefficient:

$$RL_{\text{floor}} \approx -20 \log_{10} \left| \frac{Z_0^{2x} - Z_0^{fix}}{Z_0^{2x} + Z_0^{fix}} \right|, \qquad (1)$$

where Z_0^{2x} and Z_0^{fix} are the characteristic impedance of the 2x Thru and test fixture, respectively, and RL_{floor} is the value of the return loss peaks induced by 2x Thru impedance mismatch. If the DUT is matched to below this value, the de-embedding process cannot recover the DUT behavior.

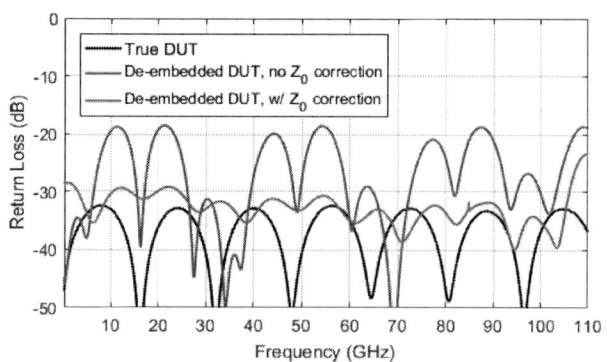

Figure 5. Return loss comparison between true DUT and DUT de-embedded with and without impedance correction.

978-1-7281-6162-4/20 $31.00 © 2020 IEEE

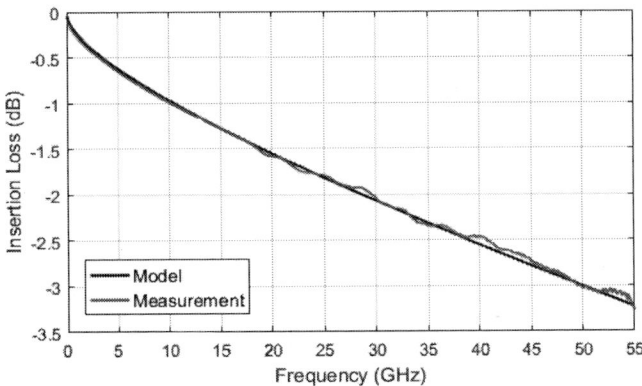

Figure 6. Insertion loss correlation for de-embedded model and measurement.

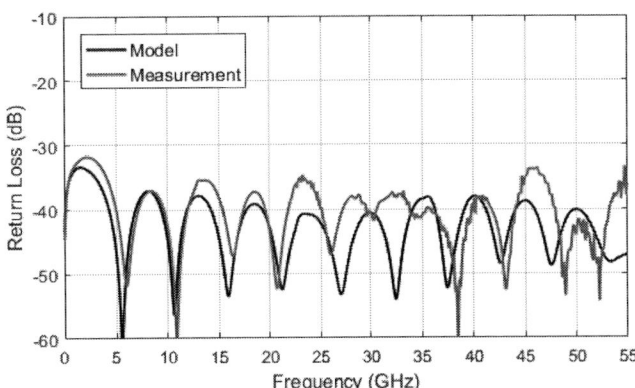

Figure 7. Return loss correlation for de-embedded model and measurement.

In sum, by removing one source of error in the 2x Thru algorithm, impedance correction can significantly improve the de-embedded return loss accuracy in the presence of 2x Thru impedance mismatch, and therefore improves the reliability of the 2x Thru method. The insertion loss is also somewhat improved when mismatch loss becomes appreciable. It is therefore recommended that impedance correction should always be used. But it is no panacea. Mismatch still limits the accuracy of the de-embedded result, and reasonable efforts should not be spared to ensure the 2x Thru is as well-matched to the fixture as possible. The fixture electrical requirements in the IEEE P370 standard include reasonable limits for mismatch along with the resulting data quality that can be expected [3].

IV. DE-EMBEDDING ACCURACY FOR MEASURED DATA

By accounting for the limitations already discussed, one can consistently achieve very good model-to-measurement correlation for package transmission line DUTs, which serves to confirm the accuracy of the de-embedded result.

The achievable post-de-embedding correlation quality is illustrated in Fig. 6 and Fig. 7 for 50 Ω single-ended stripline structures fabricated on a test package. The fixtured DUT is 20 mm long, while the 2x Thru is 5 mm long. These structures were measured with VNA up to 67 GHz at controlled temperature and humidity, and de-embedded using the impedance-corrected algorithm. In post-processing, the data were truncated at 55 GHz and macromodeled to suppress trace noise amplified during the de-embedding process.

The model is constructed using cross-sectional dimensions taken from the measured unit, and measurement-based dielectric material models using the metrology described in [7]. Both fixtured DUT and 2x Thru are modeled, and the DUT is de-embedded using the modeled 2x Thru. Using de-embedded model data for comparison with measurement circumvents the TDR offset issue described in Section II, since both datasets include the impedance increase induced by the fixture loss.

Fig. 6 and Fig. 7 show the overlaid insertion loss and return loss results, respectively. The insertion loss agrees to within $\Delta S < 0.009$ up to 55 GHz, while the return loss agrees to within $\Delta S < 0.012$, neglecting a noisy spike near 55 GHz. These results are encouraging, as the level of agreement approaches the convergence standard for the EM simulation, and is within the error bounds of the dimensional and material measurements used to construct the model. Further, the level of agreement after de-embedding is noticeably better than the agreement before de-embedding, because the de-embedded DUT geometry is simpler and lacks elements such as vias that are difficult to cross-section. Note that the non-impedance-corrected de-embedded data are not shown, as the fabricated structures are very well-matched and the data are therefore qualitatively very similar to those shown.

V. CONCLUSION

The accuracy of 2x Thru de-embedding for package transmission line DUTs has been evaluated through simulation and confirmed through measurement. A small systematic offset in TDR impedance leading to return loss inaccuracy has been highlighted and addressed. The impact of 2x Thru–fixture mismatch on de-embedding accuracy has been assessed and impedance correction has been recommended as a satisfactory, though not perfect, solution. By accounting for these effects, excellent model-to-measurement correlation has been achieved for a de-embedded package transmission line, confirming the expected accuracy of the method.

REFERENCES

[1] S. Moon, X. Ye and R. Smith, "Comparison of TRL calibration vs. 2x thru de-embedding methods," *2015 IEEE Symposium on Electromagnetic Compatibility and Signal Integrity*, Santa Clara, CA, 2015, pp. 176-180.

[2] C. Wu, B. Chen, T. Mikheil, J. Fan and X. Ye, "Error bounds analysis of de-embedded results in 2x thru de-embedding methods," *2017 IEEE International Symposium on Electromagnetic Compatibility & Signal/Power Integrity (EMCSI)*, Washington, DC, 2017, pp. 532-536.

[3] *Electrical Characterization of Printed Circuit Board and Related Interconnects at Frequencies up to 50 GHz*, IEEE Standard P370, unpublished.

[4] B. Chen *et al.*, "Analytical and numerical sensitivity analyses of fixtures de-embedding," *2016 IEEE International Symposium on Electromagnetic Compatibility (EMC)*, Ottawa, ON, 2016, pp. 440-444.

[5] M. Resso, E. Bogatin and A. Vatsyayan, "A new method to verify the accuracy of de-embedding algorithms," *2016 IEEE MTT-S Latin America Microwave Conference (LAMC)*, Puerto Vallarta, 2016, pp. 1-4.

[6] H. Barnes, E. Bogatin and J. Moreira, "Development of a PCB kit for s-parameter de-embedding algorithms verification," *2017 IEEE International Symposium on Electromagnetic Compatibility & Signal/Power Integrity (EMCSI)*, Washington, DC, 2017, pp. 510-515.

[7] C. S. Geyik, Y. S. Mekonnen, Z. Zhang and K. Aygün, "Impact of Use Conditions on Dielectric and Conductor Material Models for High-Speed Package Interconnects," in *IEEE Transactions on Components, Packaging and Manufacturing Technology*, vol. 9, no. 10, pp. 1942-1951, Oct. 2019.

Accurate BGA Package Solder Joint Modeling for High Speed SerDes Interfaces

Jiwei Sun
Assembly and Test Technology
Development
Intel Corporation
Chandler, USA
jiwei.sun@intel.com

Zhiguo Qian
Assembly and Test Technology
Development
Intel Corporation
Chandler, USA
zhiguo.qian@intel.com

Cemil S. Geyik
Assembly and Test Technology
Development
Intel Corporation
Chandler, USA
cemil.s.geyik@intel.com

Kemal Aygün
Assembly and Test Technology
Development
Intel Corporation
Chandler, USA
kemal.aygun@intel.com

Abstract— **The solder joint of the ball grid array (BGA) package becomes a performance bottleneck as the serializer/deserializer (SerDes) speeds increase beyond 112 Gbps. This paper presents a comprehensive methodology to correctly model the volume and shape variations of solder joints for both nominal and worst-case electrical models. It is found that it is indispensable to accurately predict the performance of vertical transitions from package to board beyond 30 GHz. Construction of a worst-case solder joint model is also critical to ensure adequate electrical performance for a high-volume manufacturing solution.**

Keywords—BGA package, SerDes, solder joint modeling

I. INTRODUCTION

High speed digital standards are evolving towards terabit Ethernet speeds to keep pace with emerging technologies such as 5G, Internet of Things, artificial intelligence, and autonomous vehicles [1]. Over the past ten years, SerDes speeds have dramatically increased from 10 Gbps to 112+ Gbps. Such an aggressive speed scaling has demanded corresponding improvements in channel bandwidth, as the signaling is approaching the Shannon limit [2].

As a key component in input/output channels, BGA packages have been widely used in SerDes solutions. Besides the lateral routing loss in package substrates [3], the vertical transition from package to board also plays an important role [4]-[5]. As SerDes speeds scale beyond 112 Gbps, the solder joint connecting the package substrate to the board creates a big discontinuity in the signaling path and becomes a major bandwidth limiter. Consequently, the assumptions used for past designs that either overlook or oversimplify solder joint modeling are also no longer valid.

This paper focuses on the characteristics of the solder joint formed after the package is mounted to the board. Section II explains the challenging bandwidth demand for the SerDes speed scaling. Section III highlights the importance of shape and volume modeling of solder joints. Section IV introduces an accurate solder joint model for full-wave electromagnetic field solvers. Finally, section V outlines a method to construct a worst-case ball model which is used to investigate the impact of solder joint manufacturing variations on electrical performance.

II. SOLDER JOINT IMPACT ON SPEED SCALING

As seen in Fig. 1 (a), the solder joint that connects the package substrate to the board is a major portion of the overall vertical transition in a typical package. It has a substantial impact on multiple electrical performance metrics including impedance, return loss, insertion loss, and crosstalk. This paper focuses on the insertion loss as one of the key performance metrics. Fig. 1 (b) shows the modeled differential insertion loss of a typical vertical transition of a SerDes package designed for 112 Gbps, including micro- vias, plated through holes (PTHs) and solder joints. This result demonstrates that the insertion loss up to 28 GHz, the Nyquist frequency of the Pulse-Amplitude Modulation 4-Level (PAM-4) coding scheme, is less than ~0.5 dB. This accounts for a small portion of the total package loss budget. However, as the speed scales beyond 112 Gbps, the corresponding Nyquist frequency is above 28 GHz with the PAM-4 coding scheme. It can be observed in Fig. 1(b) that the significant roll-off beyond 30 GHz has the slope of ~0.083 dB/GHz, much higher than 0.017 dB/GHz below 30 GHz, resulting in the rapid insertion loss increase. That is too high to meet the package loss budget, and the strong dispersion makes the performance worse. Hence, a higher bandwidth solder joint solution is required to reduce the loss of the vertical transition with low dispersion. In addition, the insertion loss above 30 GHz with a rapid roll-off is more susceptible to solder joint shape variations.

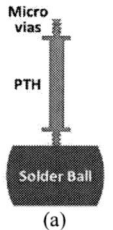 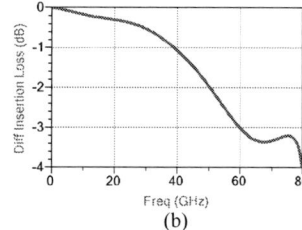

Fig. 1. An example of a typical BGA package vertical transition (a) geometry illustration (b) insertion loss for 112 Gbps SerDes.

III. SOLDER JOINT MODELING CHALLENGES

Full-wave electromagnetic field solvers are widely used to predict the electrical performance of packages. In the past, solder joints of BGA packages are usually approximated as cylinders or spheres with the volume of the incoming solder ball [5]-[6]. This assumption underestimates the loss, especially at frequencies beyond 30 GHz for two reasons. First reason is that the amount of solder volume coming from the solder paste on the package and the board is not considered in this approach. The real solder joint is bigger than what is modeled and therefore has higher parasitics. The total solder volume needs to be correctly calculated to avoid underestimation.

The second reason is that the shape variation after the surface mount process is ignored. As an example, the cross

978-1-7281-6162-4/20 $31.00 © 2020 IEEE

sections of solder joints at different locations of an assembled BGA package are shown in Fig. 2. Despite having similar solder volume, the shape varies significantly due to warpage and board stencil design. Some joints are close to a spherical shape (Fig. 2 (a)), some are stretched to be cylindrical (Fig. 2 (b)), and others are compressed to be elliptical (Fig. 2 (c)). The stretched cylindrical solder joints have the largest height and the smallest width, while the compressed elliptical ones have the smallest height and the largest width. These 3 different shapes are all modeled and compared in Fig. 3. The spherical one is referred to as the nominal case. The stretched cylinder one increases the standoff height by 110 um and achieves better performance than the nominal case. The compressed elliptical one reduces the standoff height by 75 um and results in dramatically worse performance, as shown in Fig. 3 (a). The impedance result based on the time domain reflectometry (TDR) simulations in Fig. 3 (b) show that the compressed solder joint causes the largest capacitance leading to the biggest impedance discontinuity, which is detrimental to the overall performance. Hence, it is critical to have a solder joint model that represents its shape accurately after the surface mount process.

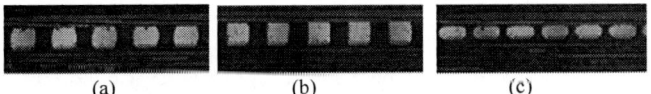

(a)   (b)   (c)

Fig. 2. (a) Spherical (b) cylindrical and (c) elliptical shape of solder joints at different locations of the same BGA package.

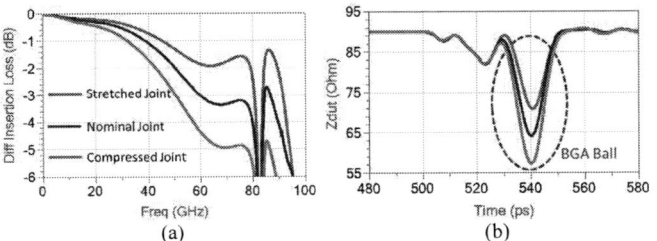

(a)   (b)

Fig. 3. Impact of solder joint shape on (a) insertion loss and (b) impedance.

IV. ACCURATE SOLDER JOINT MODELING

Both the volume and the shape of the solder joint need to be correct to accurately predict the performance. For the solder volume, the solder pastes on both the package and the board also need to be included. There may be non-negligible solder paste on package substrate for the ball attachment. The solder paste on the board for surface mount often has much larger volume, which can be 30-40% of the incoming ball volume, depending on the stencil design. It is imperative to consolidate all the information into the total volume of the solder joints.

The shape of the solder joint depends on quite a few factors, such as the package-to-board assembly warpage, the package form factor, the stencil design, the location of the joint, etc. This drives the need of two models: One is the nominal model needed for typical performance estimation before the package design starts. The other is the worst-case model required for the electrical performance verification after the package design is completed.

The nominal model is constructed with a truncated sphere [7], as illustrated in Fig. 4 (a). The geometry follows two

assumptions: First, the solder joint shape is spherical. Second, for a solder mask defined (SMD) pad, the ball landing radius is equal to the solder resist opening (SRO) radius. While for a metal defined (MD) pad, the ball landing radius is equal to the pad radius. Given the solder joint volume V and ball landing radius of package side r_1 and PCB side r_2, (1) and (2) can be used to calculate the solder joint height H and width W [7] as

$$V = \frac{\pi}{3}\left[\sqrt{R^2 - r_1^2}(2R^2 + r_1^2) + \sqrt{R^2 - r_2^2}(2R^2 + r_2^2)\right] \quad (1)$$

$$H = \sqrt{R^2 - r_1^2} + \sqrt{R^2 - r_2^2} \quad (2)$$

where $R = W/2$ is the spherical solder joint radius. The nominal model can guide the BGA ball size selection and be used in package design studies.

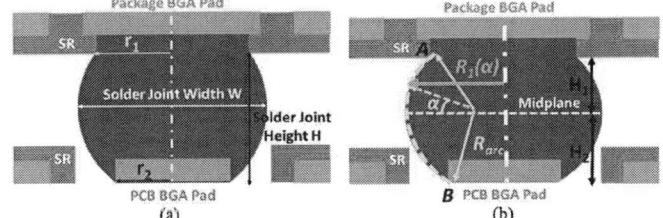

(a)   (b)

Fig. 4. (a) Nominal model and (b) worst-case model.

Once the design is completed, a two-step approach can be applied to develop the worst-case BGA ball model. The first step is to extract the solder ball geometries of high-speed SerDes signals. For the balls selected, the compressed ball with the minimum solder joint height and maximum solder joint width is identified as the worst case, since it implies the largest capacitance at BGA ball transition. The second step is to determine the surface profile of the worst-case BGA ball from any two known parameters among solder joint height (H), width (W) and volume (V), based on the following three assumptions. The first is that the solder completely covers a metal defined pad and does not flow beyond the pad. The second is that no solder paste flows out of SRO for a solder mask defined pad. The third is that the meridian defining the surface profile of the solder joint in Fig. 4 (b) is approximated by a circular arc, although the solder ball shape is no longer spherical. In addition, defining the exact worst-case solder joint model requires the midplane, as shown in Fig. 4 (b). With reference to this midplane, the worst-case model can be split into two regions, the upper half and lower half. The key to find the midplane is the arc radius determination, which can be obtained from (3) and (4) below

$$H = H_1 + H_2 = \sqrt{2d_1 R_{arc} - d_1^2} + \sqrt{2d_2 R_{arc} - d_2^2} \quad (3)$$

$$V = V_1 + V_2 = \int_0^{\theta_1} \pi[R_1(\alpha)]^2 R_{arc} \cos\alpha d\alpha + \int_0^{\theta_2} \pi[R_1(\alpha)]^2 R_{arc} \cos\alpha d\alpha \quad (4)$$

where $d_1 = W/2 - r_1$, $d_2 = W/2 - r_2$, R_{arc} is the arc radius, H_1 and V_1 are the upper half height and volume, H_2 and V_2 are the lower half height and volume, respectively. $R_1(\alpha)$ shown in Fig. 4 (b) is given by

$$R_1(\alpha) = W/2 - R_{arc}(1 - \cos \alpha) \qquad (5)$$

where α is the angle between R_{arc} and the midplane. Equation (4) can be solved for the integration variable α using the boundary conditions, where, at position A, $\alpha = \theta_1 = arctan(H_1/(R_{arc} - d_1))$, and at position B, $\alpha = \theta_2 = arctan(H_2/(R_{arc} - d_2))$. When $d_1 = d_2$ and $\theta_1 = \theta_2$, the worst-case solder joint is symmetric about the midplane. Generally, the smallest solder joint height or the largest solder joint width leads to the worst-case model.

V. APPLICATION OF THE WORST-CASE SOLDER JOINT MODEL

The worst-case solder joint model results in a larger capacitance, so it creates a low impedance discontinuity and consequently causes insertion loss increase. Primary parameters considered for the worst-case model demonstrated above include solder joint width, height, volume and shape. For SerDes speeds beyond 112 Gbps, the impact of the parameters on solder joint capacitance and insertion loss of the vertical transition becomes more notable. As a result, the utilization of the worst-case model is critical for high-volume manufacturing solutions targeting these speeds. Such models can help determine the impact of the solder joint variation on the electrical performance and optimize the assembly process accordingly.

In Fig. 5, the impact of solder joint width and height variation on solder joint capacitance is shown as an example. The capacitance increase is relative to a baseline, i.e., nominal solder joint model, which is the lower-right corner in the plot. Clearly, solder joint width is the dominant parameter in ball capacitance increase. When the width variation is less than 40 um, the capacitance increase is not sensitive to the height reduction. Also, for large width variations (> 40 um), every 10 um increase in width can increase the capacitance by ~6 fF, which is about two times than the capacitance increase for small width variations (< 30 um).

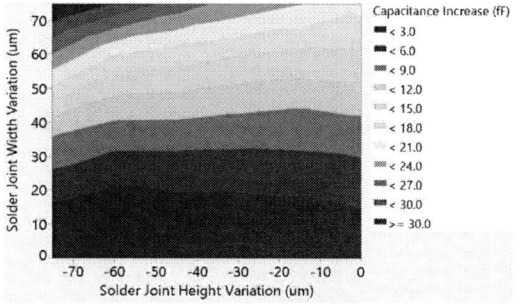

Fig. 5. Impact of solder joint width and height variation on ball capacitance.

Similarly, an example of the solder joint variation impact on insertion loss of the vertical transition beyond 30 GHz is shown in Fig. 6. The contour plot shows how the insertion loss increases as the solder joint width increases and the solder joint height decreases. A 40 um increase in joint width from the baseline causes more than 0.5 dB extra loss, corresponding to about 12 fF solder joint capacitance increase. When the width is well controlled to less than 20 um increase, the ball capacitance and loss increase are less than 6 fF and 0.3 dB, respectively. It's evident that a 0.2 dB insertion loss is attributed to every 6 fF solder joint capacitance increase. As illustrated in Fig. 6, the joint width also shows much bigger impact on the insertion loss increase than the solder joint height. Therefore, to meet the loss target, the most important parameter to control is the solder joint width. For instance, the white dash line indicates the variation limit of 0.1 dB loss increase. This sets the requirement of the maximum joint width and the minimum joint height for an example product implementation. Advancements in assembly technology may be required to achieve such a variation control.

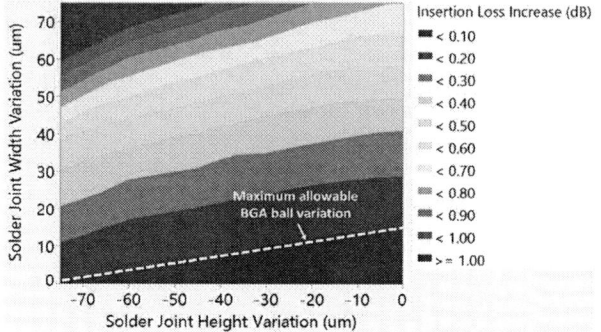

Fig. 6. Insertion loss dependence on solder joint width and height.

VI. CONCLUSIONS

Solder joints of BGA packages become a major bandwidth limiter as SerDes speed scales beyond 112 Gbps. This paper introduces a solder joint modeling methodology to accurately capture the impact of the volume and shape variation of the solder ball for electrical performance analysis. Based on this methodology, the worst-case solder joint models are used to determine the key parameters for solder joint assembly that can meet the product electrical performance requirements of future high speed SerDes interfaces.

ACKNOWLEDGMENT

The authors would like to acknowledge the following colleagues for their technical supports: Patrick Nardi, John Harper, and Hemanth Dhavaleswarapu.

REFERENCES

[1] "2020 Ethernet roadmap: the past, present and future of Ethernet." [Online]. Available: https://ethernetalliance.org/technology/2020-roadmap/

[2] C. E. Shannon, "A mathematical theory of communication," *Bell Syst. Tech. J.*, vol. 27, no. 4, pp. 623-656, July 1948.

[3] C. S. Geyik *et al.*, "Impact of Use Conditions on Dielectric and Conductor Material Models for High-Speed Package Interconnects," *IEEE Trans on Compon., Packag., Manuf. Technol.*, vol. 9, no. 10, pp. 1942-1951, Oct. 2019.

[4] Q. Zhu *et al.*, "Package Design Optimization for Intel SoC Xeon-D," *IEEE Trans on Compon., Packag., Manuf. Technol.*, vol. 8, no. 4, pp. 531-537, April 2018.

[5] J. Lim *et al.*, "ASIC package to board BGA discontinuity characterizationin >10 Gbps SerDes Links," in *Proc. IEEE Int. Symp. Electromagn. Compat.*, Denver, CO, USA, Aug. 5–9, 2013, pp. 569–574.

[6] S. Jin *et al.*, "Analytical equivalent circuit modeling for BGA in high speed package," *IEEE Trans. Electromagn. Compat.*, vol. 60, no. 1, pp. 68–76.

[7] K. Chiang and C. Yuan, "An overview of solder bump shape prediction algorithms with validation," *IEEE Trans. Adv. Packag.*, vol. 24, no. 2, pp. 158–162, May 2001.

3D Integration of Ka-band RFIC by Inductive Inter-chip Wireless Communication Using Figure-8 Coils

Masairo Usui[1], Kota Shiba[2], Mototsugu Hamada[3], Tadahiro Kuroda[3]

[1]Department of Electronics and Electrical Engineering, Keio University, Yokohama, Japan
[2]Department of Electrical Engineering and Information Systems, The University of Tokyo, Tokyo, Japan
[3]System Design Lab, The University of Tokyo, Tokyo, Japan
usui@kuroda.t.u-tokyo.ac.jp

Abstract— Inter-chip communications using inductive coupling with figure-eight coils for a Ka-band RFIC is presented. Compared with a conventional rectangular coil, the figure-8 coil achieves 15dB better interference rejection while the signal transmission is inferior only by 1 dB. The minimum coil pitch can be set to 250 μm, which is 44 % smaller than that of the conventional rectangular coil, contributing to the miniaturization of Ka-band RFIC's.

Keywords—3D integration, inductive-coupling, mm-wave

I. INTRODUCTION

Pitches between the elements of an array antenna is the half wavelength of the communication signal, and they become shorter as the communication frequency becomes higher. Therefore, RFICs must be designed to be smaller than the pitch, and the demand for miniaturization of RFICs is increasing. Stacking IC chips three-dimensionally and improving area efficiency are proposed to meet this demand.

For three-dimensional integration, it is necessary to connect signals between chips, and through-silicon vias (TSVs) drives up cost because of requiring additional manufacturing processes. A method of inter-chip wireless communication technology using inductive-coupling by on-chip coils, known as a ThruChip Interface (TCI), has been proposed. The coils are formed by standard CMOS process without additional processes unlike TSV. Therefore, TCI can suppress the increase in cost. TCI has been extensively researched for high-speed digital communications [1], [2].

In this paper, we propose to use TCI technology to transmit Ka-band analog signal and make the shape of the communication coil into figure-eight shape. It can improve the carrier-to-interference ratio better than a rectangular coil. We focus on 40 GHz in Ka-band. In this paper, Performance evaluation is performed using the maximum available gain, because the loss due to impedance matching of a figure-eight coil and a rectangular coil is considered to be about the same. In Section II, the theory of TCI using conventional rectangular coils and the concept of figure-eight coils are explained. In Section III, the performance of the Ka-band inter-chip wireless communications using figure-eight coils proposed in this paper is explained as compared with conventional rectangular coils. In Section IV, the conclusions are given.

II. THURUCHIP INTERFACE

A. Theory of ThuruChip Interface

TCI is a communication method that utilizes mutual induction of adjacent coils, as shown in Fig. 1. According to the Faraday's law of induction, an induced voltage is generated in the receiving coil by the change in the magnetic field

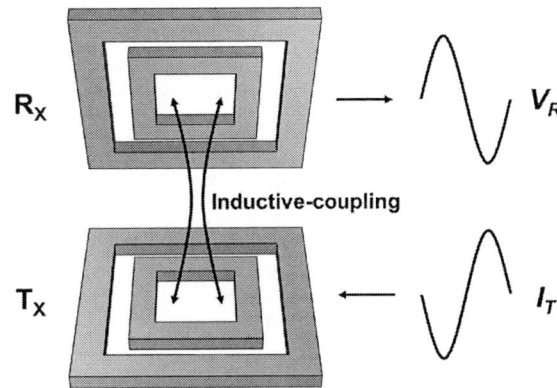

Fig. 1. The appearance of TCI.

generated from the transmit coil, and the relationship between the receiver voltage V_R and the transmit current I_T can be expressed as

$$V_R = k\sqrt{L_T L_R}\frac{dI_T}{dt} \tag{1}$$

where k is the coupling coefficient between transmit and receive coil, L_T and L_R are their inductance. However, in reality, the coil does not function as an inductor at a frequency higher than the self-resonant frequency, because a self-resonance phenomenon occurs due to the coil parasitic element. Fig. 2 shows the equivalent circuit of TCI when the same coil is used for transmit and receive coil. Inductance L and capacitance C, which are modeled on a rectangular coil, are given by

$$L = 1.62\times10^{-3}D_{out}^{-1.21}w^{-0.147}D_{avg}^{2.4}s^{-0.03} \tag{2}$$

$$C = C_G(D_{out}^2 - Din^2) + \frac{4C_c nD_{out}}{s} \tag{3}$$

[3] where D_{out} is the outer diameter of the coil, D_{in} is the inner diameter of the coil, D_{avg} is the average of the inner and outer diameter of coils, equal to $(D_{out}+D_{in})/2$, C_G is the bottom-plate capacitance, C_c is inter-winding capacitance, w is the line width, s is the line space, and n is the number of turns of coil. The self-resonant frequency of the transmit and receive coil that takes the effect of parasitic elements is given by

$$f_{SR} = \frac{1}{2\pi\sqrt{LC}} \tag{4}$$

978-1-7281-6162-4/20 $31.00 © 2020 IEEE

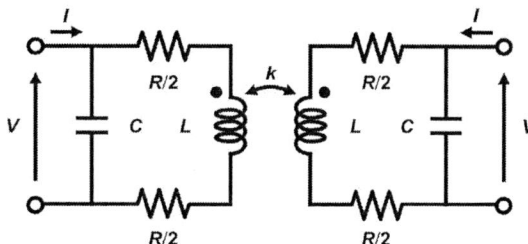

Fig. 2. The equivalent circuit of TCI.

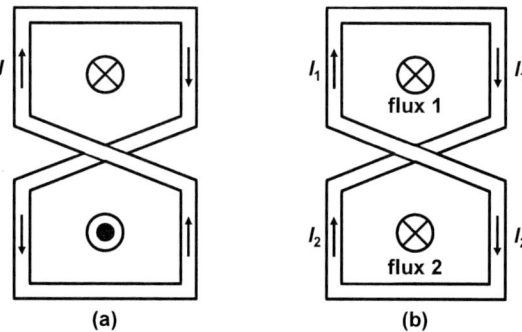

(a) **(b)**

Fig. 3 The appearance of figure 8 coil and interference reduction function.

In order to communicate using inductive coupling of coils, it is necessary to design the coils so that the following relation is satisfied.

$$f_R < f_{SR} \qquad (5)$$

Where f_R is the communication frequency.

B. Figure-eight coil

Using figure-eight coils for TCI has been proposed as a method of suppressing crosstalk interference between channels to improve channel density [4], [5]. Figure-eight coils obtain two rectangular coils turned inversely. As shown in Fig. 3 (a), magnetic fields in the opposite directions are generated, when source current I flows in a figure-eight coil. It is considered that the magnetic fields in opposite directions cancel each other at a distant place, and crosstalk interference becomes small. A figure-eight coil also has resistance to common mode noise. This is because, as shown in Fig.3 (b), currents in opposite directions I_1 and I_2 induced by identical flux 1 and flux 2 respectively cancel each other, when the magnetic flux in the same direction passes through a figure-eight coil.

III. SIMULAITON RESULT

A. Insertion loss of TCI

Fig. 4 shows the appearance of TCI using figure-eight coils. The outer diameter of coils is 100 μm, the line width is 10 μm, the line space is 4 μm, and the length of the lead wiring is 10 μm. The communication distance X is changed to 10 μm, 20 μm, 30 μm, 40 μm, 50 μm, and the insertion loss is evaluated with the maximum available gain, and the simulation result is shown in Fig. 5. When X is 10 μm, the magnitude of impedance is shown in Fig.6, and the self-resonant frequency is 43.7 GHz. The maximum available gain

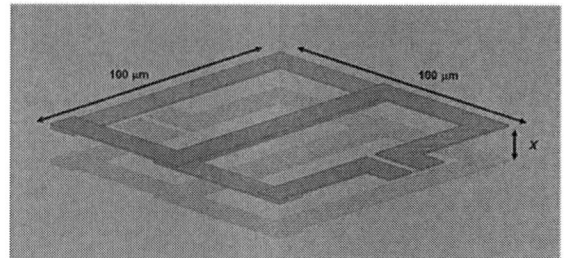

Fig. 4. The appearance of TCI using figure-eight coils.

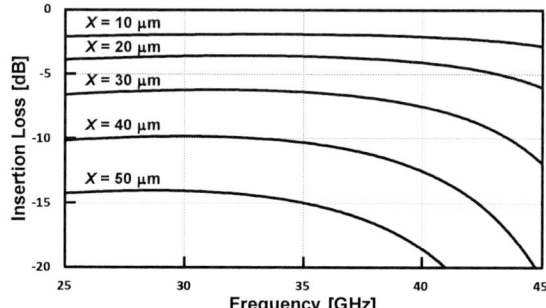

Fig. 5. The relationship between the communication distance X and the insertion loss using figure-eight coils.

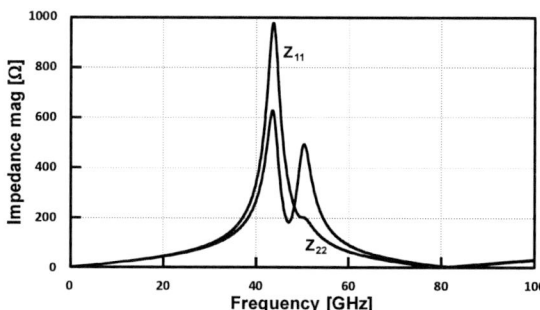

Fig. 6. The magnitude of impedance of figure-eight coils.

of the insertion loss is -2.1 dB at 40 GHz when the communication distance is 10 μm, and The insertion loss improves as the communication distance becomes shorter.

In [6], digital signal inter-chip communication using TCI was confirmed between stacked, 10 μm-thick chips. Therefore, we employed the communication distance of 10 μm in this paper.

Fig. 7 shows the appearance of TCI using conventional rectangular coils. In order to match the conditions with the figure-eight coil, the outer diameter of coils is 100 μm, the line width is 10 μm, the line space is 4 μm, and the length of lead wiring is 10 μm. Fig. 8 compares the maximum available gains of the insertion loss when the communication distance is 10 μm. The insertion loss of the rectangular coils is about 1 dB better than that of the figure-eight coils. It is considered this is because the area inside the coil is reduced due to the overlap of wiring in the figure-eight coils, and the opposite magnetic fields generated in the figure-eight transmit coil weaken each other before reaching the receive coil.

978-1-7281-6162-4/20 $31.00 © 2020 IEEE   114

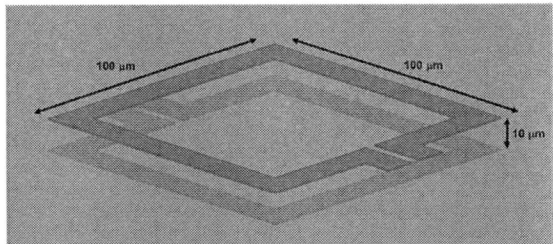

Fig. 7. The appearance of TCI using rectangular coils.

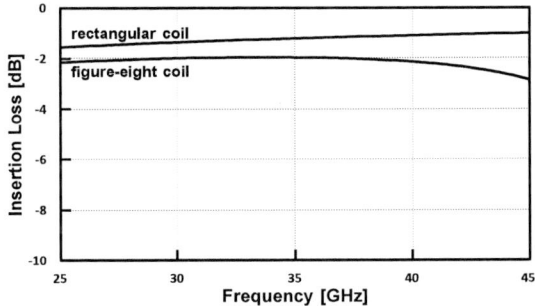

Fig. 8. The insertion loss of TCI using the figure-eight coils and the rectangular coils.

B. Crosstalk interference between channels

The communication distance is fixed at 10 μm and the interference between adjacent channels is simulated. The distance D between the two sets of TCIs is varied from 50 μm to 400 μm at intervals of 50 μm, and the crosstalk to the coil of the adjacent channel is simulated. Fig. 9 shows the appearance using figure-eight coils and Fig. 10 shows the appearance using rectangular coils.

Fig. 11 shows the results of the carrier-to-interference ratio using the maximum available gain with respect to the pitch, where the pitch of the TCI is the summation of the length of the distance D and the outer diameter of the coil. The carrier-to-interference ratio improves as the pitch becomes larger, because the influence of the magnetic fields from the coil of the adjacent channel is reduced by increasing the distance. Moreover, using figure-eight coils improves the carrier-to-interference ratio by about 15 dB. It is considered that this is because the opposite magnetic fields generated by the figure-eight coils cancel out each other at a distance. Furthermore, assuming a system that requires the carrier-to-interference ratio of 60 dB, the pitch of TCI using the figure-eight coils is 250 μm, while that of the rectangular coils is 450 μm, so the pitch can be 56 %.

IV. Conclusion

We propose TCI using figure-eight coils for the Ka-band analog signal, and the evaluation was performed using the maximum available gain. Compared to the conventional rectangular coils, the insertion loss is inferior by about 1 dB, but the carrier-to-interference ratio improves by about 15 dB. Assuming a system that requires the carrier-to-interference ratio of 60 dB, the pitch can be set to 250 μm, which is 56 % of the conventional rectangular coils. Therefore, TCI using figure-eight coils can contribute to downsizing of the RFIC.

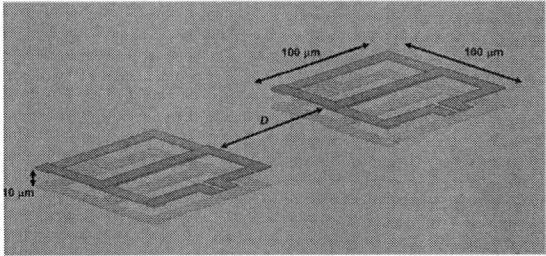

Fig. 9. The appearance of the simulation of crosstalk interference between two channels using figure-eight coils.

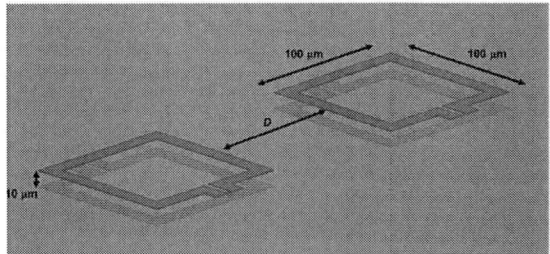

Fig. 10. The appearance of the simulation of crosstalk interference between two channels using rectangular coils.

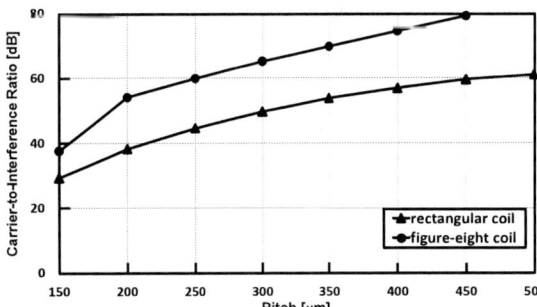

Fig. 11. The relationship between the pitch of TCIs and the carrier-to-interference ratio.

References

[1] M. Saito, N. Miura, and T. Kuroda, "A 2Gb/s 1.8pJ/b/chip Inductive-Coupling Through-Chip Bus for 128-Die NAND-Flash Memory Stacking," IEEE International Solid-State Circuits Conference (ISSCC'10), Dig. Tech. Papers, pp. 440-441, Feb. 2010.

[2] D. Ditzel, T. Kuroda, and S. Lee, "Low-cost 3D chip stacking with ThruChip wireless connections," in Proc. IEEE Hot Chips Symp. (HCS), Aug. 2014, pp. 1–37.

[3] L. Hsu, J. Kadomoto, S. Hasegawa, A. Kosuge, Y. Take, and T. Kuroda, "Analytical ThruChip Inductive Coupling Channel Design Optimization," 2016 21st Asia and South Pacific Design Automation Conference (ASP-DAC), pp. 731-736, Jan. 2016.

[4] Y. Yoshida, N. Miura, and T. Kuroda, "A 2 Gb/s bi-directional inter-chip data transceiver with differential inductors for high density inductive channel array," IEEE Journal of Solid-State Circuits (JSSC), vol. 43, no.11, pp. 2363-2369, Nov. 2008.

[5] Y. Yuan, A. Radecki, N. Miura, I. Aikawa, Y. Take, H. Ishikuro, and T. Kuroda, "Simultaneous 6Gb/s Data and 10mW Power Transmission using Nested Clover Coils for Non-Contact Memory Card," IEEE Symposium on VLSI Circuits, Dig. Tech. Papers, pp. 199-200, June 2010.

[6] K. Ueyoshi, et al., "QUEST: Multi-Purpose Log-Quantized DNN Inference Engine Stacked on 96-MB 3D SRAM Using Inductive Coupling Technology in 40-nm CMOS," IEEE Journal of Solid-State Circuits (JSSC), vol. 54, no. 1, pp. 186-196, Jan. 2019.

Analysis of the Influence of Roughness on the Propagation Constant of a Waveguide via Two Sparse Stochastic Methods

Ruben Waeytens, Dries Bosman, Martijn Huynen, Michiel Gossye, Hendrik Rogier, Dries Vande Ginste

Electromagnetics Group/IDLab, Department of Information Technology, Ghent University/imec, Gent, Belgium

Dries.VandeGinste@UGent.be

Abstract—The aim of this contribution is to study the effect of roughness on the propagation constant in interconnect structures. For this purpose, a stochastic framework is constructed around a full-wave electromagnetic field solver. To reduce the number of repeated calls to the full-wave simulator, two sparse stochastic techniques have been implemented and tested. A balance between calculation time and accuracy is sought for and found, which is demonstrated for a rough rectangular waveguide.

Index Terms—Interconnect structures, line edge roughness, stochastic testing, sparse polynomial chaos

I. INTRODUCTION

The need for smaller and faster electronics imposes serious challenges to their design, as previously negligible phenomena now influence their performance. One such example is line edge roughness (LER) [1], where the edges of electronic structures have a random, rough profile. This can be induced on purpose for improved adhesion or unintentionally by the nature of the production process. No matter its origin, the ever-increasing skin effect, pushing the current towards the edges of a conductor, only reinforces the influence of the LER, necessitating a rigorous study. A full-wave approach imposes itself to capture all emerging phenomena, entailing computational challenges. Full-wave electromagnetic simulations tend to be very time-consuming, while stochastic methods typically demand repeated runs of the electromagnetic field solver.

In this contribution, we investigate two stochastic methods to model the influence of roughness on the propagation constant of a waveguide. To keep the computational burden as small as possible, whilst still accurately capturing the relevant features, the two stochastics methods are sparse. More specifically, a novel sparse grid version of the well-known Stochastic Testing (ST), originally proposed by Zheng Zhang *et al.* in 2013 [2], is constructed and the sparse Polynomial Chaos (SPC) method, proposed by Blatman and Sudret in 2009 [3], is adapted to our waveguide problem. Both stochastic approaches require (a few) calls to a full-wave simulator, which in this contribution is a finite-element method (FEM) based solver. The obtained stochastic framework is applied to the variability analysis of the TE_{10} mode's propagation constant of a rectangular waveguide. The validation of the methods is provided by means of comparison with a brute-force Monte Carlo (MC) analysis.

II. DETERMINISTIC PROBLEM AND FULL-WAVE SOLVER

We start from Maxwell's curl equations and assume, based on the longitudinal invariance encountered in waveguides (e.g., Fig. 1), a z-dependency of $e^{-j\beta z}$ for the electromagnetic fields, where the z-axis is the direction of invariance and β the propagation constant of the mode. A finite-element

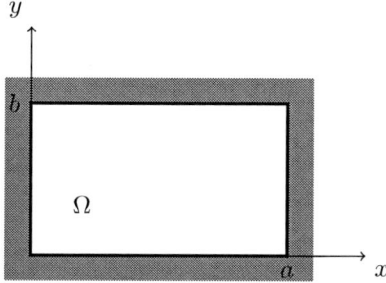

Fig. 1. An air-filled rectangular waveguide with perfectly conducting edges and dimensions a and b.

method (FEM) is employed to solve Maxwell's equations numerically. Approximating the electric field in the domain Ω by an appropriate set of hierarchical vector basis functions and applying a Galerkin weighting procedure eventually yields a quadratic eigenvalue problem

$$\left[\bar{\bar{M}}\beta^2 + \bar{\bar{C}}\beta + \bar{\bar{K}} \right] \bar{v} = 0. \qquad (1)$$

The expressions for the matrices $\bar{\bar{M}}$, $\bar{\bar{C}}$, and $\bar{\bar{K}}$ are not detailed, as they are readily derived from the standard FEM procedure. Solving (1) yields the propagation constants β for the given geometry (and also the corresponding eigenvector $\bar{v}$).

III. RANDOM PROBLEM AND SPARSE TECHNIQUES

A. Roughness and Karhunen-Loève transform (KLT)

Fig. 2. Generation of a rough edge (red) by shifting k points along the nominal edge (black) over a random distance h_i ($i = 1, \ldots, k$) (green) along the outward pointing normal to that edge.

978-1-7281-6162-4/20 $31.00 © 2020 IEEE

As shown in Fig. 2, we model the rough edge by a multivariate Gaussian distribution, similar as in [4]:

$$P(\bar{h}) = \frac{1}{\sqrt{|(2\pi)^k \bar{\bar{\Sigma}}|}} \exp(-\frac{1}{2}\bar{h}^T \bar{\bar{\Sigma}}^{-1} \bar{h}). \qquad (2)$$

The $\bar{h}$-vector collects the deviations of k points on the edge $\partial\Omega$, measured along the outward pointing normal to the edge. $\bar{\bar{\Sigma}}$ is the correlation matrix, for which a Gaussian relationship on the distance between two points, measured along the edge, is chosen:

$$\bar{\bar{\Sigma}}_{ij} = \sigma_r^2 \exp\left(-\frac{\|\vec{r}_i - \vec{r}_j\|_{\partial\Omega}^2}{L_c^2}\right). \qquad (3)$$

The parameter σ_r determines by how much points on the edge can deviate from their nominal position, while the correlation length, L_c, is a measure for how much the heights of neighbouring nodes depend on each other. The Karhunen-Loève transform (KLT) approximates the distribution of $\bar{h}$ by

$$\bar{h}(\bar{\xi}) \approx \bar{\bar{V}} \bar{\bar{\Lambda}}_{N_{RV}}^{1/2} \bar{\xi}, \qquad (4)$$

where $\bar{\xi}$ contains a set of N_{RV} stochastically independent standard normal random variables (RVs). $\bar{\bar{\Lambda}}_{N_{RV}}$ has on its diagonal the first N_{RV} eigenvalues, λ_n, of $\bar{\bar{\Sigma}}$, in descending order. The columns of $\bar{\bar{V}}$ are the corresponding eigenvectors. A measure for the fraction of variability that is captured by this approximation is given by:

$$\Theta = \frac{\sum_{n=1}^{N_{RV}} \lambda_n}{\sum_{n=1}^{k} \lambda_n}. \qquad (5)$$

B. Polynomial chaos expansion (PCE)

A relationship between the rough edge and a set of RVs has been established by the KLT. The distribution of the propagation constant β will be provided through a spectral decomposition into a set of $P+1$ orthonormal polynomials, $\phi_i(\bar{\xi})$, given by:

$$\beta(\bar{\xi}) \approx \sum_{i=0}^{P} a_i \phi_i(\bar{\xi}). \qquad (6)$$

The focus now shifts towards determining the expansion coefficients a_i $(i = 0, \ldots, P)$ in this polynomial chaos expansion (PCE). These can be found in several ways.

C. Sparse grid Stochastic Testing (ST) method

A popular technique, able to deal with a large number of RVs, is stochastic testing (ST) [2]. Within ST, (6) is tested for a certain number of 'interesting' values of $\bar{\xi}_i \in \{\bar{\xi}_i\}_{\text{test}}$:

$$\beta(\bar{\xi}_j) = \sum_{i=0}^{P} a_i \phi_i(\bar{\xi}_j) \quad \forall j \in \{0, \ldots, P\}$$
$$\Rightarrow \bar{\beta} = \bar{\bar{\Phi}} \cdot \bar{a}, \qquad (7)$$

where the number of testing nodes is equal to the number of basis functions. Calculating the expansion coefficients is now equivalent to solving a matrix equation. The selection of 'interesting' testing nodes results from a node picking algorithm [2]. In this algorithm, first a tensor grid of nodes is constructed, based on quadrature rules. Then, nodes with a higher weight are favored and selected, in the meantime guaranteeing the well-conditionedness of (7).

When applying this approach to the analysis of roughness in waveguides, the standard ST procedure still entails too many realizations $\beta(\bar{\xi}_j)$, and thus calls to the full-wave FEM solver. The exponential growth in the number of function evaluations is due to the tensor product of univariate quadratures to integrate a multivariate function. Whereas in integration routines such a procedure is said to be second-order accurate, albeit component-wise, it turns out, however, that nodes can be omitted without sacrificing on accuracy. Removing these redundant nodes from the tensor grid, is the crux of so-called sparse grid techniques [5].

In this contribution, we apply a novel sparse grid ST algorithm, which is conceptually quite simple, yet effective. The procedure is similar to the one proposed by Gossye et al. [6] to analyze statistically spatially varying dielectric-property profiles. Starting from a sparse grid, we apply the rules of the standard ST algorithm to pick the interesting nodes, i.e., favoring nodes with a higher weight and guaranteeing that the matrix (7) remains well-conditioned. This reduces the number of calls to the FEM solver drastically, whilst maintaining accuracy, as demonstrated in Section IV.

D. Sparse Polynomial Chaos (SPC)

It can be shown that the standard ST and the sparse grid ST techniques boil down to the observation that not all basis functions $\phi_i(\bar{\xi})$ in the polynomial chaos expansion (PCE) (6) are equally relevant. Therefore, we will also compare our sparse grid ST technique with another sparse technique, i.e., sparse polynomial chaos (SPC). SPC is explored as a possible technique to exploit the aforementioned observation to the fullest. The SPC algorithm further reduces the required number of full-wave runs, while extracting the most important terms in the expansion. Several algorithms, attempting to construct a sparse polynomial chaos expansion, exist in literature. In this contribution, the algorithm described by Blatman and Sudret [3] is adapted to our needs. The basic idea is to perform a fixed number of runs once and for all. Afterwards, an adaptive sparse set of basis functions is selected to construct a PCE. The decision whether or not to add or remove a basis function to this sparse set, is based on solving subsequent overdetermined systems. After several iterations of adding and removing basis functions to a temporary set of basis functions, a sparse set of relevant basis functions is retained.

IV. APPLICATION TO A ROUGH RECTANGULAR WAVEGUIDE

We now apply the two sparse stochastic techniques to the rectangular waveguide of Fig. 1, where we focus on the variability of the TE_{10} lowest-order propagation constant, further denoted β. As values for a and b, we choose $2l$ and $1l$, respectively, with l an arbitrary length unit. The free-space wavenumber is given by $k_0 = 10l^{-1}$, leading to a discretization of Ω into 1094 triangles. A second-order expansion leads to a finite-element linear system in 5745 unknowns. The roughness profile is generated by varying $k = 82$ uniformly spaced nodes along the edge $\partial\Omega$. Additionally, the following constants were assigned for the KLT: $L_c = 0.5l$, $\sigma_r = 0.01l$, and $\Theta = 0.85$, yielding $N_{RV} = 8$ RVs collected in the vector $\bar{\xi}$.

We compare the results of the two sparse algorithms to a standard Monte Carlo (MC) approach. MC is here considered as the golden standard, given its guaranteed albeit slow $1/\sqrt{N_{MC}}$-convergence, with N_{MC} the number of samples drawn. For the particular example, it is observed that for $N_{MC} = 50000$, the average of the propagation constant β has converged up to a relative error of $6 \cdot 10^{-6}$ and its standard deviation up to a relative error of $2 \cdot 10^{-2}$. Specifically, the following values are obtained: $\mu_\beta = \left(9.87589 \pm 6 \cdot 10^{-5}\right) l^{-1}$ and $\sigma_\beta = \left(0.00192 \pm 4 \cdot 10^{-5}\right) l^{-1}$.

Employing the novel advocated sparse grid ST technique, combined with a PCE, demands only 45 runs of the full-wave solver. This is a speed-up by a factor of more than 1000 compared to the brute force MC method. Note that the time needed for node picking is negligible compared to one full-wave simulation. The sparse grid ST predicts $\mu_\beta = 9.87590\, l^{-1}$ and $\sigma_\beta = 0.00196\, l^{-1}$. Hence, both values are within the bounds determined by the MC simulation. Furthermore, Fig. 3 compares the distribution of β obtained by the MC method and sparse grid ST. To construct the distribution by means of sparse grid ST, we sample our surrogate model (6) through a MC simulation, which only requires simple arithmetic. Thus, constructing such a distribution is also expedited tremendously. To quantify the good correspondence, we also performed a Cramèr-von Mises (CvM) test [7], which delivers a p-value of 0.18. This value shows that both datasets indeed originate from the same distribution.

We also apply the aforementioned SPC algorithm to the rough rectangular waveguide. For this simulation, the number of full-wave calculations could be further reduced from 45 to 30, cutting off a third from the calculation time. The computational cost of the iterative procedure to select the pertinent basis functions for the PCE is negligible compared to one full-wave run. SPC predicts $\mu_\beta = 9.87589\, l^{-1}$ and $\sigma_\beta = 0.00196\, l^{-1}$, which is again within the bounds set by the MC simulation. These results are also summarized in Table I. Moreover, the obtained distribution for β is added to the plot in Fig. 3. This distribution predicted by SPC is again plotted by employing the corresponding PCE as a surrogate model. A good qualitative resemblance is clearly visible and a CvM test was carried out as well, resulting in a p-value of 0.1241. It is therefore safe to assume that both the MC and SPC simulation also originate from the same distribution.

V. Conclusions

The goal of this contribution was to evaluate the influence of roughness on the propagation characteristics of interconnect structures. Given the computational burden of a single full-wave simulation, two sparse stochastic methods were implemented, in conjunction with a finite-elements full-wave solver.

	$\mu_\beta\,[l^{-1}]$	$\sigma_\beta\,[l^{-1}]$
MC	$9.87589 \pm 6 \cdot 10^{-5}$	$0.00192 \pm 4 \cdot 10^{-5}$
ST	9.87590	0.00196
SPCE	9.87589	0.00196

TABLE I

STATISTICS OF THE TE_{10} PROPAGATION CONSTANT β, OBTAINED BY MC SIMULATION, SPARSE GRID ST AND SPC.

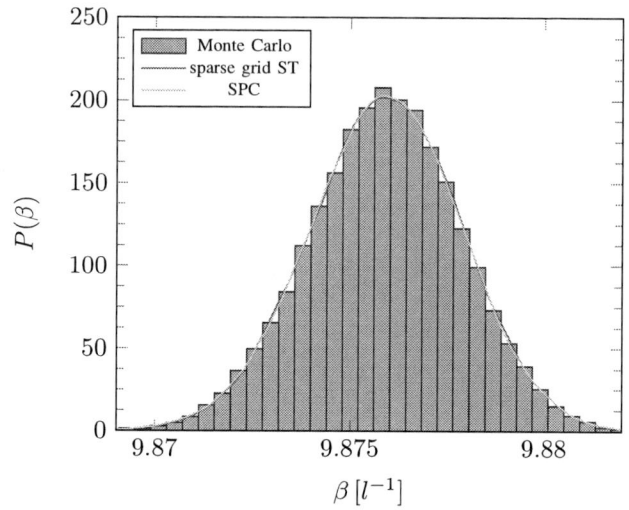

Fig. 3. Comparison of the distribution of the TE_{10} propagation constant β of a rectangular waveguide, obtained by MC, sparse grid ST and SPC.

The methods were applied to the computation of the TE_{10} propagation constant of a rectangular waveguide. It was shown that the methods reduce the computational burden, whilst still accurately predicting the influence of the roughness profile.

In future research, both sparse techniques will be further explored and improved, pushing the number of RVs to the limit. Other roughness profiles and waveguide structures, including, e.g., lossy microstrip interconnects, will be investigated too.

References

[1] H. Braunisch, X. Gu, A. Camacho-Bragado, and L. Tsang, "Off-chip rough-metal-surface propagation loss modeling and correlation with measurements," *Proceedings - Electronic Components and Technology Conference*, pp. 785–791, Feb. 2014.

[2] Z. Zhang, T. A. El-Moselhy, I. M. Elfadel, and L. Daniel, "Stochastic testing method for transistor-level uncertainty quantification based on generalized polynomial chaos," *IEEE Transactions on Computer-Aided Design of Integrated Circuits and Systems*, vol. 32, no. 10, pp. 1533–1545, 2013.

[3] G. Blatman and B. Sudret, "An adaptive algorithm to build up sparse polynomial chaos expansions for stochastic finite element analysis," *Probabilistic Engineering Mechanics*, vol. 25, pp. 183–197, Apr. 2009.

[4] Z. Zubac, D. De Zutter, and D. Vande Ginste, "Scattering from two-dimensional objects of varying shape combining the method of moments with the stochastic Galerkin method," *IEEE Transactions on Antennas and Propagation*, vol. 62, pp. 4852–4856, Sep. 2014.

[5] F. Heiss and V. Winschel, "Likelihood approximation by numerical integration on sparse grids," *Journal of Econometrics*, vol. 144, no. 1, pp. 62–80, 2008.

[6] M. Gossye, G. Gordebeke, K. Y. Kapusuz, D. Vande Ginste, and H. Rogier, "Uncertainty quantification of waveguide dispersion using sparse grid stochastic testing," *IEEE Transactions on Microwave Theory and Techniques*, vol. 68, no. 7, pp. 2485–2494, 2020.

[7] T. W. Anderson, "On the Distribution of the Two-Sample Cramer-von Mises Criterion," *Ann. Math. Statist.*, vol. 33, pp. 1148–1159, Sept. 1962.

Determine Socket's Inductance and Contact Resistance by Using PRF Method

1st Tao Wang

Qualcomm Technologies, Inc.
San Diego,CA, USA
wangtao@qti.qualcomm.com

2nd Jun Fan
Electrical and Computer Engineering
Missouri University of Science and Technology
Rolla, USA
jfan@umsystem.edu

Abstract—In this paper, an accurate socket modeling methodology is proposed for signal integrity and power integrity applications using a quasi-static EM solver and considering the contact resistance of the socket which is a non-negligible component in socket modeling. The contact resistance characterization of a specific pogo-pin socket has been conducted by measuring the socket mounted on a specially designed test vehicle. A Parallel Resonance Frequency (PRF) method [1] is used to validate the EM model and to extract the contact resistance accurately.

Index Terms—pogo pin socket, contact resistance, signal integrity, power integrity, quasi-static, PRF methodology.

I. INTRODUCTION

Nowadays, a test socket is often used to electrically link the testing PCB and semiconductor chips and to eliminate the soldering process as shown in Fig.1.

To accurately evaluate the performance of high-speed interfaces and power delivery networks (PDN), high-quality electromagnetic (EM) socket models are needed in system level simulation for both signal integrity (SI) and power integrity (PI) applications to capture crosstalk and Vdroop. Socket were characterized by using a first order equivalent circuit model or building socket model with H type of equivalent circuit. Unfortunately, direct micro probe landing is still used to measure the socket which will introduce unavoidable errors.

The resistance behavior of the socket is a function of the force applied to the socket by the socket clamp. The contact design, material, and mechanical assembly will impact the contact resistance range. In this study, a socket fixture with a force-control clamp is designed and used so the compression is kept constant with a mechanical stop. In applications where the socket can be treated as electrically small, it is adequate to add a series contact resistance component to model all the contact resistances that may exist at several points along the socket pin. Sockets with contact resistance included are characterized using the PRF method that utilizes a parallel LC resonance to characterize the socket's loop resistance and inductance over a broad frequency range. Through further analysis, the socket pin's contact resistance can be extracted. The measurements were done with a specially designed test-vehicle that enables the creation of accurate models by avoiding probe-to-socket pin contact and instead utilizing the PRF methodology.

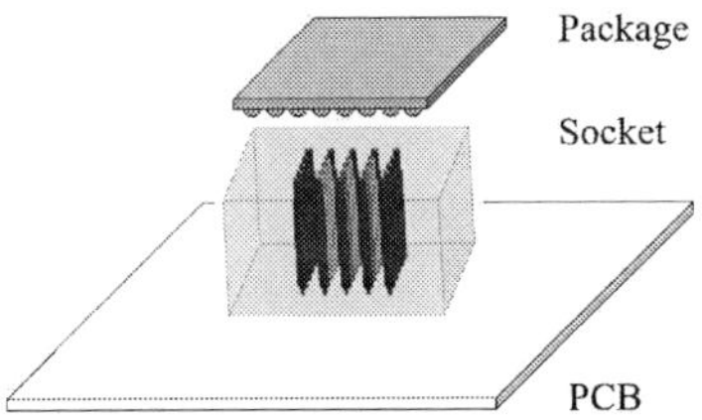

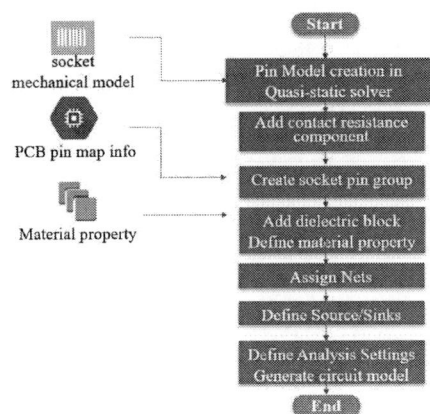

Fig. 1: Diagram of a pogo pin socket with PKG and PCB.

II. SOCKET MODELING METHODOLOGY

A. Socket Modeling Flow

This paper provides insights into extracting EM socket models for SI and PI applications using a quasi-static EM solver by extracting RLGC parameters of socket pins and export an equivalent circuit model that can be used in system level simulations to characterize channel or PDN performance. As shown in Fig. 2, a complete socket modeling flow is established to extract a socket model that includes contact resistance. In this flow, the first step is to create a pin model. After the individual pin model is completed, a socket pin array is created using the pin pattern of interest.

Fig. 2: Socket modeling flow

978-1-7281-6162-4/20 $31.00 © 2020 IEEE

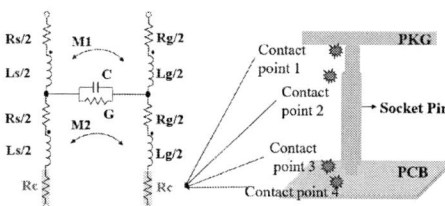

Fig. 3: Schematic diagram of a two pin socket with contact resistance

B. Contact Resistance

In a pogo-pin socket there may be several metal-to-metal contact points along the pin that contribute to the contact resistance as shown in Fig.3. The contact resistance also includes the locations where the socket pin touches the PKG and PCB in addition to the metal to metal contact points along the socket pin. In most cases, a socket pin is electrically small, so it is valid to sum the contact resistances at different points into a total contact resistance element as shown in the equivalent circuit in Fig. 3

Contact resistance of single socket pin ranges from 10 mOhm to 100 mOhm for commonly used sockets, with a representative average of contact resistance of 40 mOhm. In PI applications, such resistances can cause specification compliance failure so the contact resistance should be carefully evaluated. The PRF method captures contact resistance as will be discussed in the following section.

III. SOCKET MODELING VALIDATION

A. PRF Methodology

In validating the representative socket EM model, good correlation between the simulation and measurements have been achieved using the PRF method. A frequency domain measurement utilizing a vector network analyzer was performed to measure the impedance of a fixtured socket. The PRF technique is applied by designing a fixture that has a well characterized, fixed capacitance which interacts with the socket inductance to form a parallel resonance. The parallel resonance peak offers insight into the DUT and fixture since the Q factor of the peak is a function of the resistance, inductance, and capacitance of the DUT. The parallel resonance peak is an eigenvalue function of the fixture, thus, it is robust and not prone to parasitic inductance error that can be introduced from traditional de-embedding techniques. With the PRF method, the fixture inductance and resistance can be separated from the socket inductance and resistance, effectively de-embedding the fixture.

The DUT impedance in relation to the 2-port S-parameter is defined in (1). Since the two ports at the probing pads in Fig. 5 provide an input wave from one port while the second port senses the received wave, what is really measured is a transfer-impedance between the two ports, which is related to the DUT S-parameters as:

$$Z_{DUT} = \frac{Z_0 S_{21}}{2(1 - S_{21})} \tag{1}$$

where Z_0 is the port reference impedance (50 Ω).

B. Fixture Design

The fixtures designed for mounting the DUT socket are intended for de-embedding utilizing the parallel resonance peak. The fixture provides an accurate and easily measured capacitance source while loop inductance comprises the socket pins and the metal shorting mechanism. The PCB plane capacitance and the mounted socket pin inductance form the parallel resonance. The PCB dielectric material relative permittivity and plane dimensions and probe location are designed to have the capacitance resonate at a frequency where it is distinguishable from the cavity resonances [1].

Key parameters to the fixture are the amount of fixture capacitance and inductance needed to achieve a desired peak resonance at the desired frequency. Once the desired frequency is known, the following equation can help determine the fixture key electrical parameters.

The PRFs from the Fixture Short measurement and Socket Mounted from Fig. 4 can be used to calculate L_{DUT} to yield $L_{fixture}$ and $L_{fixture+socket}$ by subtracting the two extracted inductance values, effectively de-embedding the fixture, so that the inductance of the socket can be derived as L_{Socket}. The socket resistance, R_{DC} can be derived from subtracting the fixture resistance from the fixture and socket resistance.

As a first step, $L_{fixture}$ can be derived from (2):

$$f_{PRF_{fixture}} = \frac{1}{2\pi\sqrt{L_{fixture}C_{fixture}}} \tag{2}$$

$C_{fixture}$ can be calculated from impedance curve obtained in the Fixture Open scenario as shown in Fig. 6. After mounting the socket on top of test fixture, PRF becomes:

$$f_{PRF_{fixture+socket}} = \frac{1}{2\pi\sqrt{L_{loop}C_{loop}}}$$

$$= \frac{1}{2\pi\sqrt{L_{fixture+socket} \cdot \frac{1}{\frac{1}{C_{socket}} + \frac{1}{C_{fixture}}}}} \tag{3}$$

where:

$$L_{loop} = L_{fixture+socket} \tag{4}$$

$$C_{loop} = \frac{1}{\frac{1}{C_{socket}} + \frac{1}{C_{fixture}}} \tag{5}$$

Since C_{socket} is very small compared to $C_{fixture}$, the C_{socket} term can be ignored. $L_{fixture+socket}$ then can be derived from (3)

The fixture on the PCB comprises parallel planes with probe contact pads and the socket mounting footprint. Fig. 4 shows the 2D cross-section representation of the fixture. The same fixture design is used for the Fixture Open, Fixture Short, and Socket Mounted cases. A shorting metal is utilized to allow

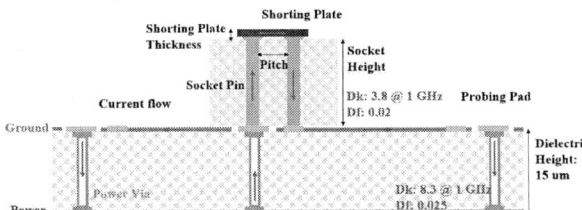

Fig. 4: Fixture + Socket 2D Cross-section diagram. (Dimension not to scale)

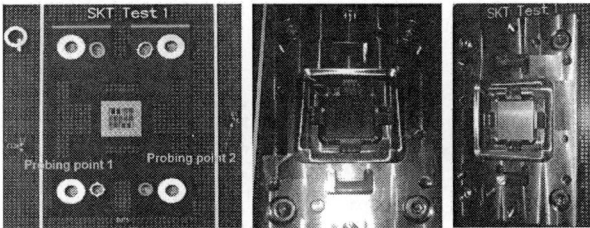

Fig. 5: Fixture design. left: top view of fixture design with two probing points located at edge of the ficture. middle: PCB fixture with socket mounted on top. right: PCB fixture with socket mounted on top and shorting plate is added on top of socket to make a short.

for a loop impedance measurement for the Fixture Short and Socket Mounted. A real fixture design is shown in Fig. 5

In order to correlate measured data to EM simulation results and be able to de-embed the fixture from the socket, three measurements must be made.

1) Fixture Open circuit. The fixture is referred as Open due to the open circuit at the socket footprint when the DUT socket is not mounted. The key design parameters for the Open fixture are governing the size of the planes, finding the minimum thickness, and finding the suitable dielectric with a permittivity value that achieves right amount of capacitance to generate a desired PRF.

2) Fixture Short. The socket footprint will now have a metal short between the positive and negative pins. With the size, thickness, and dielectric material fixed from the Open Fixture design, determining the location of the port landing pad will affect the effectiveness of the PRF. By determining the desired probe landing location, the PRF peak should be at a lower frequency than the cavity mode resonances.

3) Fixture with socket (Socket Mounted). The DUT fixture is essentially the exact copy of the Open fixture. Measurement is done while the socket is mounted and the shorting plate on the top of the socket is applied. The PRF formed by this measurement will include the fixture and the DUT.

Fig. 6 shows the measured impedance of the fixture in the open and short configurations. The impedance measurement of the open fixture allows characterization of the fixture's capacitance. Impedance measurement of the Socket Mounted captures the combined loop inductance and resistance of the fixture and the socket. DC resistance increase dramatically in the presence of the socket pins. The contact resistance of the socket pins dominates the resistance at low frequencies and dampens the Q of the resonance at higher frequencies.

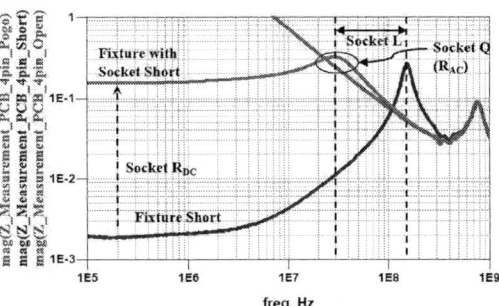

Fig. 6: Measured impedance of the Socket Mounted comparing with fixture.

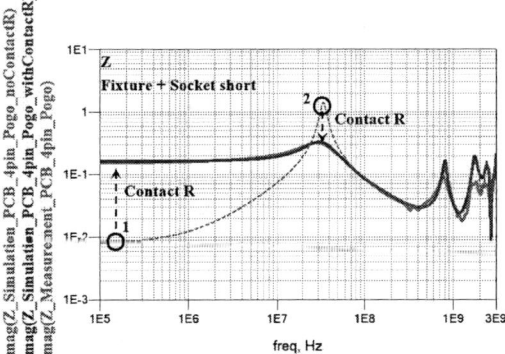

Fig. 7: Fixture Open EM model well correlated with measured data after PCB cross-section. PCB cross-section is used to get the real layer thickness of fixture PCB.

C. Pogo-pin Socket Model Correlation

There are no concerns of port prasitic elements during correlation as the socket was mounted directly on top of the fixture and a shorting plate was used to form the current loop. The solid blue curve in Fig. 7 shows the impedance of the mounted socket model with contact resistance added from simulation and it is well correlated to the measurement. By achieving this close correlation, the socket model is concluded to be a good representation of the socket's electrical performance.

Fig. 7 highlights the importance of contact resistance in the socket. The dashed curve shows the simulation model of the mounted socket pin without modeling additional contact resistance. After tuning the additional representative contact resistance, the simulation model correlates well with measured data at both low and high frequencies. The parallel resonance peak correlates well, which indicates that the AC loop inductance and resistance are modeled accurately. The Q factor is much dampened due to the contact resistance.

REFERENCES

[1] V. Sriboonlue, J. Shin and T. Michalka, "Novel Parallel Resonance Peak Measurement and Lossy Transmission Line Modeling of 2-T and 3-T MLLC Capacitors for PDN Application," *IEEE 67th Electronic Components and Technology Conference*, Lake Buena Vista, 2017.

978-1-7281-6162-4/20 $31.00 © 2020 IEEE   121

RX Equalization for a High-Speed Channel based on Bayesian Active Learning using Dropout

Xianbo Yang*, Junyan Tang*, Hakki M. Torun**, Wiren D. Becker*, Jose A. Hejase***# and Madhavan Swaminathan**

*POWER Series Servers Hardware Development Group, IBM Corporation, Austin, TX, 78758
**3D Systems Packaging Research Center (PRC), Georgia Institute of Technology, Atlanta, GA, 30332
*** Mixed Signal Development Group, Nvidia Corporation, Austin, TX, 78717

 # Contributions to this work completed prior to April 2020 while with IBM Corporation

Abstract—Determining optimal equalization settings in high-speed bus design simulations is becoming more important due to increased complexity and data rates of current server systems, but it is also time and resource consuming. In this paper, a probabilistic machine learning technique, Bayesian Active Learning using Dropout (BAL-DO), is utilized to perform RX equalization and optimization to address this issue. Largest HEYE opening and corresponding equalization settings are obtained with high prediction accuracy without performing extensive time-domain analysis, thereby significantly reducing the cost of engineering time and computational resources.

I. INTRODUCTION

The demand for faster data rates and more versatile functionality in computing systems results in more complex systems with a large number of components and interconnects. A key part of high-performance systems are board-to-board interconnects for the high-speed links which transmit signals between components on different boards. Such links can exist in between processors and other modules such as FPGAs, memory controllers and expanders. Due to the increase in complexity, signal integrity (SI) properties such as insertion loss (IL) and crosstalk (XT) can vary significantly from channel to channel, which complicates confirming the channel compliance for sign-off and production. To maximize and stabilize the performance of channels that contain cables, connectors and PCBs, channel equalization designed in I/O circuitry is required to have stronger capability and robustness. Various equalization schemes such as feed-forward equalizer (FFE), decision feed-back equalizer (DFE) and continuous time linear equalizer (CTLE) can be implemented on the transmitter (TX) side and the receiver (RX) side. A direct consequence of adding more equalization capabilities is the increased complexity of the I/O circuitry and larger design space for SI simulations. Normally, SI engineers need to perform time-domain simulations by sweeping most of the equalization settings to obtain desired eye sizes for different channel configurations. However, as the combinations of the equalization settings increase exponentially, sweeping through the entire design space becomes extremely time and resource consuming and, ultimately, impractical. It is critical for SI analysis to find an efficient method that can reduce the optimization time as well as resources and allow the designer to understand the preferred equalization settings for various channel configurations.

An intuitive idea is to apply machine learning (ML) techniques to these problems [1][2]. However, conventional ML methods are data hungry, and difficult to capture domain expertise. Equalization optimization with time-domain analysis requires intensive computational resources, limiting the amount of data being available in many cases. A probabilistic method that predicts a posterior distribution rather than deterministic predictions is a better candidate for limited training data scenarios. Recently, a novel ML technique based on Bayesian Active Learning (BAL) has been developed [3]. A new algorithm, Bayesian Active Learning using Dropout (BAL-DO) was proposed to achieve accurate data space exploration when learning data is limited. This has been tested on IBM's POWER9 channel, and successfully acquired the worst-case horizontal eye (HEYE) opening with high accuracy among large number of channels in minimum amount of time-domain simulations. The main advantage is the significant reduction of computational costs while maintaining high accuracy, and this has been demonstrated in [3] by comparing with other machine learning methods. In this work, the BAL-DO algorithm is extended to include RX equalization for a high-speed memory channel [4] to determine the equalization setting that provides the largest HEYE along with the confidence bounds of the resulting eye which is a key feature of BAL-DO.

The BAL-DO technique mainly consists of two parts, the optimization stage and the active learning stage, and they are combined associatively in the algorithm [3]. As the code starts with no training data, one set of initial training data is generated and fed into the program. In the optimization stage, the next sampling point is selected using Bayesian Optimization (BO) method with self-learning acquisition function strategy obtained using a Gaussian Process (GP) model. The goal for the active learning part is to minimize the uncertainty of the GP predictions for non-simulated equalization settings by selecting the setting that maximizes entropy. The information from both stages is combined in a single GP while using a dropout technique to prioritize optimization over learning [3]. The next sample is then evaluated using the simulation framework to get the corresponding HEYE, followed by re-training the GP and proceeding to the optimization stage again.

II. SIMULATION PROCEDURE AND SETUP

The simulation framework in this paper is based on a high-speed differential channel passing signal from CPU to the memory buffer running at 32Gb/s NRZ. S-Parameters for the whole channel are first generated, time-domain simulations are then performed under various equalization setting combinations using an in-house tool to obtain the corresponding HEYE opening results. BAL-DO uses this framework to determine next equalization settings to be simulated in an automated fashion. At each simulation iteration, the GP is trained by using all the data obtained in previous iterations. The largest HEYE opening, the probability density function (PDF) of HEYE and a sensitivity

978-1-7281-6162-4/20 $31.00 © 2020 IEEE

analysis that ranks the equalization settings in terms of creating a variation in HEYE is then derived through the learned GP after a certain number of simulations.

In this work, the optimum eye opening is searched by varying the receiver equalizations which includes 4 different settings with their varied combinations defining certain frequency dependent RX peaking curves. These 4 different settings are: long tail equalizer (LTE) gain, LTE zero, and 2 peaks of CTLE. LTE gain and LTE zero have 8 setting options each, and 2 CTLE peaks have 16 setting options each. This results in total number of 16384 possible combinations of equalization settings for the RX equalization. Each combination contains 4 variables, and is defined as a single input vector, to be fed into the BAL-DO algorithm and call the in-house simulator, HSSCDR, to run the time-domain simulation. The main objective is to find the largest HEYE opening and its corresponding equalization settings, while the aforementioned sensitivity analysis and HEYE PDF are considered as secondary objectives. In addition, FFE and DFE are also included as part of TX and RX equalization, but their settings are auto-adapted in HSSCDR. The data rate in the simulation is set as 32Gb/s and the HEYE is found at BER=10^{-15}.

III. SIMULATION RESULTS

Since the number of equalization combinations is large, 700 iterations are used in BAL-DO to ensure convergence, which is about 4.3% of total numbers of combinations. One single simulation iteration takes approx. 20 min, leading to the whole run to be completed in about 10 days. It will be seen later that significantly less numbers of iterations can be considered to arrive at convergence, which leads to considerably less time.

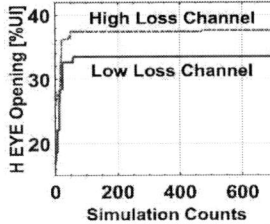

Fig. 1. Largest HEYE opening vs. simulation counts for high and low loss channels.

Different parameters from channel components such as lengths, manufacturing/packaging corners and impedance can be picked to form channels with different properties. To understand the performance of BAL-DO algorithm for channels with different properties, the optimization is performed for two channels with high loss and low loss (22.7 dB and 10.4 dB loss at 16GHz, respectively) for the same bus topology and results are compared. Fig. 1 shows the largest HEYEs derived from BAL-DO method for both high and low loss channels as simulation counts increase. HEYEs for both channels increase at early counts towards the optimal, reaching 33.5%UI at 59th simulation and 37.4%UI at 49th simulation for low and high loss channels, respectively, which corresponds to less than 1 day of CPU time. After that, largest HEYE for low loss channel remains unchanged till 700 simulations are done. For high loss channel, HEYE increased by 0.2%UI to 37.6%UI at 471st simulation, and then no change till finishing all 700 simulations. Note that in order to obtain this extra 0.2%UI, the algorithm takes approx. 6 extra days, thus the "trade-off" between

excessive extra simulation time along with resources, and small amount of HEYE opening improvement needs to be considered and balanced. One may not need to do additional intensive simulations to only receive a tiny margin of improvement. It is worth noting that the largest HEYE opening for low loss channel is smaller than that of high loss channel. Possible reasons for that are: (1) increased insertion loss deviation -reflections- for the low loss channel and (2) RX peaking circuit design targeting high loss channel equalization thus causing over-equalization for the low loss channel.

The converging criteria of BAL-DO is when the maximum value of upper confidence boundary (UCB) is within a small margin of the largest HEYE found after each iteration. This guarantees the optimal value found has 95% confidence level to be the global optimum. Convergence curves with respect to number of simulations for low and high loss channels are shown in Fig. 2(a) and 2(b), respectively. At early iterations, max of UCBs are very large, meaning the prediction has large uncertainty and goal has not been met yet. As BAL-DO progresses to find the largest HEYE opening, two curves get closer, and both shows relatively good convergence after 100 iterations. Note that for the high loss channel, the gap between max UCB and Largest HEYE is larger than that of low loss channel after converging. This is due to equalization having more influence on HEYE opening of high loss channel than low loss one. As simulation counts increase, the gap becomes smaller, indicating prediction gets more accurate. To further verify, intensive simulations were done through HSSCDR and confirms that 37.6% UI is the best HEYE for the high loss channel.

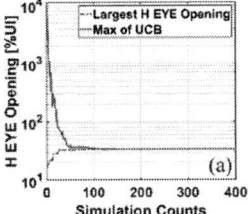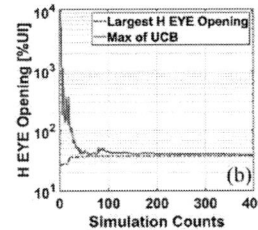

Fig. 2. Convergence for low (a) and high (b) loss channels

Values for RX equalization variables rendering optimal HEYE opening for different channels obtained from BAL-DO are listed in Table I.

TABLE I.   RX EQUALIZATION VALUES FOR OPTIMAL HEYE OPENING

	LTE zero	LTE gain	Peak 1	Peak 2	Largest HEYE	Simulation Needed
low loss channel	2	2	1	1	33.5%UI	59
high loss channel	8	1	10	14	37.4%UI	49
	3	2	7	5	37.6%UI	471

The number of variables, and the number of values being swept in the pool of each variable, are the two main factors that affect the minimum iterations for BAL-DO to arrive at desired results though the former has higher impact. In this case, 4 RX equalizers are defined as variables, making this a 4-dimensional problem. As mentioned above, if the extra 0.2%UI improvement can be neglected for the high loss channel, 100 counts (0.6% of overall combinations) is enough to derive the largest HEYE with good accuracy and low uncertainty. As comparison, a previous BAL-DO test with 9 variables required to simulate 4-5% out of

978-1-7281-6162-4/20 $31.00 © 2020 IEEE   123

the overall combination pool [3]. For brevity, the rest of the paper deals with the high loss channel.

The PDFs of the high loss channel after 100 & 400 simulations are plotted and compared in Fig. 3. In both cases, the uncertainty values (blue histogram) at 37.4%UI are very low, and they are very close (0.0027 vs. 0.0022, as shown in the inset images), indicating that 400 iterations do not improve the accuracy too much as compared to 100 iterations. In addition, both PDF curves show that the distributions are left-skewed. especially after 400 iterations as in Fig. 3(b), meaning that the samples selected by BAL-DO are focused more on the larger EYE region, and this is consistent with the goal of finding largest HEYE. The uncertainty given as blue shading on the large HEYE side is steeper in Fig. 3(b) as compared to Fig. 3(a). This is because as the number of iterations increase, more training data are available for BAL-DO to improve the overall accuracy, especially in the large HEYE region.

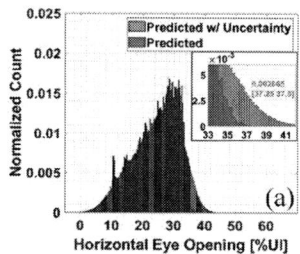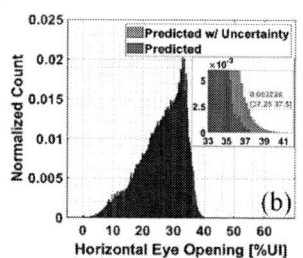

Fig. 3. PDF after 100 (a) and 400 (b) iterations.

BAL-DO algorithm starts with no training data initially, thus requires an initial input vector to generate a corresponding HEYE opening and form one set of training data. The variable combination pool including all combinations is generated by BAL-DO and remains unchanged, and the starting input vector is randomly selected by the algorithm from it. To understand the influence of the initial point selection, three different input vectors are manually picked and fed into BAL-DO for the high loss channel. Fig. 4 shows the largest HEYE openings with respect to simulation iterations by applying 3 different initial inputs. Low, mid and high index initial points denote the 10th, 8010th, and 16010th combinations in the combination array, respectively. Their corresponding values are listed in Table II.

TABLE II.   EQUALIZATION VALUES FOR EACH INITIAL INPUT POINT

	LTE zero	LTE gain	Peak 1	Peak 2
10th point	1	1	1	10
8010th point	4	8	5	16
16010th point	8	7	13	9

As demonstrated in the left inset image in Fig. 4, when reaching 37.4%UI, simulation with low index initial point takes 49 iterations, mid index initial points run takes 59 iterations, and the one with high index initial point takes 79 iterations. This again indicates that 100 simulations are enough to achieve the objective regardless of initial point used. As the simulation count increase above 450, as shown in the right inset image of Fig. 4, all three runs eventually reach 37.6%UI. Again, the trade-off between tiny margin of improvement and excessive simulation time needs to be taken into consideration.

Sensitivity analysis is also performed after different simulation counts for the high loss channel. As shown in Fig. 5, CTLE peak 1 has the most influence on the HEYE opening as

simulation counts increase. Meanwhile, as more training data becomes available as simulation counts increase, the area between UCB and lower confidence boundary (LCB) for each variable decrease, indicating the uncertainty over weights becomes smaller, hence prediction becomes more confident and accurate. Note that in Fig. 5(a), peak 1 has a slightly higher weight than peak 2. In this case, more training data are needed to distinguish the differences between them, and 150 iterations is enough to draw the conclusion instead of going up to 400.

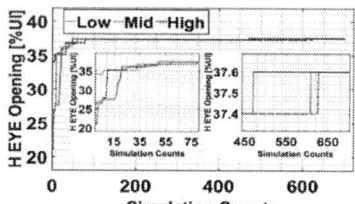

Fig. 4. Largest HEYE opening vs. simulation counts for BAL-DO different starting points.

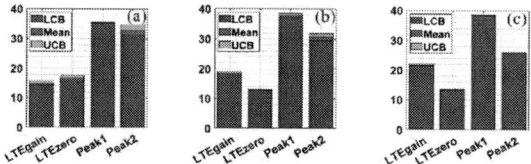

Fig. 5. Sensitivity for 100 (a), 150 (b), and 400 (c) simulation counts for high loss channel.

IV. CONCLUSION

Bayesian Active Learning using Dropout (BAL-DO) is extended to successfully perform efficient RX equalization optimization for high-speed channel design. Results show that for both high and low loss channels, main objective of finding largest HEYE openings and their corresponding equalization settings can be achieved within 100 simulations with good convergence and low uncertainly. With different initial input points, results still converge within 100 simulations, which is only 0.6% of overall equalization setting combinations. This significantly reduces engineering time and resources while maintaining high accuracy. The sensitivity analysis shows Peak 1 is the dominant variable and has higher influence on the HEYE opening. BAL-DO technique has great capability of using minimum resource to find global optima with high predictive accuracy, and it can be extended to different types of equalization problems with more variables.

REFERENCES

[1] M. Swaminathan, H.M. Torun, H. Yu, J. A. Hejase, and W. D. Becker, "Demystifying Machine Learning for Signal and Power Integrity Problems in Packaging," *IEEE Transactions on Components, Packaging and Manufacturing Technology*, vol. 10, no. 8, pp. 1276-1295, Aug. 2020.

[2] J. A. Hejase, P. R. Paladhi, R. S. Krabbenhoft, Z. Chen, J. Tang and D. J. Boday, "A neural network based method for predicting PCB glass weave induced skew," in Proc. IEEE 25th Conf. Elect. Perform. Electron. Packag. Syst. (EPEPS), Oct. 2016, pp. 151-154.

[3] H. M. Torun, J. A. Hejase, J. Tang, W. D. Becker and M. Swaminathan, "Bayesian Active Learning for Uncertainty Quantification of High Speed Channel Signaling," in Proc. IEEE 27th Conf. Elect. Perform. Electron. Packag. Syst. (EPEPS), Oct. 2018, pp. 311–313.

[4] B. Cai *et al.*, "OpenCAPI Memory Interface Signal Integrity Study for High-Speed DDR5 Differential DIMM Channel with Standard Loss FR-4 Material and SNIA SFF-TA-1002 Connector," *2019 IEEE 69th Electronic Components and Technology Conference (ECTC)*, Las Vegas, NV, USA, 2019, pp. 1200-1207.

Hyperparameter determination in multivariate macromodeling based on radial basis functions

Alessandro Zanco, *Student Member, IEEE*, Stefano Grivet-Talocia, *Fellow, IEEE*

Abstract—This paper introduces a simple and effective algorithm for the automated selection of Radial Basis Function hyperparameters in the context of high-dimensional multivariate macromodeling. Numerical results show an average speedup of at least one order of magnitude with respect to direct hyperparameter optimization.

I. INTRODUCTION

Multivariate (parameterized) macromodels aim at reproducing the frequency behavior of complex structures through surrogate compact dynamical systems [1], whose coefficients depend on a number of free variables related to geometry, material, processes, temperature, etc. Such models are usually identified from sampled responses through a multivariate approximation, whose accuracy and compactness depend on the specific choice of some basis function sets embedded in the model equations.

It has been shown that Radial Basis Functions (RBFs) provide an excellent candidate for scalability to high parameter dimension, thanks to their mesh-free structure. RBFs are usually defined in terms of one or more *hyperparameters*, which should be carefully optimized for best performance. Various techniques exist for select appropriate values, usually based on optimization of cross validation and maximum likelihood estimators [2], [3] or direct optimization or grid search [4]–[6]. Here, we propose a simple yet very effective algorithm that, with negligible loss of accuracy and no overhead, allows a suboptimal determination of hyperparameters related to Gaussian RBFs.

II. BACKGROUND AND FORMULATION

We consider a generic P-port electronic, electrical or electromagnetic structure, whose behavior depends on ρ geometric or physical parameters collected in vector $\boldsymbol{\vartheta} = [\vartheta_1, \dots, \vartheta_\rho] \in \Theta$. The structure is characterized through its frequency responses $\breve{\mathbf{H}}_{k,m} = \breve{\mathbf{H}}(s_k, \boldsymbol{\vartheta}_m)$ known at a discrete set of $\bar{k}$ frequency $s_k = \mathrm{j}\omega_k$ and $\bar{m}$ parameter samples $\boldsymbol{\vartheta}_m$. We seek for a surrogate parametric model $\mathbf{H}(s, \boldsymbol{\vartheta})$ such that

$$\mathbf{H}(s_k, \boldsymbol{\vartheta}_m) \approx \breve{\mathbf{H}}_{k,m} \quad \forall k, m \tag{1}$$

in order to approximate with a controlled error both the parametric and the broadband frequency behavior of the data.

The model structure we use is well established [1]

$$\mathbf{H}(s; \boldsymbol{\vartheta}) = \frac{\sum_{n=0}^{\bar{n}} \sum_{\ell=1}^{\bar{\ell}} \mathbf{R}_{n,\ell} \, \xi_\ell(\boldsymbol{\vartheta}) \, \varphi_n(s)}{\sum_{n=0}^{\bar{n}} \sum_{\ell=1}^{\bar{\ell}} r_{n,\ell} \, \xi_\ell(\boldsymbol{\vartheta}) \, \varphi_n(s)}, \tag{2}$$

where frequency dependence is captured by the basis functions $\varphi(s)$, which correspond to the standard partial fraction basis associated with Vector Fitting (VF) "basis" poles, and parameter variability is captured by the basis functions $\xi_\ell(\boldsymbol{\vartheta})$. The model coefficients $\mathbf{R}_{n,\ell}$, $r_{n,\ell}$ are computed through the so-called *Parameterized Sanathanan Koerner* (PSK) algorithm [1], [7], which is an iterative linear relaxation which converts (1) into a sequence of linear least squares problems

$$\boldsymbol{\Gamma}^\mu \mathbf{c}^\mu = \mathbf{b}, \quad \mu = 1, 2, \dots \tag{3}$$

where $\boldsymbol{\Gamma}^\mu \in \mathbb{C}^{\bar{k}\bar{m}, (P+1)(\bar{n}+1)\bar{\ell}}$ contains, suitably ordered, products of basis functions $\varphi_n(s)$ and $\xi_\ell(\boldsymbol{\vartheta})$, see [5]. The iterations stop when the model coefficients $\mathbf{R}_{n,\ell}$, $r_{n,\ell}$, stored in unknown vector $\mathbf{c}$, stabilize.

III. PROBLEM STATEMENT

Several choices have been documented for the selection of the parameter basis functions $\xi_\ell(\boldsymbol{\vartheta})$, including orthogonal polynomials, Bernstein polynomials, spectral (trigonometric) expansions, and Radial Basis Functions (RBF) [4], [5], [8]. Only the latter enable mesh-free (unstructured) expansions providing a good scalability to high dimension [4], [5]. Therefore, this work focuses on Gaussian RBFs, defined as

$$\xi_\ell(\boldsymbol{\vartheta}) = e^{-\varepsilon \|\boldsymbol{\vartheta} - \boldsymbol{\vartheta}_\ell\|^2} \tag{4}$$

where $\boldsymbol{\vartheta}_\ell$ denotes the center and $\varepsilon > 0$ is the *shape parameter*. Figure 1 depicts through a projection onto a single dimension the effect of ε, which determines how "fat" (ε small) or "thin" (ε large) is the RBF.

The approximation capability of the macromodel is strongly dependent on ε, as well as the numerical steps needed to construct the model (the regressor matrix $\boldsymbol{\Gamma}^\mu = \boldsymbol{\Gamma}^\mu(\varepsilon)$ has a strong dependence on ε). The main objective of this work is to find a sub-optimal value of ε that provides accurate macromodels, with limited overhead required for its determination. In particular, we would like to avoid a direct search on the shape parameter, which would require repeated model construction for different values of ε within an optimization loop.

IV. SUB-OPTIMAL HYPER-PARAMETER SELECTION

Let us consider the two asymptotic cases $\varepsilon \to 0$ and $\varepsilon \to \infty$.

- For $\varepsilon \to 0$, Gaussian RBFs become increasingly flat (see Fig. 1). It is well known that, under suitable conditions, RBF interpolation in this limit approaches a polynomial accuracy [9], [10], so that the model is expected to be very accurate. Unfortunately, as ε decreases, the condition number of $\boldsymbol{\Gamma}^\mu(\varepsilon)$ grows exponentially fast, making the solution of (3) numerically unstable. Figure 2 illustrates

978-1-7281-6162-4/20 $31.00 © 2020 IEEE

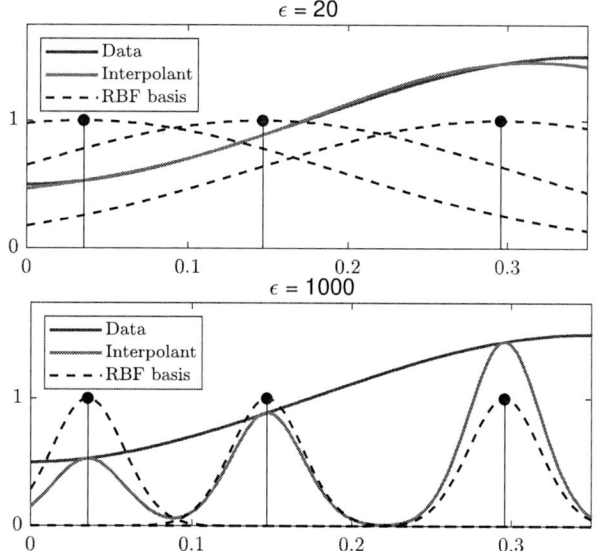

Fig. 1. Comparison of Gaussian RBF interpolants for different shape parameter values. Top panel (ε small): the interpolant accurately approximates the data. Bottom panel (ε large): the basis functions are too narrow to capture the data variability.

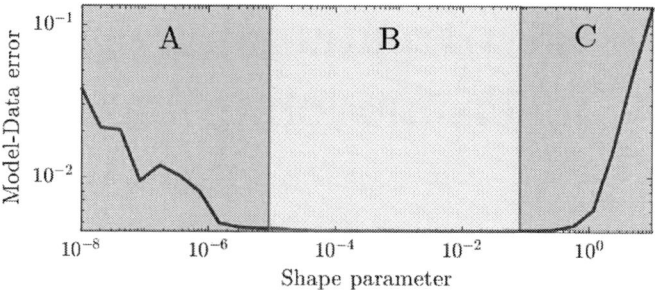

Fig. 2. Model-data error vs shape parameter ε. Region A: numerical instabilities associated with small ε. Region B: candidate sub-optimal shape parameter values. Region C: loss of parameterization capabilities.

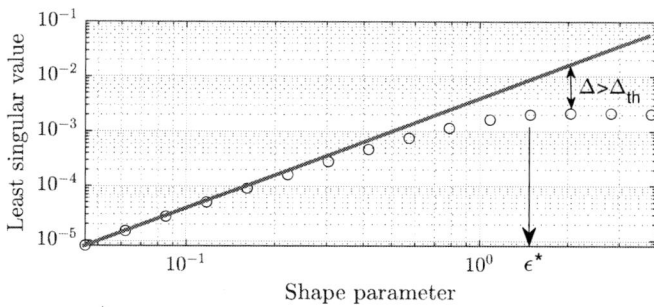

Fig. 3. Black dots: least singular values $\underline{\sigma}(\varepsilon)$ samples. Red solid line: log-log regression line evaluated on the samples $\varepsilon \leq \varepsilon^*$.

the dependence of the model-data error on ε. In case of small ε (region A), such numerical problems are evident.

- When $\varepsilon \to \infty$, the Gaussian RBF $\xi_\ell(\boldsymbol{\vartheta})$ concentrates more and more around its center $\boldsymbol{\vartheta}_\ell$, until it reaches the asymptotic limit

$$\xi_\ell(\boldsymbol{\vartheta}) = \begin{cases} 1 & \text{if } \boldsymbol{\vartheta} = \boldsymbol{\vartheta}_\ell \\ 0 & \text{if } \boldsymbol{\vartheta} \neq \boldsymbol{\vartheta}_\ell \end{cases} \qquad (5)$$

A corresponding model is thus expected to have perfect accuracy only at the RBF centers, but it will be unable to provide a continuous parameterization. As expected, the model-data error increases (Fig. 2, region C).

Candidate sub-optimal shape parameters that minimize model-data error are located in the region B of Fig. 2, where they are

1) sufficiently large to avoid numerical instabilities, and
2) sufficiently small to effectively parameterize the model.

It is also advisable to minimize the condition number $\kappa(\varepsilon)$ of $\boldsymbol{\Gamma}^\mu(\varepsilon)$ for numerical robustness, therefore, we aim at selecting the largest admissible ε located at the interface between regions B and C. This selection must be performed without explicitly computing the model-data error, which in turn would require multiple model estimation and long runtime.

As a proxy to such (inverse) condition number, we consider the least singular value $\underline{\sigma}(\varepsilon)$ of $\boldsymbol{\Gamma}^\mu(\varepsilon)$, whose typical behavior is depicted in Fig. 3. By definition, indeed, $\kappa(\varepsilon) = \bar{\sigma}(\varepsilon)/\underline{\sigma}(\varepsilon)$, where the leading singular value $\bar{\sigma}(\varepsilon)$ is proven not to change significantly with respect to ε variations. It can be proved that $\underline{\sigma}(\varepsilon)$ increases exponentially in regions A and B, until it stabilizes to a finite value for large ε in region C. It is thus sufficient to identify the corner point where the $\underline{\sigma}(\varepsilon)$ trajectory begins to flatten, in a log-log scale. To this end,

1) we precompute a set of $\bar{t}$ least singular values $\{\underline{\sigma}(\varepsilon_1), \ldots, \underline{\sigma}(\varepsilon_{\bar{t}})\}$ of matrix $\boldsymbol{\Gamma}^\mu(\varepsilon)$;

2) we iteratively build a log-log regression line on the pairs

$$\{(\varepsilon_t, \log_{10} \underline{\sigma}(\varepsilon_t)), \ t = 1, 2, \ldots, t_i\}, \quad i = 2, 3, \ldots.$$

This process stops at iteration i^* when the relative deviation Δ of the regression line with respect to the singular value sample t_{i^*+1} exceeds a predefined threshold Δ_{th}. See Fig. 3 for a graphical illustration. The sub-optimal shape parameter ε^* is thus selected as

$$\varepsilon^* = \varepsilon_{t_{i^*}} \qquad (6)$$

The above procedure still requires the construction of the regression matrix $\boldsymbol{\Gamma}^\mu(\varepsilon)$ at each iteration, leading to a potentially different $\varepsilon^* = \varepsilon^*(\mu)$ and consequently different RBF expansions at each iteration. This is not even necessary, since the spectral properties of matrix $\boldsymbol{\Gamma}^\mu(\varepsilon)$ are not expected to change significantly throughout the PSK iterations. Therefore, our proposed algorithm predetermines the sub-optimal ε^* at the first iteration using $\boldsymbol{\Gamma}^1(\varepsilon)$, thus significantly improving runtime.

A second and even more efficient implementation considers the Kernel matrix

$$\mathbf{K}(\varepsilon) = \begin{pmatrix} \xi_1(\boldsymbol{\vartheta}_1) & \cdots & \xi_{\bar{\ell}}(\boldsymbol{\vartheta}_1) \\ \vdots & & \vdots \\ \xi_1(\boldsymbol{\vartheta}_{\bar{m}}) & \cdots & \xi_{\bar{\ell}}(\boldsymbol{\vartheta}_{\bar{m}}) \end{pmatrix} \in \mathbb{R}^{\bar{m} \times \bar{\ell}} \qquad (7)$$

whose size is significantly smaller than $\boldsymbol{\Gamma}^1(\varepsilon)$, and whose construction requires negligible time. It can be shown (details will be presented in a forthcoming report) that the least singular value of $\mathbf{K}(\varepsilon)$ is strongly related to the corresponding least singular value of $\boldsymbol{\Gamma}^1(\varepsilon)$ and, in particular, has the same

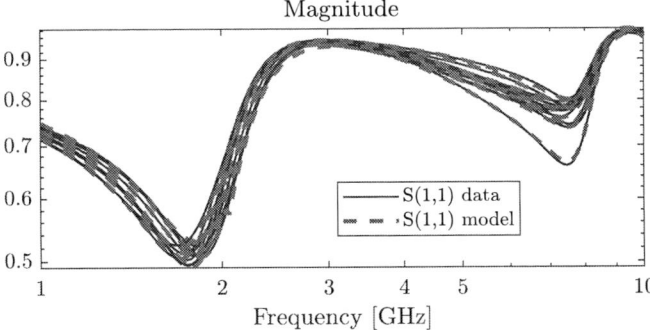

Fig. 4. Model responses (red dashed lines) compared with validation data samples (blue solid lines)

TABLE I
PERFORMANCE OF PROPOSED SHAPE PARAMETER IDENTIFICATION IN TERMS OF MODEL-DATA ERROR AND RUNTIME.

Method	ε^*	Error	Runtime(s)	Speed-up
Grid search	$2.60 \cdot 10^{-3}$	$7.9 \cdot 10^{-3}$	1243	-
Proposed with $\mathbf{\Gamma}(\varepsilon)$	$2.98 \cdot 10^{-2}$	$9.2 \cdot 10^{-3}$	134	$9.2\times$
Proposed with $\mathbf{K}(\varepsilon)$	$2.98 \cdot 10^{-2}$	$9.2 \cdot 10^{-3}$	99	$12.5\times$

dependence with ε as depicted in Fig. 3. The reason for that resides in the definition of the structured matrix $\mathbf{\Gamma}^1(\varepsilon)$, whose composing blocks depend on ε through $\mathbf{K}(\varepsilon)$, inheriting similar spectral properties. Therefore, our second proposed implementation considers matrix $\mathbf{K}(\varepsilon)$ for the identification of the sub-optimal ε^*.

V. NUMERICAL EXAMPLES

We illustrate the proposed approach on a 10-parameter benchmark structure, the Low Noise Amplifier (LNA), originally presented in [11]. For details on the parameterization see [5]. We aim at comparing the presented automated approach with a direct search on the shape parameter ε in terms of runtime and model accuracy.

The structure is known through $\bar{m} = 2000$ frequency responses uniformly covering the parameter space, each including $\bar{k} = 701$ frequency samples in the range [1 Hz, 10 GHz]. The model to be constructed has $\bar{n} = 16$ poles and is parameterized by means of $\bar{\ell} = 40$ and $\bar{\ell} = 4$ Gaussian basis functions at numerator and denominator, respectively. The RBFs centers were randomly selected among the available samples, as in [5].

The proposed algorithm was applied to find the optimal shape parameter ε in the range [0.001, 10]. A total of $\bar{t} = 20$ log-spaced candidate points were used, with a stopping threshold $\Delta_{th} = 10^{-2}$. A single model extraction required 99 s. The overhead to estimate the sub-optimal shape parameter was 35 s using matrix $\mathbf{\Gamma}^1(\varepsilon)$ and negligible using the kernel matrix $\mathbf{K}(\varepsilon)$. Both approaches led to the same $\varepsilon^* = 2.98 \cdot 10^{-2}$, with a corresponding model-data error $9.2 \cdot 10^{-3}$. Figure 4 compares the model responses (red dashed lines) with data (blue solid lines) for 10 randomly chosen validation samples. As a comparison, a standard shape parameter optimization based on direct search on the same set of candidate shape parameters required 20 minutes, leading to $\varepsilon = 2.60 \cdot 10^{-3}$ and a model-data error $7.9 \cdot 10^{-3}$. Therefore, the proposed approach was able to construct a sub-optimal model of comparable accuracy with a $9.2\times$ and $12.5\times$ speedup, see Table I.

Extensive application to a larger set of benchmark structures (10 different test cases depending on 1 up to 10 parameters) led to an average speed up of about $11\times$ using $\mathbf{\Gamma}^1(\varepsilon)$ and of $21\times$ using $\mathbf{K}(\varepsilon)$. These results will be documented in a forthcoming report.

VI. CONCLUSIONS

In the context of high-dimensional parameterized macro-modeling, this paper introduced a heuristic yet effective strategy to select a sub-optimal shape parameter ε of the Gaussian Radial Basis Functions forming the model structure. The proposed method exploits the spectral properties of a small-size kernel matrix as a criterion to optimize the shape parameter. This strategy avoids a direct search, which would require repeated model construction within an optimization loop, and leads to accurate models in significantly reduced runtime.

The presented approach effectively solves only one problem in mesh-free parameterized macromodeling based on unstructured RBF expansions, but several open problems remain to be addressed, such as the automated selection of both number and centers of Gaussian RBFs.

REFERENCES

[1] P. Triverio, S. Grivet-Talocia, and M. S. Nakhla, "A parameterized macromodeling strategy with uniform stability test," *IEEE Trans. Advanced Packaging*, vol. 32, no. 1, pp. 205–215, Feb 2009.

[2] M. Mongillo, "Choosing basis functions and shape parameters for radial basis function methods," *SIAM undergraduate research online*, vol. 4, pp. 190–209, 2011.

[3] S. Rippa, "An algorithm for selecting a good value for the parameter c in radial basis function interpolation," *Advances in Computational Mathematics*, vol. 11, no. 2-3, pp. 193–210, 1999.

[4] A. Zanco and S. Grivet-Talocia, "High-dimensional parameterized macromodeling with guaranteed stability," in *2019 IEEE 28th Conference on Electrical Performance of Electronic Packaging and Systems (EPEPS)*, 2019, pp. 1–3.

[5] A. Zanco, S. Grivet-Talocia, T. Bradde, and M. De Stefano, "Uniformly stable parameterized macromodeling through positive definite basis funtions," *IEEE Transactions on Components, Packaging and Manufacturing Technology*, 2020, submitted.

[6] A. Zanco and S. Grivet-Talocia, "A mesh-free adaptive parametric macromodeling strategy with guaranteed stability," in *2020 International Symposium on Electromagnetic Compatibility, Rome, Italy, 7–11 Sept.* IEEE, 2020, in press.

[7] C. Sanathanan and J. Koerner, "Transfer function synthesis as a ratio of two complex polynomials," *Automatic Control, IEEE Transactions on*, vol. 8, no. 1, pp. 56–58, jan 1963.

[8] S. Grivet-Talocia and E. Fevola, "Compact parameterized black-box modeling via fourier-rational approximations," *IEEE Transactions on Electromagnetic Compatibility*, vol. 59, no. 4, pp. 1133–1142, 2017.

[9] B. Fornberg and G. Wright, "Stable computation of multiquadric interpolants for all values of the shape parameter," *Computers & Mathematics with Applications*, vol. 48, no. 5-6, pp. 853–867, 2004.

[10] B. Fornberg, E. Larsson, and N. Flyer, "Stable computations with gaussian radial basis functions," *SIAM Journal on Scientific Computing*, vol. 33, no. 2, pp. 869–892, 2011.

[11] T. Buss, "2 GHz low noise amplifier with the BFG425W," Philips Semiconductors, B.V., Nijmegen, The Netherlands, Tech. Rep., 1996. [Online]. Available: http://application-notes.digchip.com/004/4-7999.pdf

978-1-7281-6162-4/20 $31.00 © 2020 IEEE

Signal Integrity Characterization of Channels With Asymmetric Via Stubs

Yanyan Zhang*, Mahesh Bohra, Nam Pham, Pavel R. Paladhi, Wiren D. Becker and Daniel M. Dreps

IBM Systems Hardware Development

Austin, TX, USA

* zhangya@us.ibm.com

Abstract—The characterization of the signal integrity (SI) performance of differential high-speed channels that have an imbalance due to mismatched via stub length is investigated. The impact of the asymmetric stubs are characterized with regards to impedance matching, differential-to-common mode conversion and intra-pair skew computation. Simulations for the short and long differential vias are carried out in both the frequency-domain and the time-domain with a back-drilling tolerance of ±10 mils for the residual stubs. It is shown that there is a 9 Ω differential impedance variation, over -40 dB mode conversion loss above 5GHz and 2~3 ps of intra-pair skew is introduced. An example is illustrated to show the impact of the stub asymmetry for the high-speed channel with a time-domain eye simulation at 40 Gbps data rate. It is shown that the short imbalanced differential via with a 20 mil stub asymmetry has the eye diagram degradation in the peak-peak jitter that doubles the jitter for the channel with vias with equal-length residual stubs.

I. INTRODUCTION

As high-speed serial link channel data rates increase, any asymmetry or imbalance in the differential signal routing significantly degrades the signal integrity of multi-gigabit interconnects [1-3]. A plated through hole (PTH) in the printed circuit board (PCB) is referred to as a via for this study. This PTH via has a current carrying part from the surface connection to an internal wiring layer that is called the active via length. The portion of the via after the internal wiring layer is removed, typically by a back-drilling process that leaves a residual via stub [4]. The via stub lengths are greater than zero due to tolerances in PCB fabrication. For the class of PCBs considered for this study, we assume that a via stub tolerance of +/- 10 mils around a so-called do-not-cut line is achieved. This 20mil stub length variation causes a mismatch of the via stub loading on the differential signaling which impacts the impedance matching and induces an intra-pair skew. Skew mitigation is very important for maintaining the signal integrity. Even picosecond skews can impact the performance of the channel and induce undesirable differential-to-common (DTC) mode noise and cause electromagnetic radiation (EMI). This study characterizes the impedance variation and the skews introduced by the asymmetric via stub. For this study, we are assuming a data rate of 25 Gbps or more.

The paper is organized as follows. Section II presents the 3D full-wave differential via modeling with asymmetrical stubs in both short and long active via lengths, respectively. Through simulation, the differential via impedance, the differential insertion loss and the DTC mode conversion are determined.

The intra-pair skew as a function of frequency is computed with a method using the modified mixed-mode S parameter [5]. Section III provides an eye simulation example at a data rate of 40 Gbps comparing the imbalanced via stub response with the balanced via stub. Finally, the conclusion is given in Section IV.

II. MODELING OF IMBALANCED DIFFERENTIAL VIA WITH ASYMMETRIC STUBS

A pair of differential vias with the via stubs is shown in Fig. 1. The active via length, *h*, is from the surface differential microstrip trace to the differential stripline traces on an internal wiring layer. The via stub lengths, b_1 and b_2, extend below the internal wiring layer. In order to compare the SI impact of the stub asymmetry, two active via lengths are chosen as shown in Fig.1(a) with *h*=52mils (~1.32mm) and Fig. 1(b) with *h*=102.7mils (~2.61mm). The stub asymmetry is studied as the stub length b_2 varies with the center at b_1=15mils with a variation of +/- 10 mils. A stubless via model is simulated as the reference for comparison. The S-parameter models are extracted using a 3D full-wave solver.

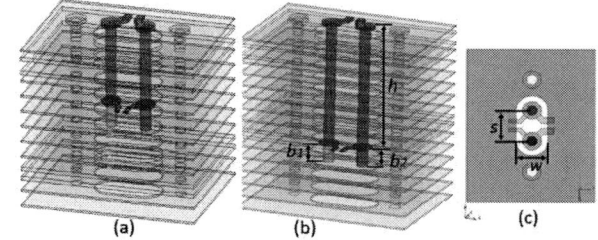

Fig. 1. Modeling with a pair of differential via with via stubs: (a) and (b) front views for short and long vias; (c) top view.

Both the microstrip and stripline traces are designed to have an 85 Ω differential impedance. The via pad is 18 mils, the drill dimeter is 10 mils, the via spacing, *S*, is 31.5 mils, the oblong antipad width, *W*, is 30 mil, and the antipad length, (*W*+*S*), is 61.5mil. The board cross-section is of a hybrid construction [4] having an effective relative dielectric constant of ε_r=3.26 and a loss tangent tanδ=0.003 at 10GHz for the low-loss internal wiring layers with the breakout wiring, and a relative dielectric constant of ε_r=4.15 and a loss tangent tanδ=0.021 at 10GHz for the layers without internal high-speed signal traces.

Figs. 2a and 2b shows the simulated differential time-domain reflectometry (TDR) results for the short *h*=52 mils and long *h*=102.7 mils differential vias, respectively. A 0-100% rise time of 20 ps is launched at the breakout stripline on the internal

978-1-7281-6162-4/20 $31.00 © 2020 IEEE

wiring layer. One via stub is fixed at the length b_1=15 mils, and the other via stub b_2 varies the length from 5 mils to 25 mils in light of the back-drilling tolerance of ±10 mils. Both the short and long differential vias show the capacitive impedance with the aforementioned board stackup, and the via impedance dip increases as via stub b_2 gets shorter by approximately 2.5 Ω per 5 mils change in b_2. For the stubless case, the short and long differential vias are around 80 Ω. As b_2 changes from 5mils to 25mils, the impedance of the short differential via varies by 9.1 Ω decreasing from 73.8 Ω to 64.7 Ω . For the long differential via, the impedance decreases from 71.0 Ω to 62.4 Ω, an 8.6 Ω change. Therefore, the back-drilling tolerance of ±10 mils results in a differential impedance uncertainty of approximately 9 Ω for the vias in this PCB.

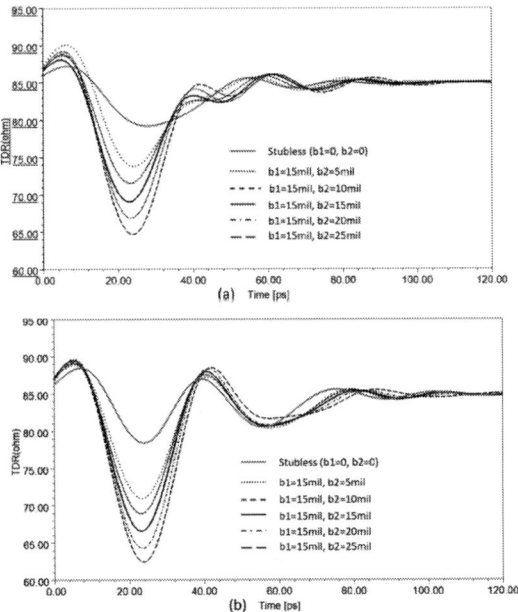

Fig. 2. Simulated differential TDR impedance of the differential via from the breakout stripline to the top microstrip with rise time of 20 ps and source reference impedance at 85 Ω: (a) active via length h=52mils and (b) active via length h=102.7mils.

The simulated differential-to-common mode conversion as denoted by Scd21 with varied stub lengths for both the short and long differential vias are shown in Fig. 3. It is clearly seen that the asymmetry of the via stubs causes the DTC loss to increase with the frequency. Scd21 rises over -40dB above 5 GHz. However, the equal-stub case (b_1=15mils, b_2=15mils) shows that Scd21 is less than -50 dB over the 35 GHz frequency range and nearly follows the ideal stubless case. The induced mode conversion increases with more asymmetry as shown in Table 1. In addition, there is nearly a 10dB variation in mode conversion over the back-drilling tolerance of ±10mils. Note that the Scd21 curves are very similar for the short and long active vias showing that it is the stub asymmetry that increases the mode conversion. Meanwhile, the Scd21 curves start to drop near 30GHz as is caused by the poor impedance matching and the worse return loss.

Also note that the differential insertion loss Sdd21 mirrors the aforementioned impedance variation by the via stubs in TDR plots. The differential via with the longest stub b_2=25 mils

has the biggest insertion loss. For instance, Sdd21 of the imbalanced short differential via drops 0.5 dB at 20 GHz and 1.4dB at 30GHz as compared with the stubless via model; while for the long differential via, it drops 0.8 dB at 20 GHz and 2.2 dB at 30 GHz.

b_2 (mils)	5	10	15	20	25
Δ(=b_2-b_1) (mils)	-10	-5	0	5	10
Scd21(dB)	-23.5	-28.8	-70.8	-27.2	-19.9

Table 1. Simulated differential-to-common mode conversion Scd21 at 16GHz for the active via length h=52mils with one stub length b_1 at 15mils.

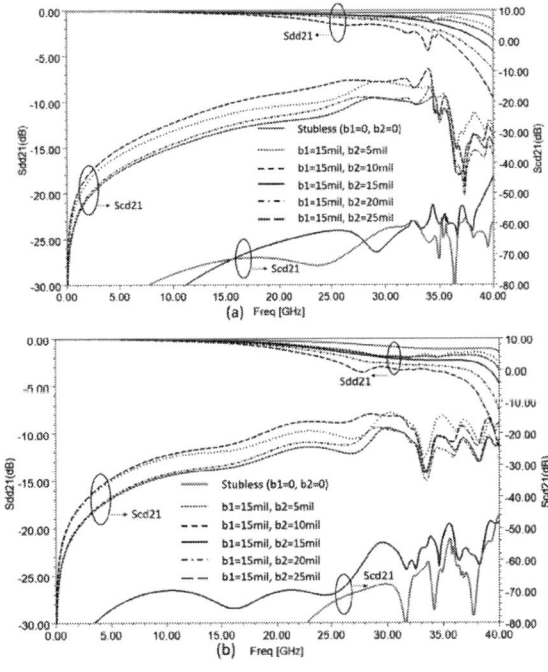

Fig. 3. Simulated differential insertion loss Sdd21and differential-to-common mode conversion Scd21 at the reference impedance 85 Ω: (a) active via length h=52mils and (b) active via length h=102.7mils.

The modified mixed-mode S parameter as proposed by the computational method in [5], takes differential-mode and common-mode into account simultaneously. Based on this method, the intra-pair skew is calculated as the function of the frequency with different stub lengths for the short and long differential vias. Fig.4 shows the calculated results of unwrapped phases and the resulting time delays for the P-side via with the stub b_1=5 mils and N-side via with the stub b_2=25 mils of the long differential via (h=102.7 mils).

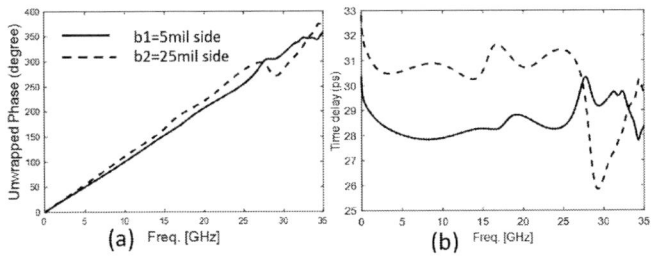

Fig. 4. Calculated unwrapped phase (a) and time delay (b) for the long differential via (h=102.7mils) with the asymmetric stubs (b_1=5mils, b_2=25mils).

978-1-7281-6162-4/20 $31.00 © 2020 IEEE   129

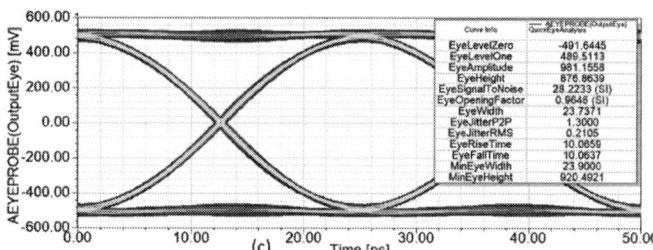

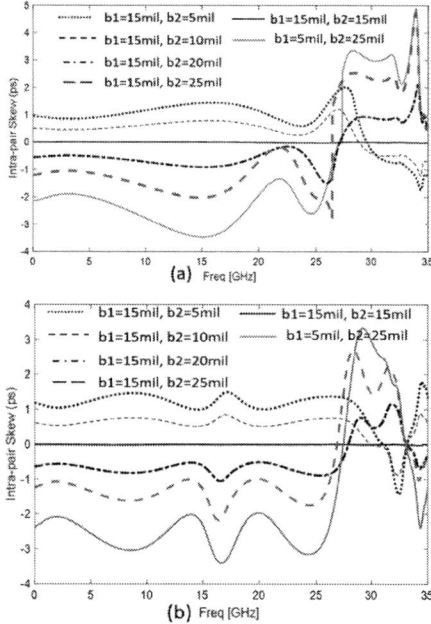

Fig. 5. Calculated intra-pair skew: (a) active via length h=52mils and (b) active via length h=102.7mils.

The intra-pair skew is defined by the difference between the time delays of P-side and N-side vias, as is derived in the following Fig. 5. As expected, the equal-stub case (b_1=15 mils, b_2=15 mils) has zero skew in the entire frequency band. At the lower frequencies less than 10 GHz, the skew changes very slowly with the frequency, and there is around a 0.5 ps step size for the skew as the stub asymmetry changes in 5 mils length steps. The case (b_1=5 mils, b_2=25 mils) has the biggest stub difference and shows the biggest skew as a result. For the short and long differential vias, the intra-pair skews are very close at a certain stub asymmetry, which agrees well with the Scd21 plots for the mode conversion.

III. TIME-DOMAIN EYE SIMULATION

Fig. 6. Eye diagrams for the short (h=52 mils) differential via at 40 Gbps: (a) stubless (b_1=0 mil, b_2=0 mil), (b) equal stubs (b_1=15 mils, b_2=15 mils) and (c) unequal stubs (b_1=5 mils, b_2=25 mils).

The short (h=52 mils) differential via is simulated in the time domain at 40 Gbps date rate. The 85 Ω Tx setting uses the peak voltage of 1 V, the bit pattern of PRBS23 and the rise time of 15 ps. No equalization is used in Tx and the eye probe in Rx. The eye width and the jitter match closely in both the ideal stubless case in Fig.6(a) and the equal 15 mils stub case in Fig. 6(b). Fig. 6(c) has the multiple traces in the eye which indicates the degraded impedance matching. In comparison with the equal stub case in Fig. 6(b), the 20 mils stub asymmetry (b_1=5 mils, b_2=25 mils) in Fig. 6(c) results in the peak-to-peak jitter doubling from 0.65 ps to 1.30 ps, and the RMS jitter increasing by a factor of 1.45 from 0.145 ps to 0.21 ps.

IV. CONCLUSION

The degradation of signal integrity properties introduced by stub asymmetry in the back-drilling process of differential vias is investigated and simulated with regards to differential impedance, the insertion loss, the differential-to-common conversion, and the intra-pair skew. The SI characterization for the imbalanced differential via with asymmetric stubs provides useful insights into the differential via designs for high-speed channel applications. At 25 Gb/s and above, the impact is significant, and steps must be taken to avoid or minimize the stub length differences between PTH vias in a differential pair by proper PCB design and fabrication process control. If sufficient control is not available, the impact of the via stub asymmetry on the signal integrity of the high-speed signal needs to be factored into the channel specifications.

REFERENCES

[1] E. Kunz, et al., "Sources and Compensation of Skew in Single-Ended and Differential Interconnects," DesignCon 2014.

[2] J. Chaves, et al., "Mode Conversion Due To Via Stubs in Differential Signaling," 2019 IEEE 23rd Workshop on Signal and Power Integrty (SPI), Chambéry, France, 2019, pp. 1-4.

[3] K. J. Han et al., "Modeling On-Board Via Stubs and Traces in High-Speed Channels for Achieving Higher Data Bandwidth," IEEE Transactions on Components, Packaging and Manufacturing Technology, vol. 4, no. 2, pp. 268-278, Feb. 2014.

[4] C. Méndez Ruiz, et al., "Improve signal integrity performance by using hybrid PCB stackup," IEEE International Symposium on Electromagnetic Compatibility, Denver, CO, 2013, pp. 317-321.

[5] S. Baek ,et al., "Compution of Intra-pair Skew for Imbalance Differential Line using Modified Mixed-mode S-parameter," 2007 IEEE Electrical Performance of Electronic Packaging, Atlanta, GA, 2007, pp. 179-182.

AUTHOR INDEX

Aberle, James T. 22
Achar, Ramachandra..101
Achar, Ram..19
Avula, Venkatesh..1
Aygün, Kemal 13, 22, 107, 110
Barnes, Heidi 4, 68
Beaman, Brian 83
Becker, Wiren D. 122, 128
Becker, Wiren 83
Beyene, Wendemagegnehu..41
Bianchi, Giovanni 68
Bianconi, Giacomo 53
Bogner, Werner..89
Bohra, Mahesh 128
Bosman, Dries 116
Bradde, T. 7
Calafiore, G. C. 7
Cangellaris, Andreas C. 62
Carrel, Jack 4
Chakraborty, Swagato..53
Chaudhary, Muhammad Waqas..10
Chen, Ching-Huei..47
Chen, Guang 41
Chen, Xu..62
Cheng, Chung-Kuan 31
Choi, Seonguk..44
Choubey, Bhaskar..10
Christ, Sean R. 13
Chun, Sungjun 83
Daniel, L. 7
Dong, Xiaoqing 16
Dreps, Daniel M...128
Dreps, Daniel 83
Durgun, Ahmet C...13
Erdin, Ihsan..19
Fan, Jun..119
Geyik, Cemil S. 22, 110
Ginste, Dries Vande 116
Gossye, Michiel..116
Grivet-Talocia, S...7
Grivet-Talocia, Stefano..125
Guglani, Surila..80
Hamada, Mototsugu..65, 113
Han, Seunghyup..25
Hashemi, Ashkan..41
He, Zichang 28
Heinig, Andy 10
Hejase, Jose A...122

Hejase, Jose..83
Hill, Michael J...13, 22
Ho, Chia-Tung 31
Hong, Seokwoo..34
Huang, Chunxing 16
Huynen, Martijn..116
Jakob, Johannes..89
Jeong, Seungtaek..34
Jeong, Yi-Ru..38
Jiao, Chao..31
Joshi, Yogendra..1
Kang, Hyo-Soon..41
Kaushik, Brajesh K...101
Keck, Florian..89
Kim, Joungho 34, 44
Kim, Minsu..44
Kim, Seongguk..44
Kim, Subin 34, 44
Kim, Youngwoo..34
Kumar, Rahul..101
Kumar, Sanjay..47
Kumar, Somesh..101
Kumar, Vijender..92
Kunze, Kevin 89
Kuroda, Tadahiro..65, 113
Lee, Jaehak..34
Lee, Seongsoo..34
Li, Dongwei..53
Li, Er-Ping..62
Li, Xinbo..56
Liao, Chun-Lin..47
Liu, Chang..50
Liu, Xiaoping..41
Ma, Hanzhi..62
Mahmood, Z...7
Marek, Damian..59
Miura, Reiji..65
Moreira, Jose..68
Muthusamy, Sukumar..92
Mutnury, Bhyrav..47, 92
Narayan, S. S. Likith..101
Nguyen, Thong..71, 104
Okhmatovski, Vladimir..56
Ong, Chong-Jin..74
Paladhi, Pavel R...128
Paladhi, Pavel Roy..83
Park, Gapyeol..44
Park, Hyunwook..34, 44

Pathania, Sunil 92
Peterson, Zachariah M. 77
Pham, Nam 128
Proskurnikov, A. V. 7
Qian, Zhiguo 110
Rodriguez, Daniel 83
Rogier, Hendrik 116
Roy, Sourajeet 80, 101
Röhrl, Franz Xaver 89
Sandler, Steve 4
Scharff, Katharina 86
Scharl, Andreas 89
Schierholz, Christian Morten 86
Schuster, Christian 86
Schutt-Aine, Jose 71, 104
Seema, P K 92
Sepaintner, Felix 89
Sharma, Rohit 92, 101
Sharma, Shashwat 59, 98
Sharma, Vijender Kumar 95
Shi, Bobi 104
Shiba, Kota 113
Shrimali, Hitesh 95
Sim, Boogyo 34
Smet, Vanessa 1
Smith, Stephen A. 107
Son, Keeyeong 34
Son, Keeyoung 44
Son, Kyunjune 44
Song, Junyeop 34
Sun, Jiwei 110
Swaminathan, Madhavan 1, 25, 122
Tang, Junyan 83, 122
Torun, Hakki M. 122
Tripathi, Jai Narayan 95
Triverio, Piero 59, 98
Usui, Masairo 113
Vardapetyan, Armen 74
Vasa, Mallikarjun 47, 92
Waeytens, Ruben 116
Wang, Tao 119
Wang, Xinyuan 31
Yang, Cheng 86
Yang, Xianbo 122
Yilmaz, Ali E. 38, 50
Zanco, Alessandro 125
Zen, Zhiyu 31
Zha, Xin 31
Zhang, Yanyan 83, 128
Zhang, Zheng 28
Zhang, Zhichao 22, 107
Zorn, Stefan 89

IEEE
445 Hoes Lane
Piscataway, NJ 08854-4141

ISBN 978-1-7281-6162-4